Wind Turbine Maintenance Technician

Level One, Volume One

Trainee Guide

nccer

Prentice Hall

Boston   Columbus   Indianapolis   New York   San Francisco   Upper Saddle River
Amsterdam   CapeTown   Dubai   London   Madrid   Milan   Munich   Paris   Montreal   Toronto
Delhi   Mexico City   São Paulo   Sydney   Hong Kong   Seoul   Singapore   Taipei   Tokyo

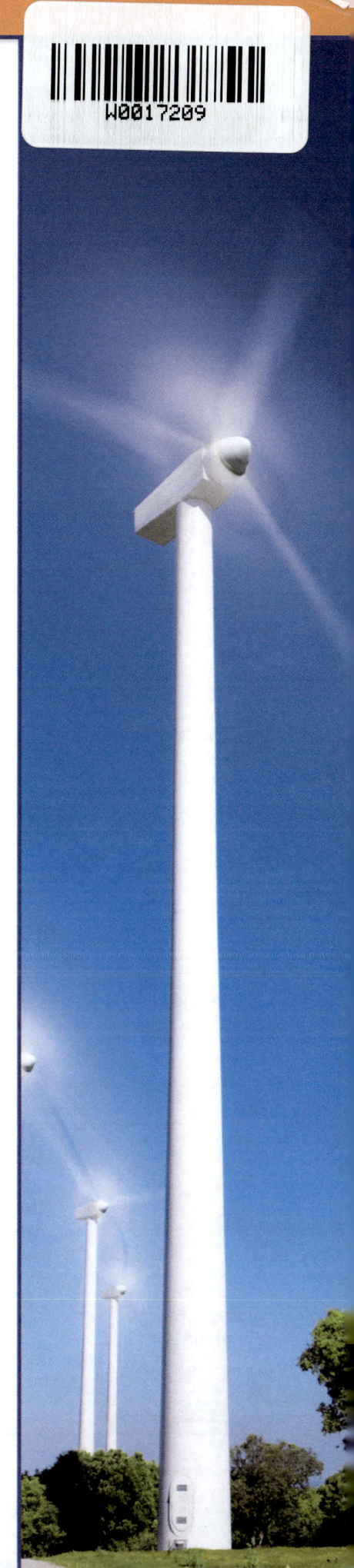

National Center for Construction Education and Research
President: Don Whyte
Director of Product Development: Daniele Stacey
Wind Turbine Maintenance Technician
  Project Manager: Rob Richardson
Production Manager: Tim Davis

Quality Assurance Coordinator: Debie Ness
Desktop Publishing Coordinator: James McKay
Production Specialist: Laura Wright
Editor: Chris Wilson

Writing and development services provided by Topaz Publications, Liverpool, NY
Lead Writer/Project Manager: Troy Staton
Desktop Publisher: Joanne Hart
Art Director: Megan Paye

Permissions Editors: Tonia Burke, Alison Richmond
Writers: Troy Staton, Thomas Burke, Charles Rogers

Pearson Education, Inc.
Editorial Director: Vernon R. Anthony
Executive Editor: Alli Gentile
Senior Product Manager: Lori Cowen
Operations Supervisor: Deidra M. Skahill
Art Director: Jayne Conte

Director of Marketing: David Gesell
Executive Marketing Manager: Derril Trakalo
Marketing Manager: Brian Hoehl
Marketing Coordinator: Crystal Gonzalez
Cover Photo: Courtesy of Buckner Companies

Composition: NCCER
Printer/Binder: Courier/Kendallville, Inc.
Cover Printer: Lehigh-Phoenix Color, Hagerstown
Text Fonts: Palatino and Univers

10 9 8 7 6 5 4 3 2 1

Prentice Hall
is an imprint of

www.pearsonhighered.com

Perfect bound   ISBN-13: 978-0-13-271895-0
                ISBN-10:   0-13-271895-2

Preface

To the Trainee

Welcome to an emerging industry and a truly exciting field of renewable energy—wind power. While harnessing the wind to power machinery and transportation is nothing new (humans have been doing so for thousands of years) the interest in developing this and other renewable technologies to reduce our dependence on fossil fuels has never been more fervent.

The first quarter of 2011 saw over 1,100 megawatts (MW) of wind power capacity installed—more than double the capacity installed in the first quarter of 2010. In addition, the U.S. wind industry has added over 35% of all new generating capacity over the past 4 years, more than nuclear and coal combined. Today, U.S. wind power capacity represents more than 20% of the world's installed wind power. The U.S. Department of Energy has set a goal of providing 20 percent of our electrical needs through wind power by 2030.

The wind industry's impact is not limited to energy production, however. As the construction and ongoing maintenance of turbines and wind farms increases, so do the employment possibilities and the need for educated and well-trained technicians.

The career field of wind turbine technicians is so new that the U.S. Bureau of Labor Statistics recently developed a category for this career to facilitate tracking of employment and wage data. At this writing, shortages already exist, and as development of both onshore and offshore wind farms continues, the technician shortage is expected to get worse before it gets better. The laws of supply and demand dictate that wages and benefits for technicians will rise as a result.

Successful wind turbine technicians possess a wide variety of skills. A mechanical and electrical system aptitude, computer literacy, an understanding of meteorology, and above-average math skills are some of the most important requirements. The specific training students receive during the educational process will prepare them for entry- and mid-level positions as wind turbine service technicians.

And if you've started with this course, you're already on your way.

Wind Turbine Maintenance Technician Level One

Prior to beginning this level, we highly recommend completion of *Power Industry Fundamentals*, a course that combines NCCER's fundamental *Core Curriculum* with the *Introduction to the Power Industry* module. If you're training through an NCCER Accredited Training Program Sponsor, *Introduction to the Power Industry* is required before any credentials for *Wind Turbine Maintenance Technician* can be obtained.

Wind Turbine Maintenance Technician Level One is the first level of training in NCCER's three-year wind turbine maintenance technician program. To satisfy the demand for this title in time for Fall 2011-2012 classes, we are releasing *Wind Turbine Maintenance Technician Level One* in two volumes. *Wind Turbine Maintenance Technician Level One, Volume Two* will be released in late 2011. Trainees must successfully complete both Volumes One and Two to receive *Wind Turbine Maintenance Technician Level One* credentials.

We wish you success as you progress through this training program. If you have any comments on how NCCER might improve upon this textbook, please complete the User Update form located at the back of each module and send it to us. We will always consider and respond to input from our customers.

We invite you to visit the NCCER website at **www. nccer.org** for information on the latest product releases and training, as well as online versions of the *Cornerstone* newsletter and Pearson's product catalog.

Your feedback is welcome. You may email your comments to **curriculum@nccer.org** or send general comments and inquiries to **info@nccer.org**.

NCCER Curricula

NCCER is a not-for-profit 501(c)(3) education foundation established in 1995 by the world's largest and most progressive construction companies and national construction associations. It was founded to address the severe workforce shortage facing the industry and to develop a standardized training process and curricula. Today, NCCER is supported by hundreds of leading construction and maintenance companies, manufacturers, and national associations. NCCER's curricula are developed in partnership with Pearson Education, Inc., the world's largest educational publisher.

Some features of NCCER's curricula are as follows:

- An industry-proven record of success
- Curricula developed by the industry for the industry
- National standardization providing portability of learned job skills and educational credits
- Compliance with the Office of Apprenticeship requirements for related classroom training (*CFR 29:29*)
- Well-illustrated, up-to-date, and practical information

NCCER also maintains a National Registry that provides transcripts, certificates, and wallet cards to individuals who have successfully completed modules of NCCER's curricula. *Training programs must be delivered by an NCCER Accredited Training Sponsor in order to receive these credentials.*

Special Features

In an effort to provide a comprehensive, user-friendly training resource, we have incorporated many different features for your use. Whether you are a visual or hands-on learner, this book will provide you with the proper tools to get started as a wind turbine maintenance technician.

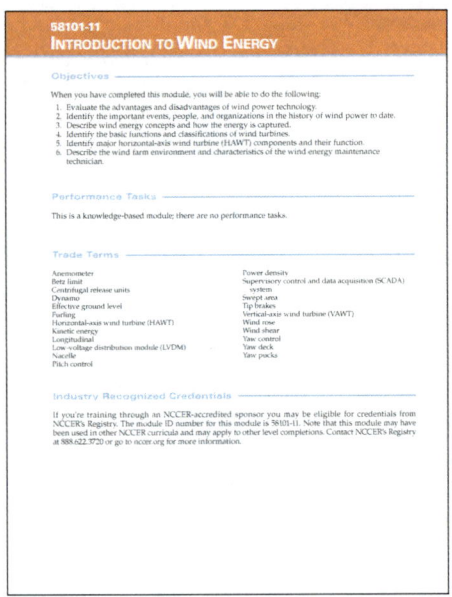

Introduction

This page is found at the beginning of each module and lists the Objectives, Performance Tasks, Trade Terms, and Required Trainee Materials for that module. The Objectives list the skills and knowledge you will need in order to complete the module successfully. The Performance Tasks give you an opportunity to apply your knowledge to the real-world duties that electricians perform. The list of Trade Terms identifies important terms you will need to know by the end of the module. Required Trainee Materials list the materials and supplies needed for the module.

On Site

On Site features provide a head start for those entering the wind turbine maintenance technician field by presenting technical tips and professional practices from master technicians in a variety of disciplines. The On Site features often include real-life scenarios similar to those you might encounter on the job site.

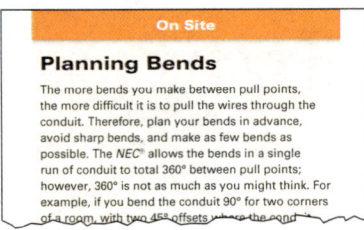

Color Illustrations and Photographs

Full-color illustrations and photographs are used throughout each module to provide vivid detail. These figures highlight important concepts from the text and provide clarity for complex instructions. Each figure reference is denoted in the text in *italic type* for easy reference.

Figure 2 Pushing down on the bender to complete the bend.

Notes, Cautions, and Warnings

Safety features are set off from the main text in highlighted boxes and are organized into three categories based on the potential danger of the issue being addressed. Notes simply provide additional information on the topic area. Cautions alert you of a danger that does not present potential injury but may cause damage to equipment. Warnings stress a potentially dangerous situation that may cause injury to you or a co-worker.

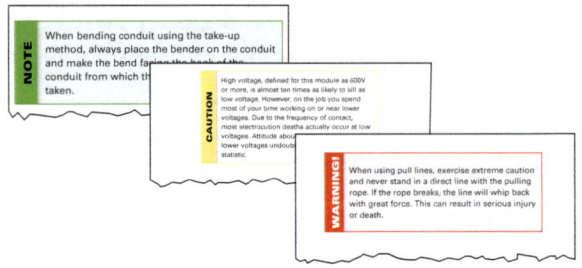

Going Green

Going Green looks at ways to preserve the environment, save energy, and make good choices regarding the health of the planet. Through the introduction of new construction practices and products, you will see how the "greening of America" has already taken root.

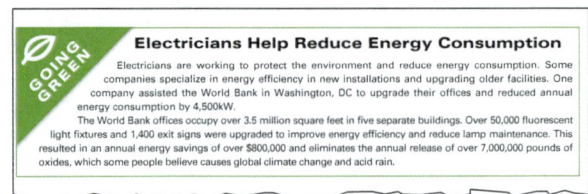

Case History

Case History features emphasize the importance of safety by citing examples of the costly (and often devastating) consequences of ignoring codes, standards, or OSHA regulations.

What's wrong with this picture?

What's wrong with this picture? features include photos of actual code violations for identification and encourage you to approach each task with a critical eye.

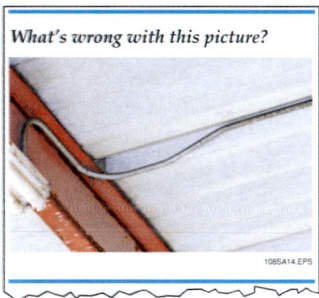

What's wrong with this picture?

108SA14.EPS

Think About It

Think About It features use "What if?" questions to help you apply theory to real-world experiences and put your ideas into action.

Step-by-Step Instructions

Step-by-step instructions are used throughout to guide you through technical procedures and tasks from start to finish. These steps show you not only how to perform a task but also how to do it safely and efficiently.

are energized until you have verified that the circuit is de-energized. This is called a live-dead-live test. Follow these steps to verify that a circuit is de-energized:

Step 1 Ensure that the circuit is properly tagged and locked out *(CFR 1910.333/1926.417)*.

Step 2 Verify the test instrument operation on a known source using the appropriately rated tester.

Step 3 Using the test instrument, check the circuit to be de-energized. The voltage should be zero.

Step 4 Verify the test instrument operation, once again on a known power source.

Trade Terms

Each module presents a list of Trade Terms that are discussed within the text and defined in the Glossary at the end of the module. These terms are denoted in the text with **blue bold type** upon their first occurrence. To make searches for key information easier, a comprehensive Glossary of Trade Terms from all modules is located at the back of this book.

Electricity is all about cause and effect. The presence of voltage (volts) in a closed circuit will cause current (amps) to flow. The more voltage you apply, the more current will flow. However, the amount of current flow is also determined by how much resistance (ohms) the load offers to the flow of current. In order to convert electrical energy into work, the load consumes energy. The amount of energy a device consumes is called power, and is expressed in watts (W). Volts (V), amps, ohms, and watts are related in such a way that if any one of them changes, the others are proportionally affected. This relationship can be seen using basic math principles that you will learn in this module. You will also learn how electricity is produced and how test instruments are used to measure electricity.

Review Questions

Review Questions are provided to reinforce the knowledge you have gained. This makes them a useful tool for measuring what you have learned.

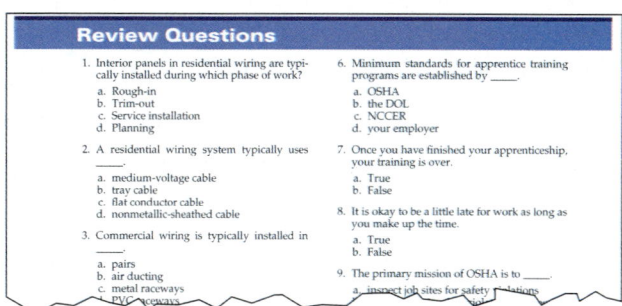

NCCER Curricula

NCCER's training programs comprise more than 80 construction, maintenance, pipeline, and utility areas and include skills assessments, safety training, and management education.

Boilermaking
Cabinetmaking
Carpentry
Concrete Finishing
Construction Craft Laborer
Construction Technology
Core Curriculum:
    Introductory Craft Skills
Drywall
Electrical
Electronic Systems Technician
Heating, Ventilating, and
    Air Conditioning
Heavy Equipment Operations
Highway/Heavy Construction
Hydroblasting
Industrial Coating and Lining
    Application Specialist
Industrial Maintenance
    Electrical and
    Instrumentation Technician
Industrial Maintenance
    Mechanic
Instrumentation
Insulating
Ironworking
Masonry
Millwright
Mobile Crane Operations
Painting
Painting, Industrial
Pipefitting
Pipelayer
Plumbing
Reinforcing Ironwork
Rigging
Scaffolding
Sheet Metal
Signal Person
Site Layout
Sprinkler Fitting
Tower Crane Operator
Welding

Green/Sustainable Construction

Building Auditor
Fundamentals of
    Weatherization
Introduction to Weatherization
Sustainable Construction
    Supervisor
Weatherization Crew Chief
Weatherization Technician
Your Role in the Green
    Environment

Energy

Introduction to the Power
    Industry
Introduction to Solar
    Photovoltaics
Introduction to Wind Energy
Power Industry Fundamentals
Power Generation Maintenance
    Electrician
Power Generation I&C
    Maintenance Technician
Power Generation Maintenance
    Mechanic
Power Line Worker
Solar Photovoltaic Systems
    Installer

Pipeline

Control Center Operations,
    Liquid
Corrosion Control
Electrical and Instrumentation
Field Operations, Liquid
Field Operations, Gas
Maintenance
Mechanical

Safety

Field Safety
Safety Orientation
Safety Technology

Management

Fundamentals of Crew
    Leadership
Project Management
Project Supervision

Supplemental Titles

Applied Construction Math
Careers in Construction
Tools for Success

Spanish Translations

Basic Rigging
    (Principios Básicos de
    Maniobras)
Carpentry Fundamentals
    (Introducción a la
    Carpintería, Nivel Uno)
Carpentry Forms
    (Formas para Carpintería,
    Nivel Trés)
Concrete Finishing, Level One
    (Acabado de Concreto,
    Nivel Uno)
Core Curriculum:
    Introductory Craft Skills
    (Currículo Básico:
    Habilidades Introductorias
    del Oficio)
Drywall, Level One
    (Paneles de Yeso, Nivel Uno)
Electrical, Level One
    (Electricidad, Nivel Uno)
Field Safety
    (Seguridad de Campo)
Insulating, Level One
    (Aislamiento, Nivel Uno)
Ironworking, Level One
    (Herrería, Nivel Uno)
Masonry, Level One
    (Albañilería, Nivel Uno)
Pipefitting, Level One
    (Instalación de Tubería
    Industrial, Nivel Uno)
Reinforcing Ironwork, Level One
    (Herreria de Refuerzo,
    Nivel Uno)
Safety Orientation
    (Orientación de Seguridad)
Scaffolding
    (Andamios)
Sprinkler Fitting, Level One
    (Instalación de Rociadores,
    Nivel Uno)

Acknowledgments

This curriculum was revised as a result of the farsightedness and leadership of the following sponsors:

Aeroflex Test Systems
ATI Career Training Center
Cianbro Corporation
Columbia Gorge Community College
Francis Tuttle Technology Center
Grand Rapids Community College
Highland Community College

Madison Comprehensive High School
Mesabi Range Community & Technical College
Michigan Institute of Aviation
Miller-Motte Technical College
Northeastern Junior College
Snap-on Inc.

This curriculum would not exist were it not for the dedication and unselfish energy of those volunteers who served on the Authoring Team. A sincere thanks is extended to the following:

Earl Bailey
Neil Browne
Tim Dean
Robert E. DeGraw
Ron Gardner
Dan Janisch

Tom Lieurance
Robert Malliet
Joe Rakoczy
Jonathan Sacks
Brian Shultz
David Vrtol

A final note: This book is the result of a collaborative effort involving the production, editorial, and development staff at Pearson Education, Inc., and the National Center for Construction Education and Research. Thanks to all of the dedicated people involved in the many stages of this project.

NCCER Partners

American Fire Sprinkler Association
Associated Builders and Contractors, Inc.
Associated General Contractors of America
Association for Career and Technical Education
Association for Skilled and Technical Sciences
Carolinas AGC, Inc.
Carolinas Electrical Contractors Association
Center for the Improvement of Construction
  Management and Processes
Construction Industry Institute
Construction Users Roundtable
Construction Workforce Development Center
Design Build Institute of America
Merit Contractors Association of Canada
Metal Building Manufacturers Association
NACE International
National Association of Minority Contractors
National Association of Women in Construction
National Insulation Association
National Ready Mixed Concrete Association
National Technical Honor Society
National Utility Contractors Association
NAWIC Education Foundation
North American Technician Excellence

Painting & Decorating Contractors of America
Portland Cement Association
SkillsUSA
Steel Erectors Association of America
The Manufacturers Institute
U.S. Army Corps of Engineers
University of Florida, M. E. Rinker School of
  Building Construction
Women Construction Owners & Executives, USA

Contents

Module One
Introduction to Wind Energy

Introduces the fundamentals of generating electrical power from wind energy. A brief history of wind energy is included as well as wind science, the interception of wind energy through a rotor, and an identification of major wind turbine generator components. (Module ID number 58101-11; 15 Hours)

Module Two
Introduction to Wind Turbine Safety

Introduces trainees to the unique safety concerns of working inside the wind turbine and in the wind farm environment. Expands on earlier safety training and provides coverage of electrical arc flash safety. (Module ID number 58102-11; 12.5 Hours)

Module Three
Climbing Wind Towers

Covers all aspects of climbing wind turbine lattice towers and tubular towers. Includes coverage of proper climbing equipment and equipment inspection, environmental hazards, proper climbing techniques, and common wind turbine safe climbing guidelines. (Module ID number 58103-11; 40 Hours)

Module Four
Introduction to Electrical Circuits

Offers a general introduction to the electrical concepts used in Ohm's law applied to DC series circuits. Includes atomic theory, electromotive force, resistance, and electric power equations. (Module ID number 26103-11, from *Electrical Level One*; 7.5 Hours)

Module Five
Electrical Theory

Introduces series, parallel, and series-parallel circuits. Covers resistive circuits, Kirchhoff's voltage and current laws, and circuit analysis. (Module ID number 26104-11, from *Electrical Level One*; 7.5 Hours)

Module Six
Electrical Test Equipment

Focuses on proper selection, inspection, and use of common electrical test equipment, including voltage testers, clamp-on ammeters, ohmmeters, multimeters, phase/motor rotation testers, and data recording equipment. Also covers safety precautions and meter category ratings. (Module ID number 26112-11, from *Electrical Level One*; 5 Hours)

Note: *NFPA 70*®, *National Electrical Code*® and *NEC*® are registered trademarks of the National Fire Protection Association, Inc., Quincy, MA 02269. All *National Electrical Code*® and *NEC*® references in this textbook refer to the 2011 edition of the *National Electrical Code*®.

Module Seven

Electrical Wiring

Provides coverage of the proper types and applications of conductors as well as their installation techniques. Also describes the technique and components used for terminating and splicing conductors. (Module ID number 58104-11; 10 Hours)

Glossary

Index

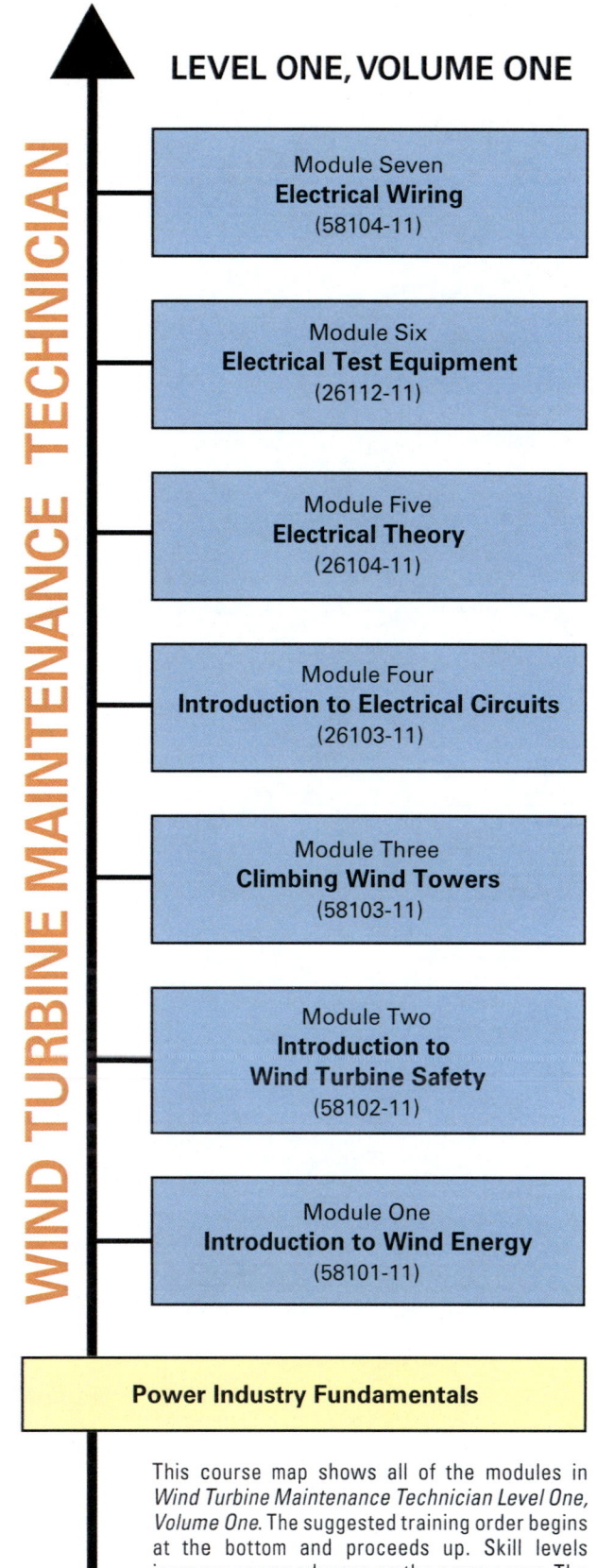

LEVEL ONE, VOLUME ONE

WIND TURBINE MAINTENANCE TECHNICIAN

Module Seven
Electrical Wiring
(58104-11)

Module Six
Electrical Test Equipment
(26112-11)

Module Five
Electrical Theory
(26104-11)

Module Four
Introduction to Electrical Circuits
(26103-11)

Module Three
Climbing Wind Towers
(58103-11)

Module Two
**Introduction to
Wind Turbine Safety**
(58102-11)

Module One
Introduction to Wind Energy
(58101-11)

Power Industry Fundamentals

This course map shows all of the modules in *Wind Turbine Maintenance Technician Level One, Volume One.* The suggested training order begins at the bottom and proceeds up. Skill levels increase as you advance on the course map. The local Training Program Sponsor may adjust the training order.

58101-11
Introduction to Wind Energy

Module One

Trainees with successful module completions may be eligible for credentialing through NCCER's National Registry. To learn more, go to **www.nccer.org** or contact us at **1.888.622.3720**. Our website has information on the latest product releases and training, as well as online versions of our *Cornerstone* newsletter and Pearson's Contren® product catalog.

Your feedback is welcome. You may email your comments to **curriculum@nccer.org,** send general comments and inquiries to **info@nccer.org**, or use the User Update form at the back of this module.

INTRODUCTION TO WIND ENERGY

Objectives

When you have completed this module, you will be able to do the following:

1. Evaluate the advantages and disadvantages of wind power technology.
2. Identify the important events, people, and organizations in the history of wind power to date.
3. Describe wind energy concepts and how the energy is captured.
4. Identify the basic functions and classifications of wind turbines.
5. Identify major horizontal-axis wind turbine (HAWT) components and their function.
6. Describe the wind farm environment and characteristics of the wind energy maintenance technician.

Performance Tasks

This is a knowledge-based module; there are no performance tasks.

Trade Terms

Anemometer
Betz limit
Centrifugal release units
Dynamo
Effective ground level
Furling
Horizontal-axis wind turbine (HAWT)
Kinetic energy
Longitudinal
Low-voltage distribution module (LVDM)
Nacelle
Pitch control

Power density
Supervisory control and data acquisition (SCADA) system
Swept area
Tip brakes
Vertical-axis wind turbine (VAWT)
Wind rose
Wind shear
Yaw control
Yaw deck
Yaw pucks

Industry Recognized Credentials

If you're training through an NCCER-accredited sponsor you may be eligible for credentials from NCCER's Registry. The module ID number for this module is 58101-11. Note that this module may have been used in other NCCER curricula and may apply to other level completions. Contact NCCER's Registry at 888.622.3720 or go to nccer.org for more information.

Contents

Topics to be presented in this module include:

1.0.0 Introduction 1
2.0.0 The History of Wind Power 1
3.0.0 The Wind Industry Today 3
4.0.0 A Study in Wind Energy 5
  4.1.0 The Power of the Wind 5
  4.2.0 More About Wind Speed 6
  4.3.0 Wind Speed and Height 6
  4.4.0 Wind Data Acquisition and Use 8
    4.4.1 Met Towers 8
    4.4.2 Wind Maps and Charts 8
5.0.0 Intercepting Wind Energy 8
  5.1.0 The Betz Limit 11
6.0.0 Wind Turbines 12
  6.1.0 HAWTs and VAWTs 12
  6.2.0 Blade Count 14
  6.3.0 Blade Size and Construction 14
  6.4.0 HAWT Yaw and Pitch 15
  6.5.0 Supervisory Control and Data Acquisition (SCADA) 16
7.0.0 HAWT Systems 16
  7.1.0 Wind Turbine Towers 17
    7.1.1 Types of Towers 17
    7.1.2 Offshore Installations 19
  7.2.0 Nacelles 19
  7.3.0 Electric Power Components 19
    7.3.1 Generator 19
    7.3.2 Pad-Mounted Transformer 21
    7.3.3 Converter 21
    7.3.4 Low-Voltage Distribution Module (LVDM) 22
  7.4.0 Drive System Components 22
    7.4.1 Gearboxes 22
    7.4.2 Braking and Hydraulic Systems 23
8.0.0 The Wind Farm 24
  8.1.0 Wind Farm Maintenance 25
9.0.0 The Wind Energy Technician 26
10.0.0 Standardized Training by NCCER 28
Appendix NCCER Credentials 36-40

Figures and Tables

Figure 1   Wind turbine 1
Figure 2   Tower mill 2
Figure 3   Charles Brush and his wind-powered electrical turbine 2
Figure 4   Wind generation 5
Figure 5   Air volume calculation illustrated 6
Figure 6   Effect on wind from ground obstructions 7
Figure 7   Average annual wind resource estimated availability for the
           contiguous U.S. 9
Figure 8   Wind rose for O'Hare Airport 11
Figure 9   Measurements that apply to circles 11
Figure 10 H-pattern VAWT 12
Figure 11 Phi, or eggbeater style VAWT 13
Figure 12 Typical HAWT 13
Figure 13 Upwind and downwind rotors 13
Figure 14 Small HAWT with tail vane 13
Figure 15 Multi-blade turbine rotor 14
Figure 16 The Enercon E126 turbine 14
Figure 17 Carbon-fiber and fiberglass turbine blades 15
Figure 18 Yaw control 15
Figure 19 Turbine blade pitch 16
Figure 20 Basic turbine system components 17
Figure 21 Guyed tower 18
Figure 22 Lattice tower 18
Figure 23 Tubular tower 18
Figure 24 Tubular tower assembly 19
Figure 25 Turbine access through the tower 19
Figure 26 Offshore tower installation 20
Figure 27 Vestas V90 nacelle 20
Figure 28 Wind turbine generator 20
Figure 29 Power converter 21
Figure 30 Wind turbine gearbox 23
Figure 31 Tip brakes deployed 24
Figure 32 Big Horn wind farm 24
Figure 33 Turbine power delivery to the utility grid 25
Figure 34 Wind turbine maintenance checklist 27
Figure 35 Wind energy technician at work 28

Table 1   Wind Shear Exponents for Various Surfaces 7

1.0.0 INTRODUCTION

Wind power is a practical alternative to fossil-fuel power (*Figure 1*). Billions are invested in wind power each year, and the U.S. is on course to produce as much as 20 percent of its total electrical power from wind by 2030. Here are a few statistics that may surprise you:

- By the end of 2009, U.S. wind power had grown to more than 35,000 megawatts (MW). This is equal to about 100 coal-fired power plants. In 2009 alone, nearly 10,000 MW of capacity was added. A megawatt is equal to 1,000,000 watts, enough to power 300 average homes.
- From 2004 to 2008, the U.S. led the world in annual additions to wind power.
- Denmark, Spain, Portugal, and Ireland all generate more than 10 percent of their total electrical power from wind turbines, with Denmark leading the way at 20 percent.
- At the end of 2009, wind power in the U.S. accounted for just over 2.5 percent of total electrical power. Iowa generates nearly 20 percent of its in-state power from wind.

Wind represents one of the best opportunities to generate power from a renewable resource. In addition to the power generated, the wind industry provides employment and economic growth. Other advantages of wind power include the following:

- There is enormous capacity available on a worldwide scale.
- Wind is a clean source of energy with no significant emissions.
- Wind energy provides an alternative to fossil fuels.

58101-11_F01.EPS

Figure 1 Wind turbine.

- Wind is free and with modern technology, it can be captured efficiently.
- Although utility-scale wind turbines can be very tall, each takes up only a small plot of land. This allows the land below to be used for agriculture or other purposes.
- Many people find that wind farms add an interesting and unique feature to the landscape.
- Remote or undeveloped areas that are not connected to the power grid can use wind turbines to produce power.

To be fair, wind also has some disadvantages. Even proponents of the technology often qualify their support by adding, "as long as it's not in my backyard!" These words have become a part of modern culture, often referred to by the acronym "NIMBY." Common disadvantages include the following:

- Wind farms do not appeal to everyone. Many people prefer the landscape to retain its natural form.
- The whirring noise from the rotor can be disturbing. Some minor noise can sometimes be heard from other components as well, and worn **yaw pucks** (much like brake pads) can emit sounds as the equipment housing, or **nacelle**, pivots on the tower.
- Shadow flickering and the reflection of light from the blades as they rotate can be a nuisance.
- Ice formation on blades can be dangerous when it breaks loose.
- Birds may be injured or killed, although the evidence to date suggests that this threat is small.
- The wind is not consistent in direction or intensity, varying from zero to storm force. As a result, there are times when turbines produce no electricity.

Every aspect of wind power will be debated for years to come. However, the commitment to its use on a global scale is very real and well underway. This module introduces the wind industry as a whole, from its history to the mechanical and electrical components that make it possible.

2.0.0 THE HISTORY OF WIND POWER

Just as early humans discovered uses for the sun, the wind proved to be a valuable resource as well. Wind power was used to propel boats as early as 5000 BC. More than 2,000 years ago, the Chinese used wind power to drive water pumps. During the same period in other areas of the world, wind power was harnessed to grind grain.

History suggests that in Western Europe, the water wheels used to grind grain inspired early windmills. Post mills in Europe seemed more advanced than earlier wind-powered systems. Hand-made wooden gears were used to transmit power from the horizontal shaft to a vertical shaft. The mill structure itself was balanced on a center post; this allowed the entire assembly to be manually rotated into the wind by the miller, or by using animal power for larger structures.

Tower mills (*Figure 2*) used the post mill concept in a more practical way. The lower portion of a tower mill remained fixed in position, often providing housing for the operator. The upper portion functioned as a post mill, with the ability to rotate into the wind as required. These structures were built primarily of brick or stone instead of wood.

Over the years, wind power was used in many applications, including sawing, spice and cocoa processing, and the mixing of paints and dyes. Industrialization and the development of steam power slowed the use of wind power for quite some time. There were obvious advantages to power production on demand, rather than power production at the mercy of the wind and weather. However, advancements in wind power continued. This included new blade designs and increased efficiency.

Although an advancing industrial society may have stalled the growth of wind power, it created a need for more electrical power. The first large windmill for power generation was built in 1888 by Charles Brush (*Figure 3*). Brush founded the Brush Electric Company, which later became the General Electric Company.

The Brush turbine used 144 blades, each with a diameter of over 17 meters (56 feet). The tail was more than 18 meters (60 feet) long and about 6 meters (20 feet) wide. At its best, the turbine was able to spin a **dynamo** at about 500 revolutions per minute (rpm) through a 50:1 step-up gearbox and produced 12 kW of electrical power. Rather than using the power as it was generated, it was sent to a bank of 408 batteries instead. The loads included 350 lights and three electric motors. This was quite a feat for a first attempt at power generation through wind power using late 19th century technology.

58101-11_F02.EPS

Figure 2 Tower mill.

58101-11_F03.EPS

Figure 3 Charles Brush and his wind-powered electrical turbine.

In 1891, a Danish inventor by the name of Poul la Cour developed the first power-generating turbine to use four-bladed rotors with an old-fashioned airfoil shape. This design continued to advance through World War I, resulting in a number of such systems across Denmark. These systems produced up to about 25 kW of power. Industrialization again began to take its toll though, as fossil fuel plants producing large volumes of power sprouted across the land.

Over the next 40 years, the U.S. electrical grid continued to expand. However, those who remained beyond the grid's reach often used wind turbines to generate local power. Two American wind pioneers, Marcellus and Joe Jacobs, installed a wind turbine at their Montana ranch in 1922. Soon they were organizing other installations at the request of neighbors. The Jacobs Wind Electric Company continues in operation today as the oldest active wind turbine company in the U.S. The company now provides turbines primarily for rural electric cooperative grids.

The state of Vermont claims ownership of the world's first megawatt-generating wind turbine connected to the power grid. The Smith-Putnam 1.25 MW unit was built in 1941. It used a twin-blade, 53-meter (175-foot) rotor with speed control to maintain 28 rpm.

In addition to the growing demands for electrical power, the use of wind power through the 1930s and 1940s was also impacted by the Great Depression. To stimulate rural economies, the electrical grid was expanded through government investment.

Through the 1950s, power generation from wind continued to shrink in the U.S., but grew in Europe and other areas where wind and economic conditions favored its use.

German work in the field of wind power in the 1960s was led by Ulrich Hutter, who advanced lightweight plastic and fiberglass blade designs. He focused on using lighter components to reduce the loads, while the Danes focused on building structures that could withstand the loads.

The first major oil crisis in 1973 led to a renewed interest in wind power. Through the 1980s, attention turned from smaller (25 kW) turbines to commercial wind turbines that could produce 50 kW or more of power and were grid-connected. Between 1981 and 1990, 17,000 wind turbines providing 20 to 350 kW each were built in California mountain passes where consistent winds and government regulations favored their construction. Each national or world event that affects fossil fuel prices sparks renewed interest in renewable energy.

By the end of 2002, U.S. wind turbines provided roughly 4,600 MW of power, with California and Texas having the greatest number of wind farms. One year later, U.S. capacity had grown to 6,300 MW.

However, Europe's attention to wind remained more consistent, leading to the region's ownership of 70 percent of the world's wind power in 2003. The Germans led Europe with 14,000 MW of wind power, supplying 3.5 percent of the country's power at that time. During the same period, the Danes held the largest proportion of wind power, generating 20 percent of the country's power and providing nearly 40 percent of the turbines used worldwide.

3.0.0 THE WIND INDUSTRY TODAY

Today's wind industry continues to expand across the globe, both in terms of technical advancement and construction. As mentioned earlier, interest in all sources of renewable energy grows as fossil fuel issues continue to emerge. Recent events have kept those issues in the headlines. From the threat of global warming to the fouling of waterways with crude oil, the attack on fossil fuels has never been more consistent. Thus, the interest in developing wind power has followed suit.

The U.S. Department of Energy has set a goal of providing 20 percent of electrical needs via wind power by 2030. As of 2010, Texas led the way in total capacity. Adding the capacity of projects underway to the existing capacity will allow the state to exceed 10,000 MW in wind power. With one megawatt of power being enough to supply about 300 homes, enough capacity will be avail-

<table>
<tr><td>On Site</td></tr>
</table>

The Work of a Previous Wind Energy Student

Credit for designing the modern wind turbine belongs to a Danish engineer by the name of Johannes Juul. An early student of Poul la Cour's Wind Electrician Training Program, Juul designed and built the Gedser turbine in 1957. It used three blades on an upwind rotor and generated 200 kW of power.

able to power 3,000,000 Texas households. The southeast U.S. lags behind other regions due to poor wind conditions. Although this will likely continue for the near future, technology continues to improve performance. Systems that can extract more power from less wind may open up new areas to wind power.

Offshore projects represent huge opportunities to capitalize on wind energy. Benefits include reduced wind turbulence, as well as higher and more consistent wind speeds. In addition, the ocean can accommodate turbines of the largest size. Offshore installations can be placed where they are not seen from the shore. The first offshore wind farm serving the U.S. is located 16 miles off the coast of Nantucket, MA. The Cape Wind Project will incorporate 130 turbines and produce nearly 500 MW of power. A Federal lease of 28 years was signed in October 2010. Although the construction costs are much higher offshore, other advantages make them good investments.

The wind industry's impact is not limited to energy production. As the construction of wind farms continues to grow, so do the job prospects and the need for trained wind technicians.

Many organizations and government agencies are involved in wind energy. They will be referred to throughout this training program. Some of these organizations include the following:

- *U.S. Department of Energy (DOE)* – The DOE (www.energy.org) has a broad mission, which includes energy security, nuclear security, scientific discovery and innovation, environmental responsibility, and effective energy management. The DOE's Wind and Water Program supports wind energy through partnerships to advance wind energy technology. The program also provides for collaboration with the electric power industry as a whole to incorporate wind-provided power into the national grid without sacrificing reliability. As mentioned earlier, the DOE has set a goal of 20 percent national power generation through wind technology by 2030. A report released in September 2010 examines the feasibility and benefits of large scale, offshore wind power generation. The annual *Wind Technologies Market Report* can be found on the DOE web site and it is a good resource for anyone in the industry.

  The DOE's efforts benefit the entire U.S. renewable energy industry. The DOE also provides support to the American Wind Energy Association (AWEA), discussed in more detail later in this section. The AWEA's primary goal is to ensure that the U.S. has an active role and a strong voice in the creation of international consensus standards, with a large part of the DOE's support provided through the National Renewable Energy Laboratory.

- *The National Renewable Energy Laboratory (NREL)* – The NREL (www.nrel.gov) conducts research in wind energy systems at its National Wind Technology Center (NWTC) located south of Boulder, CO. A variety of testing is conducted there and at various satellite locations. Newer satellite facilities will enable testing of prop blades up to 100 meters in length. Testing and research conducted at the NWTC has contributed to the success of many large-scale wind farms. Funding in part comes from the DOE. The mission of the research conducted at NWTC is to collaborate with private industry to advance wind technology and to accelerate its use. NWTC's research has contributed to many wind industry success stories and the development of successful utility-scale wind power plants.

- *American Wind Energy Association (AWEA)* – The mission statement for AWEA (http://awea.org) is to promote wind power growth through advocacy, communication, and education. AWEA develops and publishes standards for the U.S. wind industry, following their review and adoption by the American National Standards Institute (ANSI). ANSI recognizes the organization as an Accredited Standards Developer for the U.S. wind industry. AWEA also participates in standards development at the worldwide level. To this end, they cooperate with an array of other standards-producing organizations such as the American Society of Mechanical Engineers (ASME), the American Society for Testing Materials (ASTM) International, and the Institute of Electrical and Electronic Engineers (IEEE). Internationally, the organization is an active participant in the International Energy Agency (IEA), the International Electrotechnical Committee (IEC), and the International Standards Organization (ISO). These groups are responsible for the vast majority of accepted standards in many U.S. industries.

- *World Wind Energy Association (WWEA)* – WWEA (http://wwindea.org) is an international non-profit organization that incorporates the efforts of over 90 member countries to promote the use of wind energy across the globe. Their guiding principle is to replace *all* fossil fuel and nuclear energy. Although a lofty goal, it is a desirable one. One of the most important benefits of the group is the worldwide sharing of technology and research data.

There are many other organizations making contributions to the wind industry in a variety of ways. However, the guidelines, standards, and research represented by the organizations listed here will likely have the greatest impact on the wind turbine maintenance technician, as well as the future of the industry as a whole.

4.0.0 A STUDY IN WIND ENERGY

Before discussing wind energy and how to harness it, you must first understand where it comes from. Wind is created when the sun heats the planet at various rates and locations. As the air above is heated, it rises and creates a negative pressure beneath it (*Figure 4*). Cooler air rushes in at a low level to replace the air that has risen. This movement of air masses is the wind, at varying rates of flow and direction.

Once a mass of air is in motion, it carries with it the energy that created that motion. This is known as kinetic energy. A car moving at any given speed still contains the kinetic energy required to attain that speed, even if the engine is turned off. Wind turbines convert the kinetic energy of the wind into both mechanical and electrical energy. The mechanical energy can be used for pumping and grinding operations. This module focuses on the use of wind to generate electrical power.

4.1.0 The Power of the Wind

The amount of energy in the wind is a mathematical relationship between its speed (velocity) and mass. However, there is much to learn about both factors to understand the result of the calculations.

The wind speed changes constantly, sometimes moment by moment. Wind speed is affected by many factors, including time of day (or night),

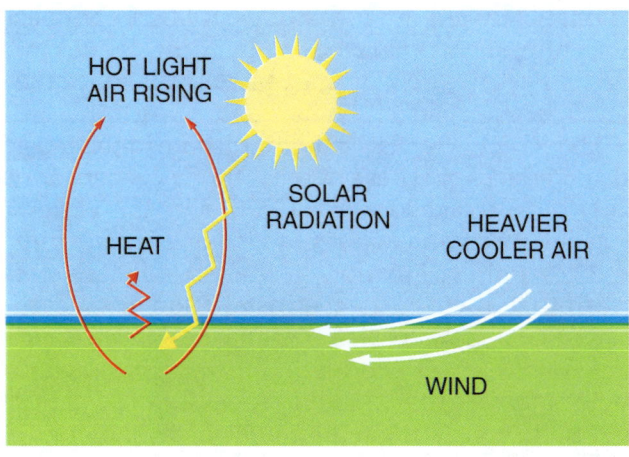

HOT LIGHT AIR RISING

SOLAR RADIATION

HEAT

HEAVIER COOLER AIR

WIND

58101-11_F04.EPS

Figure 4 Wind generation.

since the sun is the primary generator of heat-induced winds. Further, since the sun changes its position in the sky to some degree on a daily basis, its impact on a given point on the earth's surface changes accordingly. Seasonal storm patterns also affect wind speed and its consistency.

Although wind speed is intermittent by nature, some means of documenting it at a given location must be identified. Using the average wind speed over the course of a year is an accepted standard. Keep in mind that this value may change from year to year. Wind speed is important since the greater the wind speed, the more energy it contains.

The second factor in determining wind energy is its mass. At any given speed, there is more kinetic energy stored in a heavy object than in a lighter one. Although air is very light relative to most other things, it does have a measurable mass. A cubic meter of water, due to its mass, can store more energy than a cubic meter of air, but the wind has volume and velocity on its side to help make up for its low mass.

The mass (m) of air is defined as the product of the air density (represented by ρ, the Greek symbol rho) and the volume. To calculate the volume of air, multiply the wind velocity (V) by the area (A) it is passing through for a given period of time (t). See *Figure 5*. The area will be equal to the size of the circle created by the turbine rotor as it turns. This concept will be explored in more detail later in this module.

Mathematically, the complete mass equation would be:

$$\text{Mass} = \text{density} \times (\text{area} \times \text{velocity} \times \text{time})$$

or

$$m = \rho AVt$$

Once the mass of air is known, the wind energy can be found using the following equation:

$$\text{Wind energy} = \tfrac{1}{2} \times \text{mass} \times \text{velocity}^2$$

or

$$\text{Wind energy} = \tfrac{1}{2} mV^2$$

The mass calculation can be combined with the energy calculation so the math can be completed in one equation:

$$\text{Wind energy} = \tfrac{1}{2} \times \text{density} \times (\text{area} \times \text{velocity} \times \text{time}) \times \text{velocity}^2$$

or

$$\text{Wind energy} = \tfrac{1}{2}\rho AtV^3$$

The units of measure for wind energy will be discussed in a later section. They are not significant in this discussion.

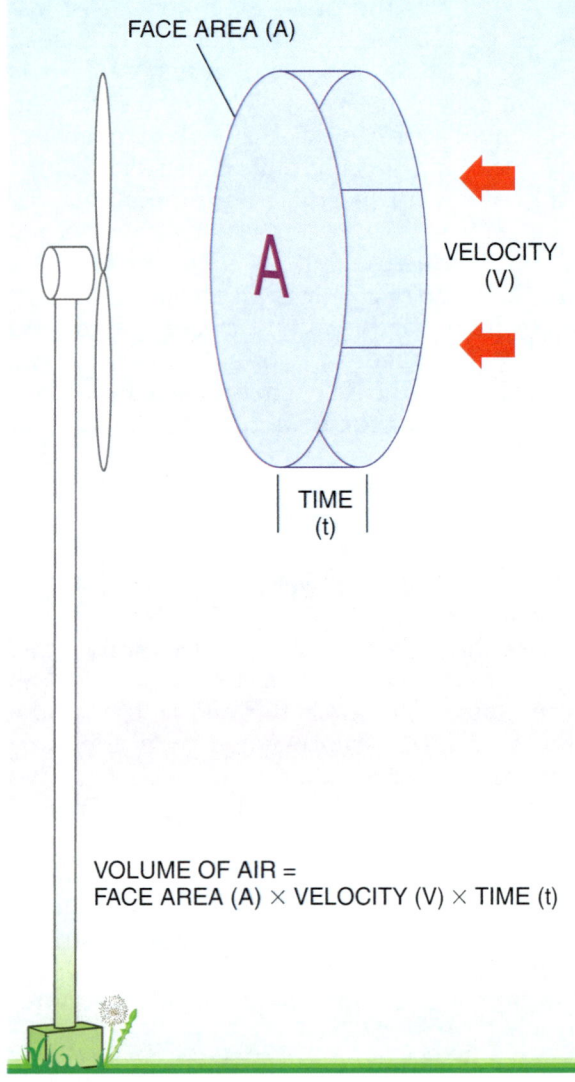

FACE AREA (A)

VELOCITY
(V)

TIME
(t)

VOLUME OF AIR =
FACE AREA (A) × VELOCITY (V) × TIME (t)

58101-11_F05.EPS

Figure 5 Air volume calculation illustrated.

- Consider a speed of 4.5 meters/sec (10+ mph). An increase in speed of 10 percent results in a 33 percent increase in available power.
- Increasing speed by 20 percent increases the available power by 73 percent.
- Doubling the speed increases the available power by 800 percent.

These effects can be calculated using the wind energy equation ($\frac{1}{2}\rho A t V^3$). These are important relationships to remember. At this point, the math is not as important as grasping the concepts.

Calculations for the effect of wind speed on available power must use either a single speed at any given moment or an average speed over a given period. Annual averages are often used. Note that an average is based on many values that are either above or below the average, and this can have a major impact on the result. To predict available power more accurately, engineers and designers often make many individual calculations and add them together. For example, power may be calculated based on the number of hours per year for one speed, then on the number of hours for another speed, and so on. This method provides a more accurate value of the available power.

Remember that increasing either the wind speed or its mass increases the available energy. However, increases in wind speed create a greater energy increase than an increase in mass, due to the cubic relationship of velocity (V^3) in the equation.

As a wind energy technician, you will not often concern yourself with these calculations. Yet there is no reason to be intimidated. Remember the concepts and the equation will make sense. Wind energy can be quantified by relating its speed and mass, the area it is passing through, and the amount of time it passes through that area.

4.2.0 More About Wind Speed

For any size of rotor, the wind speed has a major impact on the power produced. Examine the following statements about the impact of wind speed on the available power:

4.3.0 Wind Speed and Height

The wind felt at ground level may seem intense at times, but it pales in comparison to the wind speed (and power available) at higher elevations. The wind close to the Earth slows down due to the friction caused by obstructions such as trees, buildings, and hills (*Figure 6*). Some obstructions have a greater impact than others do. For example, wind blowing across a lake is subject to less friction than wind blowing through a forest.

Generally, the wind speed is greater at higher levels due to the lack of obstructions. Over mountains or hills, wind speed tends to increase more dramatically at higher altitudes. Radical changes in terrain, such as isolated tall buildings or mountain cliffs, can affect air movement well above the height of the obstruction itself.

The object is to place a wind turbine in the most ideal air stream. This not only applies geographically, but also at a local level. Although it is widely accepted that Mt. Washington in New Hampshire is the windiest location in the U.S., it does not mean that a wind turbine installation anywhere in that area is a good idea. At the local level, wind speed and power production may be best at 100 meters above ground in one spot, but even better and more consistent at another spot at 80 meters.

The concept of wind speed varying at altitude due to friction is known as **wind shear**. (Do not

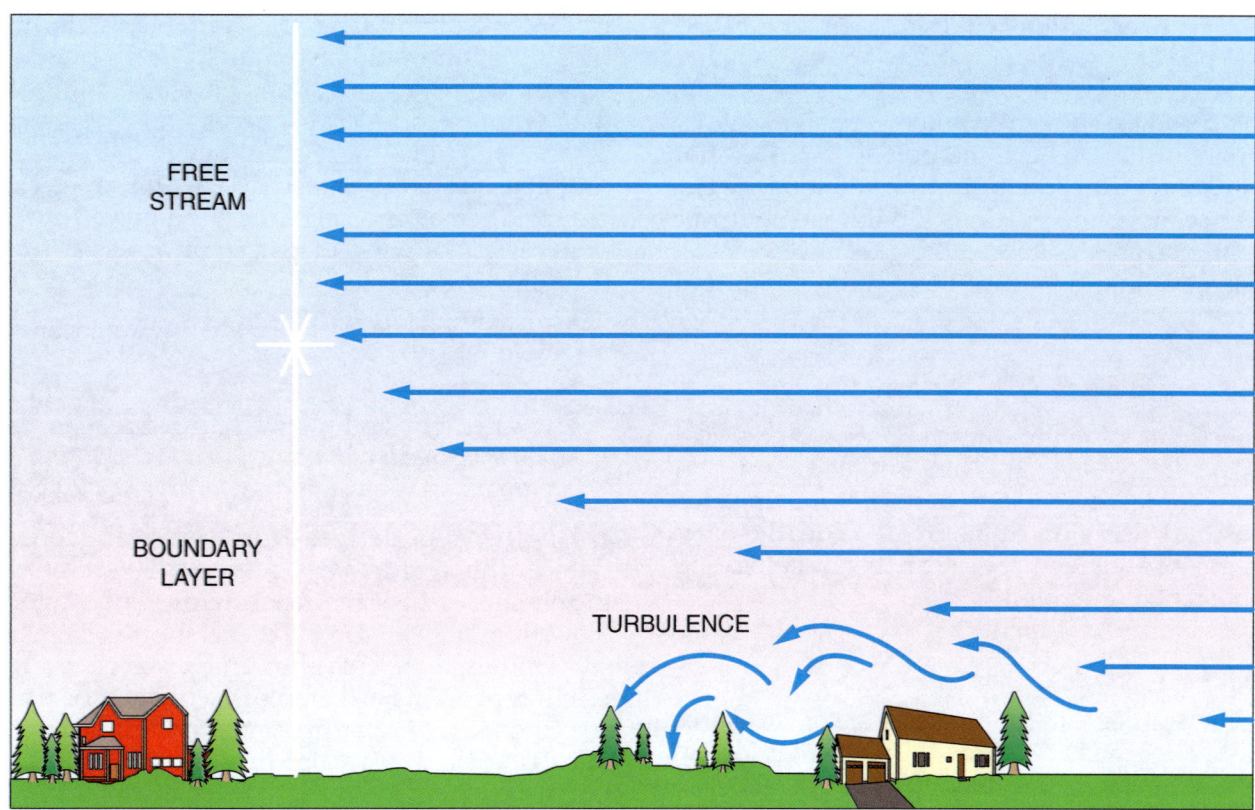

Figure 6 Effect on wind from ground obstructions.

confuse this with wind shear as it applies to aviation.) The basic theory of wind shear is true primarily for heights above the **effective ground level**. As the wind blows across a smooth lake, the effective ground level would be the surface of the water; wind shear decreases above this level and the wind moves faster. However, the effective ground level of wind blowing over a jungle would be the tops of the tree canopy, not the ground itself.

Wind velocities at a given altitude can be calculated using a value known as the wind shear exponent. These values are related to surface roughness. As the surface roughness increases, the wind shear exponent also increases. *Table 1* provides a list of different terrains, along with their surface roughness factors and wind shear exponents.

When the wind shear exponent is not identified, the value for a mown grass area is often used. It has an exponent of 0.14. Since 0.14 represents roughly

Table 1 Wind Shear Exponents for Various Surfaces

Type of Terrain	Surface Roughness Length (in meters)	Wind Shear Exponent
Ice	0.00001	0.07
Snow on flat terrain/calm water	0.0001	0.09
Snow-covered low crops	0.002	0.12
Short or mown grass	0.0007	0.14
Crops or prairies with taller grasses	0.05	0.19
Scattered tall shrubs and trees	0.15	0.24
Scattered tall shrubs and trees with a few buildings	0.3	0.29
Suburban areas with homes, etc.	0.4	0.31
Wooded areas/forests	1.0	0.43

one-seventh of one, the wind shear impact is called the ¹⁄₇ power law. The power law applies to many sites across the U.S., allowing engineers to estimate wind speed based on a common exponent.

For locations where the power law does not apply, other rules of thumb may be used. For instance, at some locations, doubling the tower height increases wind speed by about 10 percent. Remember that this is a rule of thumb and does not apply to every case.

4.4.0 Wind Data Acquisition and Use

While there are many ways to predict wind speed and available power, it is better to use real-world data recorded over a long period. Wind speed is measured using an anemometer. Ideally, measurements are taken at or near the exact location and height of the rotor.

4.4.1 Met Towers

The meteorological tower, or met tower, measures wind properties at a given location. Met towers in the 30- to 50-meter range can be set up at various locations in an area to assist in creating trend maps. Permanent met towers are sometimes built to accompany wind turbine farms as a backup source of information. Where changes in surface terrain are significant, more sites and data are likely needed to make accurate predictions. Along with speed, the wind direction and air temperature are usually recorded. Data can be stored on site through a microprocessor and memory card or transmitted via cellular or satellite technology to a central location. Not only is site-specific data needed to select the best site for the wind turbine, such data is also important to potential investors.

4.4.2 Wind Maps and Charts

In the absence of site-specific data, wind maps can be used to quickly evaluate wind conditions on a larger scale. The National Climatic Data Center (NCDC) headquartered in Asheville, NC contains weather data collected from a variety of federal resources, the most notable of these being the National Weather Service (NWS) and the Federal Aviation Administration (FAA). More than 30 years of data has been amassed for some locations in the U.S. It is highly useful data, but remember that wind properties can vary dramatically even at the local level.

Figure 7 is a map provided by the NREL of the average annual wind power available across the U.S. Many such maps are available, including those that show the estimated wind power by season. Other organizations have also compiled data from both federal sources and private enterprise to gain a more accurate picture of wind properties. This is especially true for areas that are regarded as good sites for wind power.

A variety of wind maps can be accessed from the following websites:

- *National Climatic Data Center* – www.ncdc.noaa.gov
- *National Renewable Energy Laboratory (NREL)* – www.nrel.gov and www.nrel.gov/rredc
- *U.S. Department of Energy, Wind Powering America Program* - www.windpoweringamerica.gov

Another tool derived from years of recorded data is the wind rose (*Figure 8*). This circular graph shows the frequency with which winds blow from a given direction at a specific location, reported as a percentage of time. Each concentric circle represents a different frequency of time, starting from zero at the center and progressing outward. Wind roses also display other information using color, such as how often the wind blows from each direction at a given speed. Information at the bottom helps the user to see a compilation of the data. Note that the wind rose shown in *Figure 8* for O'Hare Airport in Chicago provides the trends of wind direction and speed for a 30-year period, which is typical.

5.0.0 INTERCEPTING WIND ENERGY

The swept area is the area that the turbine rotor blades pass through. Greater energy is available to be intercepted over a larger swept area. Thus bigger is better, as long as the entire rotor remains active and is not somehow blocked.

A common horizontal-axis (propeller-style) turbine rotor sweeps a circle as it turns. Common measurements that apply to circles are shown in *Figure 9*.

The swept area of the circle is easily calculated as follows:

$$\text{Area} = \text{pi (3.1416)} \times \text{radius}^2$$

or

$$A = \pi r^2$$

Although this is a simple equation, it is easy to confuse radius with diameter since wind turbine discussions may use both terms.

Again, more area is better. As blade length is increased, the proportional increase in swept area

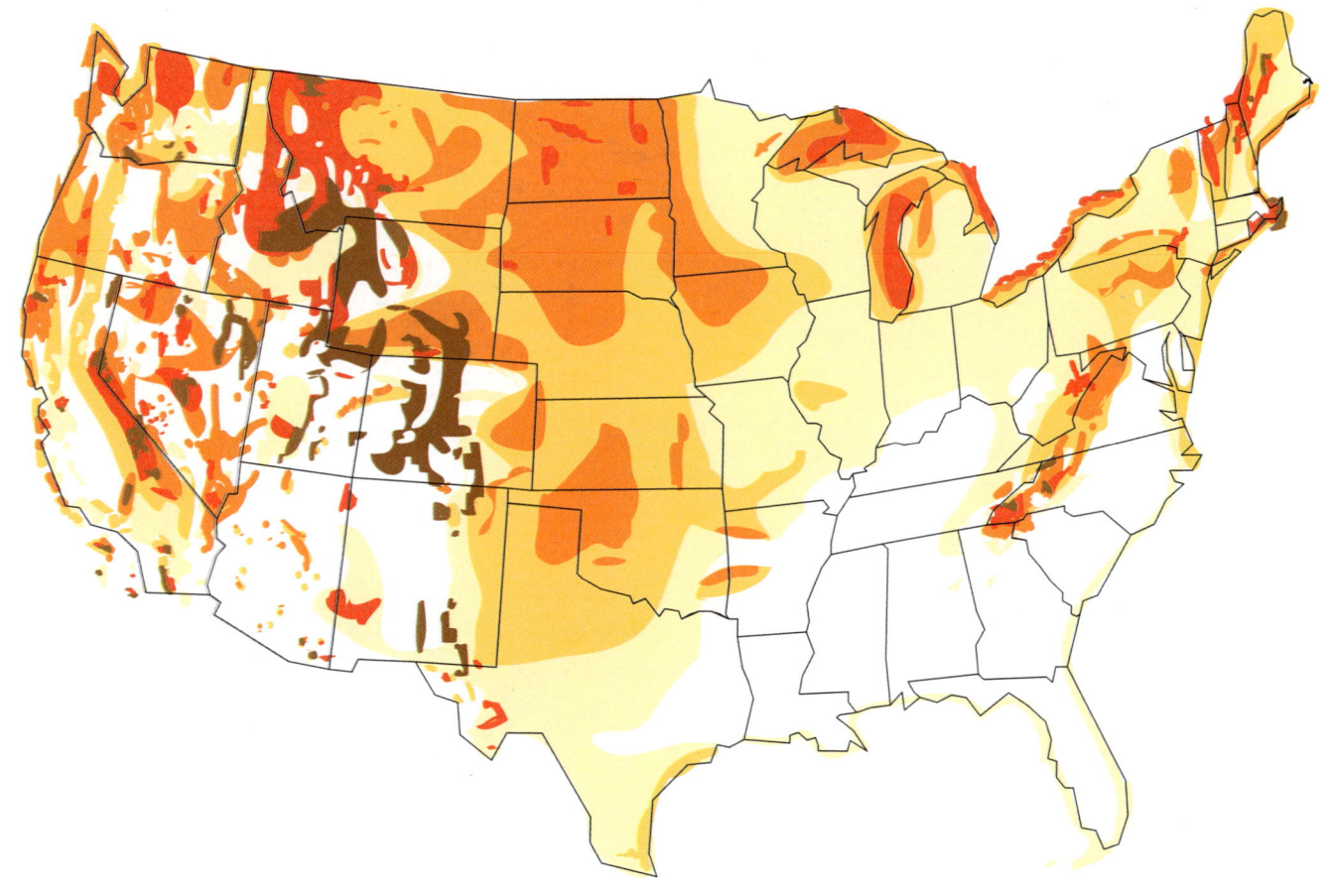

50m (164 ft)

POWER CLASS	WIND POWER (W/m²)	SPEED* (m/s)
1	<200	<5.6
2	200-300	5.6-6.4
3	300-400	6.4-7.0
4	400-500	7.0-7.5
5	500-600	7.5-8.0
6	600-800	8.0-8.5
7	>800	>8.5

*Equivalent wind speed at sea level for a Rayleigh distribution.

58101-11_F07.EPS

Figure 7 Average annual wind resource estimated availability for the contiguous U.S.

is dramatic. Doubling the length of the blade (effectively doubling the diameter) quadruples the swept area. Understanding the principle of swept area will help you to grasp the concept of how power is intercepted.

The previous section of text reviewed the concept of average wind speeds and the fact that averages often misrepresent the true characteristics of a site. For example, at any site, a given wind speed may occur 15 percent of the time, at another it occurs 6.2 percent of the time, and so on. As a result, it is often far more accurate to sum the amount of power available based on many calculations on a range of velocities than a single calculation using a time-based average.

The term power density is used to quantify the amount of power available. Power is measured in watts per square meter or w/m^2. In English units, it is expressed in watts per square foot or watts/ft^2. Power density is represented by P/A in the equation. The equation takes speed as well as air density (to quantify the air's mass) into consideration.

Assume that a site's wind has been monitored and the calculations lead to a power density of 60 w/m^2. To calculate the power available for an entire year (expressed as E/A), multiply the power density by the hours in a year (8,760), and then change the watts to kilowatt/hours:

$$E/A = P/A \times \text{hours of chosen period} \times (1 \text{ kW}/1,000 \text{ watts})$$
$$E/A = P/A \times 8,760 \text{ hours} \times (1/1,000)$$
$$E/A = 60 \times 8,760 \times (1/1,000)$$
$$E/A = 525,600 \div 1,000$$
$$E/A = 525.6 \text{ kWh/year/m}^2$$

On Site

Collecting Wind Data

Basic wind data collection requires only a few instruments. Cup anemometers are the most common. A wind directional sensor is mounted on the same tower on the left side. The propeller anemometer style is also used in some applications.

A wind data logger is needed to record the desired information over a period of time. The model shown here, which is also available with a solar power kit to eliminate utility connection or batteries, offers features such as:

- The ability to record wind speed, gusts, and wind direction from multiple anemometers simultaneously
- Logging of temperature
- A real-time clock
- Data logging interval programming of 10 seconds to 16.6 hours in one-second intervals of choice
- An RS-232 port for connection and data download to a computer
- The ability to accommodate many other sensor types, including relative humidity, rainfall, and light intensity

CUP ANEMOMETER

58101-11_SA01.EPS

DATA LOGGER

58101-11_SA03.EPS

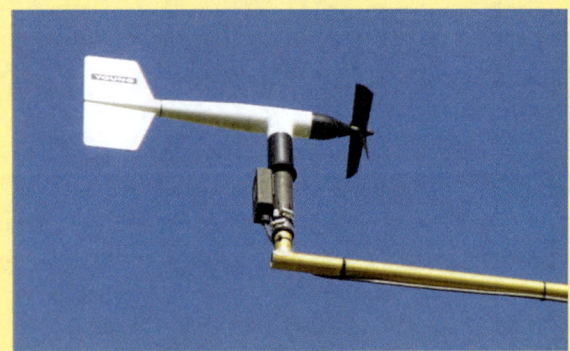

PROPELLER ANEMOMETER

58101-11_SA02.EPS

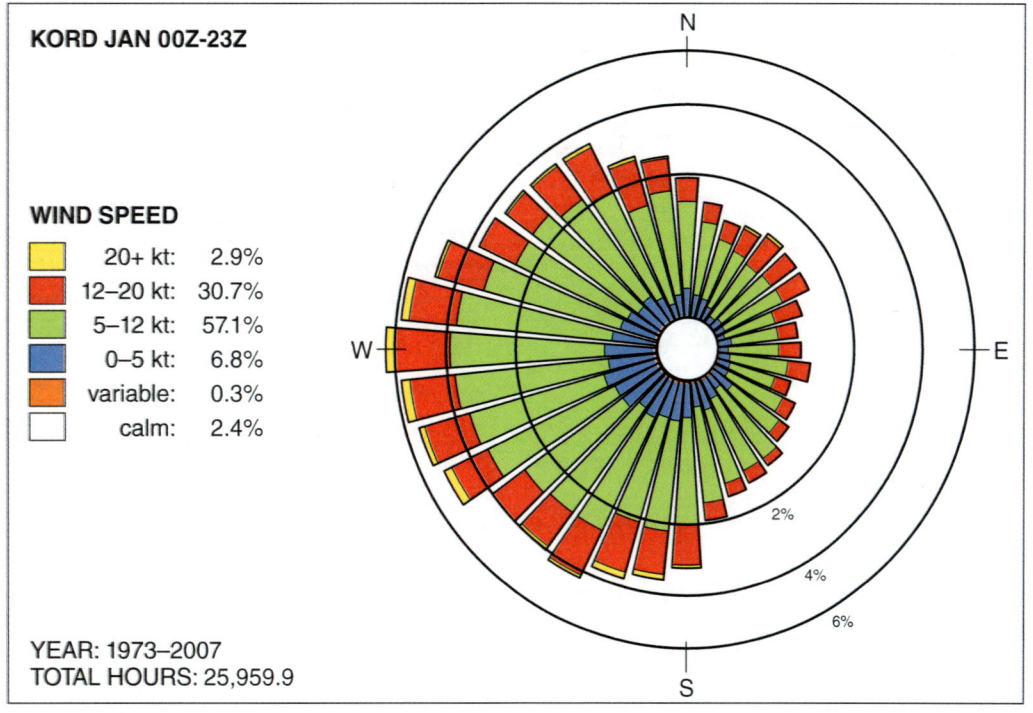

WIND SPEED

☐	20+ kt:	2.9%
☐	12–20 kt:	30.7%
☐	5–12 kt:	57.1%
☐	0–5 kt:	6.8%
☐	variable:	0.3%
☐	calm:	2.4%

KORD JAN 00Z-23Z

YEAR: 1973–2007
TOTAL HOURS: 25,959.9

58101-11_F08.EPS

Figure 8 Wind rose for O'Hare Airport.

Note that these values are the amount of power available per unit of area; in this case, square meters. This can now be related to the swept area of the rotor. For this site that provides 526 kWh/year/m², assume a rotor has a diameter of 40 meters (a radius of 20 meters). The swept area of a spinning propeller-style rotor is calculated the same as any circle:

$A = \pi r^2$
$A = 3.1416 \times 20 \text{ meters}^2$
$A = 3.1416 \times 400$
$A = 1,256.64 \text{ m}^2$

Since the example site provides 526 kWh/year/m², multiply this number by the area of the rotor to determine the amount of power the rotor will intercept over the course of a year:

Total E/A = 526 kWh/year/m² × 1,256.64 m²
Total E/A = 660,993 kWh/year intercepted by the 40-meter rotor at this set of conditions

5.1.0 The Betz Limit

In a perfect world, a wind turbine would transfer all of the available power through the rotor and make use of it. However, this is not possible. There are losses associated with the process. In order to extract all of the energy, the wind would have to come to a complete stop since it would be drained of all kinetic energy. This does not happen.

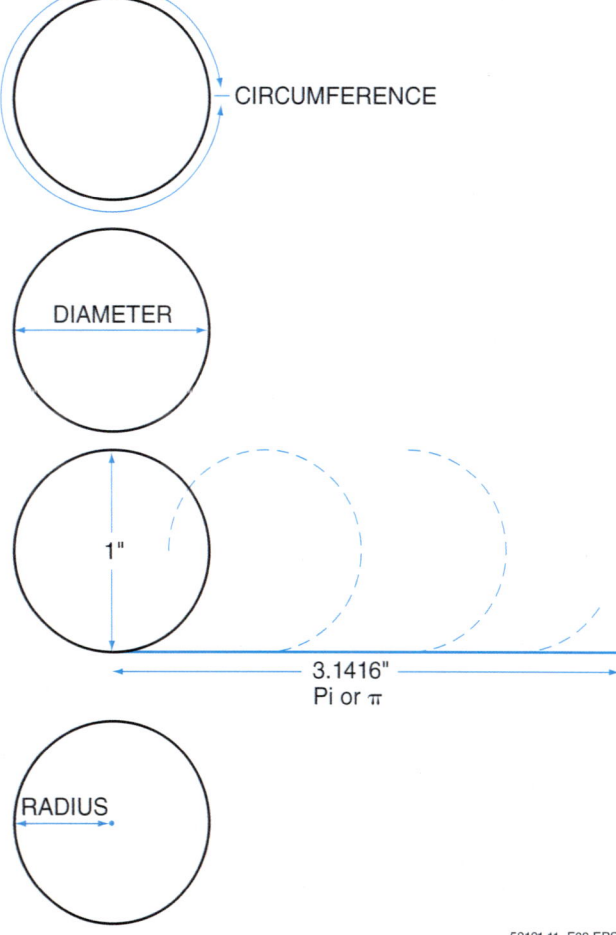

58101-11_F09.EPS

Figure 9 Measurements that apply to circles.

A German scientist named Albert Betz developed what is known as the Betz limit. This is the theoretical limit of how much power a rotor can capture. Per Betz's calculations, only 59.3 percent of the power is truly available. Again, this is the theoretical limit for all rotors, regardless of any claims that it has been exceeded. It should not be confused with how much is actually captured by every rotor. In reality, rotors rarely approach this level of efficiency. It is the target value for perfect efficiency from a rotor. Keep in mind though, that designing the perfect rotor may not necessarily be effective in the end. The perfect wind turbine provides maximum electrical power for the lowest cost in the real world. A turbine equipped with a perfect rotor may not accomplish that.

To find the Betz limit, multiply the total E/A by a value of 59.3 percent (0.593). Using the example rotor with a total E/A of 660,993 kWh/year, the Betz limit can be calculated as follows:

Betz limit = Total E/A × 0.593
Betz limit = 660,993 kWh/year × 0.593
           = 391,969 kWh/year

Large rotors can extract 40 percent or more of the wind's power. Smaller rotors capture far less than this—some as little as 12 percent. In general, smaller rotors are less efficient.

Even though 40 percent of the available power can be captured with a large rotor, not all of that is transformed into electrical power. There are still many other losses to consider, including bearing friction as well as gearbox and generator inefficiency. However, an earlier concept remains true—the rotor is the limiting factor when it comes to power capture from the wind. If the calculations show that the energy produced by a specific turbine is insufficient, there are three simple ways to increase it:

- Use a bigger rotor.
- Find a new site with better wind conditions.
- Use a turbine with higher conversion efficiencies, such as a better generator. (Note that this option does not typically net a substantial increase.)

6.0.0 WIND TURBINES

Wind turbines come in many shapes and sizes, from tiny propeller styles to wacky contraptions in the backyards of home inventors. Many different configurations are attempted. Some are successful, while others are not. What constitutes a successful design may be a matter of opinion. Many approaches have resulted in successful power generation, but not necessarily in sufficient volume or in a cost-effective manner. This section examines the basic designs and properties of common turbines.

6.1.0 HAWTs and VAWTs

Turbine designs typically fall into one of two broad categories: horizontal-axis wind turbines (HAWTs) or vertical-axis wind turbines (VAWTs).

VAWTs (*Figure 10*) have two advantages. They can accept wind power from any direction and the majority of the major elements, such as drive components and generators, can be placed at ground level for easier access and service. A number of shapes and designs have been attempted over the years, most of which were inspired in the 1920s by a French inventor named D. G. M. Darrieus. One of the more popular Darrieus-inspired designs is the phi, or eggbeater, configuration (*Figure 11*).

Due to structural and functional issues, many large VAWT designs have had limited success in commercial and utility-scale wind farms. However, variations of the H-pattern VAWT as well as

58101-11_F10.EPS

Figure 10 H-pattern VAWT.

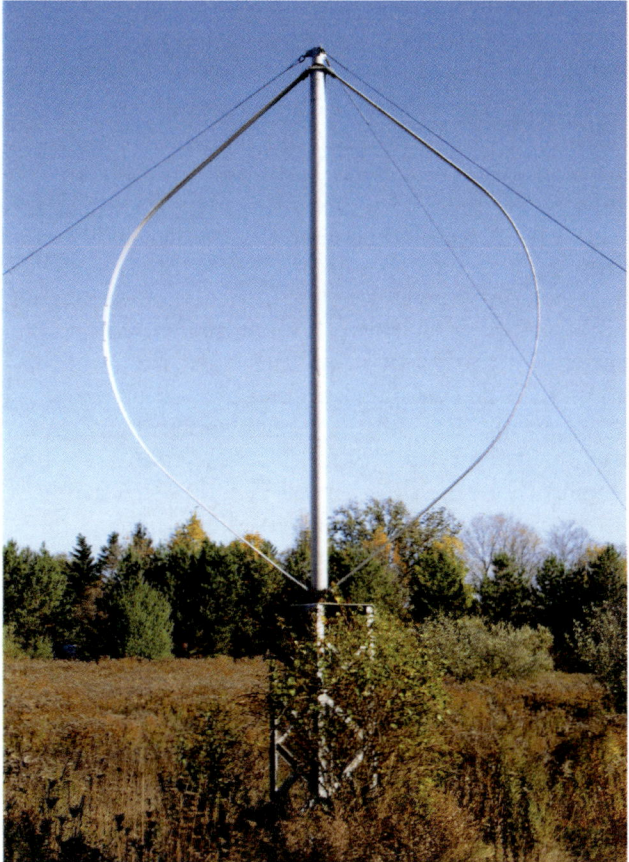

Figure 11 Phi, or eggbeater style VAWT.

Figure 12 Typical HAWT.

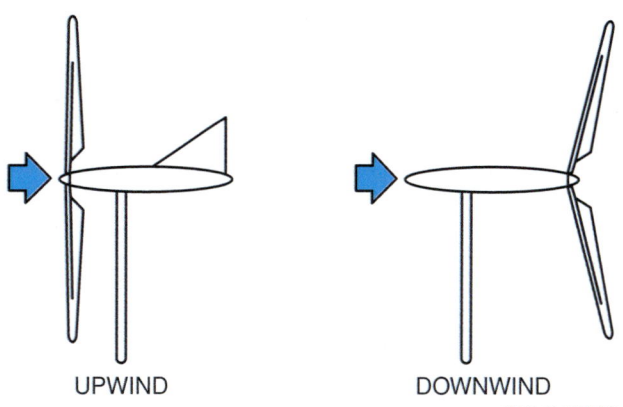

UPWIND   DOWNWIND

Figure 13 Upwind and downwind rotors.

other vertical designs used in the small turbine market have proven reliable and are relatively inexpensive. They can be used commercially when the environment allows for their installation in large numbers.

HAWTs, on the other hand, have had great success in all market sectors and are the primary focus of the wind industry today (*Figure 12*). They do require positioning of the rotor assembly into the wind and the major components are located high in the air. However, their reliability and performance are better than the VAWT in commercial and utility-scale applications.

HAWTs also vary somewhat in their design. Although most position the blade upwind (*Figure 13*), some smaller designs use a downwind rotor. Most downwind models use a very simple means of positioning the rotor into the wind, such as a tail vane. Others use a self-aligning blade design. Smaller upwind turbines also may use a tail vane (*Figure 14*). Larger, heavier turbines cannot be turned into the wind reliably by a simple tail vane and must be mechanically moved to the proper position.

Figure 14 Small HAWT with tail vane.

6.2.0 Blade Count

Another design option relates to the number of blades. Single-blade turbines do work and can be less expensive. However, efficient energy capture, strength problems, and aerodynamic issues hinder their effectiveness. Since a balancing counterweight is required that weighs as much as the single blade, some losses due to unproductive weight in motion occur. Two-blade and three-blade systems offer balance advantages, as does the multi-blade rotor (*Figure 15*). Most commercial turbines today use three-blade rotors that operate more effectively than other types.

6.3.0 Blade Size and Construction

HAWT blade sizes vary widely depending upon the application and desired output. Recall that the rotor diameter and thus the swept area of the rotor is the controlling factor in wind power capture. Typical modern commercial and utility-scale wind turbines have blade lengths of 20 to 45 meters (65 to 150 feet) and produce between 500 kW and 2 MW of power.

A number of small 50-watt turbines are available that provide minimal power for small applications such as boating and camping. Since power storage in batteries is often important to smaller turbine applications, they often produce DC power rather than AC. Turbines of this size are at the entry level of the growing small wind market. The AWEA defines the small wind market as turbines that produce less than or equal to 100 kW of power, and are designed for residential or light commercial use. HAWT's in this class have a blade length of 10 meters (32+ feet) or less. This segment continues to grow and the U.S. leads the world in production of small wind turbines.

New turbines continue to increase in size and power production as new technology encourages the development of improved materials and engineering. The world's largest capacity unit is the German-built Enercon E126 (*Figure 16*), with 24 units on line as of 2010. With a hub height of 135 meters (443 feet), a blade length of 63 meters (206+ feet), and a total height of 198 meters (650 feet), this unit can produce up to 7 MW of power. Research is now underway to produce and construct a 10-MW unit. Although turbines in this range may be initially built on land, they are generally used for offshore sites because of their size.

Many materials have been used in blade construction over the years and new materials are constantly being developed. Wood is still used on some small turbines, and wood composites have been used for blades up to roughly 43 meters (141 feet). Steel and aluminum have both been used. Steel is strong but heavy, while aluminum is light but subject to severe damage due to metal fatigue. Both have also been known to contribute to radio and TV signal interference. *Figure 17* shows carbon-fiber and fiberglass blades.

Figure 15 Multi-blade turbine rotor.

58101-11_F16.EPS

Figure 16 The Enercon E126 turbine.

Figure 17 Carbon-fiber and fiberglass turbine blades.

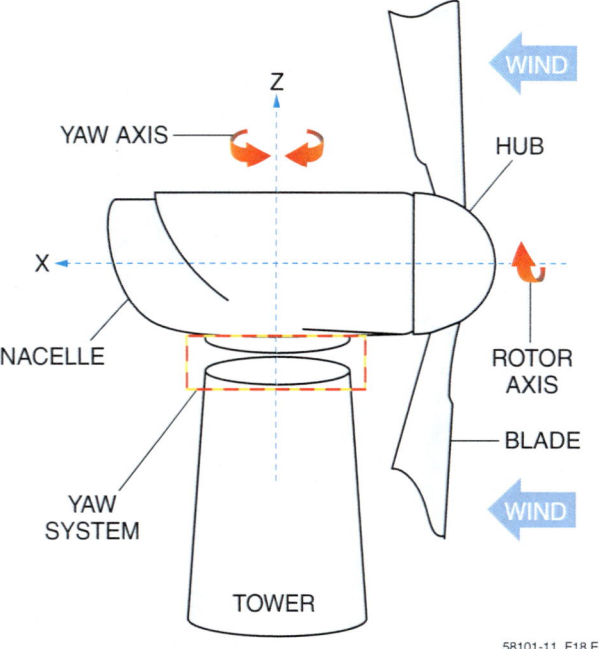

Figure 18 Yaw control.

6.4.0 HAWT Yaw and Pitch

As noted previously, HAWTs must be pointed into the prevailing wind for good performance. This means that the turbine assembly must be able to pivot on its vertical axis. This directional management is known as yaw control (*Figure 18*). Yaw control is only necessary for HAWTs. Smaller turbines may self-control their yaw angle fairly well using a simple tail vane. Larger units must be powered into position electrically or hydraulically. A small tail vane in the turbine assembly still plays a role by providing sensory input to the yaw control mechanism.

In most cases, the goal is to position the rotor perpendicular to the wind for maximum power production. However, there are occasions when the wind is simply too intense for the turbine system and the yaw control mechanism assists by furling the rotor, or positioning the rotor face away from a perpendicular wind strike. This slows the rotor speed and maintains the proper rpm for maximum power production without damaging the turbine. A fully furled rotor has

the face of the rotor parallel to the direction of the wind.

Other operating parameters may also be involved in the yaw control scheme. A minimum wind speed is often required before the yaw control system operates to avoid equipment wear and tear when there is little or no energy to gain. Yaw control schemes can also be limited by time to prevent too many movements, or by the size of the movement to avoid insignificant adjustments.

Pitch control provides another means of controlling the load and speed imposed on the rotor. The pitch of an object is generally related to its rotation around a horizontal axis, such as "pitching" the nose of an aircraft up or down, just as yaw relates to the vertical axis. For wind turbines, pitch is related to the rotation of the individual blades along their own longitudinal axis (*Figure 19*).

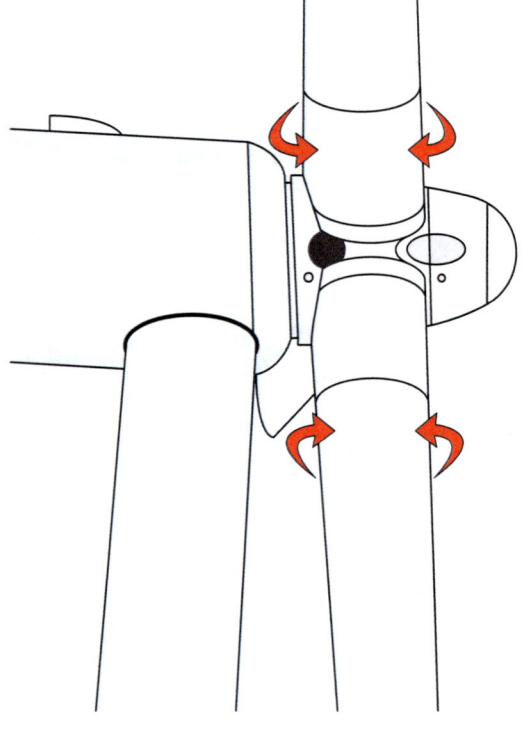

Figure 19 Turbine blade pitch.

58101-11_F19.EPS

As in yaw control, electric motors or hydraulic actuators are used to position the blades to the desired pitch. Pitch control works together with yaw control to improve performance of the turbine by optimizing the blade position to the wind. Controlling pitch also protects the blades and rotor from high winds by intentionally spoiling the airflow and inhibiting lift.

6.5.0 Supervisory Control and Data Acquisition (SCADA)

Wind turbines are often located in remote areas and must operate for long periods without on-site human intervention. Virtually every function of a turbine system must be monitored and controlled electronically, generally from a remote location for utility-scale turbines. In addition to controlling the complete turbine system to optimize power production, electronic control systems also monitor operation and help protect the turbine from damage.

Supervisory control and data acquisition (SCADA) systems are used to supervise, control, monitor, and collect historical data from an individual wind turbine or a group of turbines using real-time information and commands. A vast array of information is collected through sensor

inputs from the turbine and the site. Some of the inputs to the SCADA system include:

- Wind speed and direction
- Blade pitch
- Rotor rpm and yaw direction
- Oil and individual bearing temperatures
- Vibration levels
- Generator output

As the system uses this data input to control turbine functions in real time, vital information is collected and archived to provide trend reports. These reports assist in predictive maintenance and future programming. For example, small increases in the temperature of a bearing over time indicate a trend toward possible failure. The SCADA system can alert an operator thousands of miles away to the need for attention.

7.0.0 HAWT SYSTEMS

A complete turbine system consists of a variety of components and subsystems. Beyond the turbine assembly itself, which houses most of the mechanical and electrical components, the foundation, supporting tower, and SCADA system must also be considered part of a functioning turbine system.

Figure 20 shows the major components associated with the turbine. The components shown include the following:

- *Pitch* – The arrow indicates the rotation of the blade as the pitch is changed to accommodate the condition.
- *Low-speed shaft* – The shaft directly connected to the rotor itself, turning at the same rpm and transferring wind power to the gearbox.
- *Gearbox* – Houses a set of gears and accepts the power input from the low-speed shaft. Through gearing, the rpm is increased to a usable value for the generator to create power.
- *Generator* – Develops electrical power from mechanical energy input through the high-speed shaft.
- *Controller* – Takes in information from the anemometer, wind vane, and other sensors, and controls the pitch and yaw of the turbine. The internal controller may be interfaced with, or replaced by, a SCADA system.
- *Anemometer* – Measures wind speed.
- *Wind vane* – Monitors wind direction.
- *Nacelle* – Streamlined housing or enclosure that contains the major working components of the turbine system at the top of the tower.
- *High-speed shaft* – Transfers the higher rpm output of the gearbox to the generator.

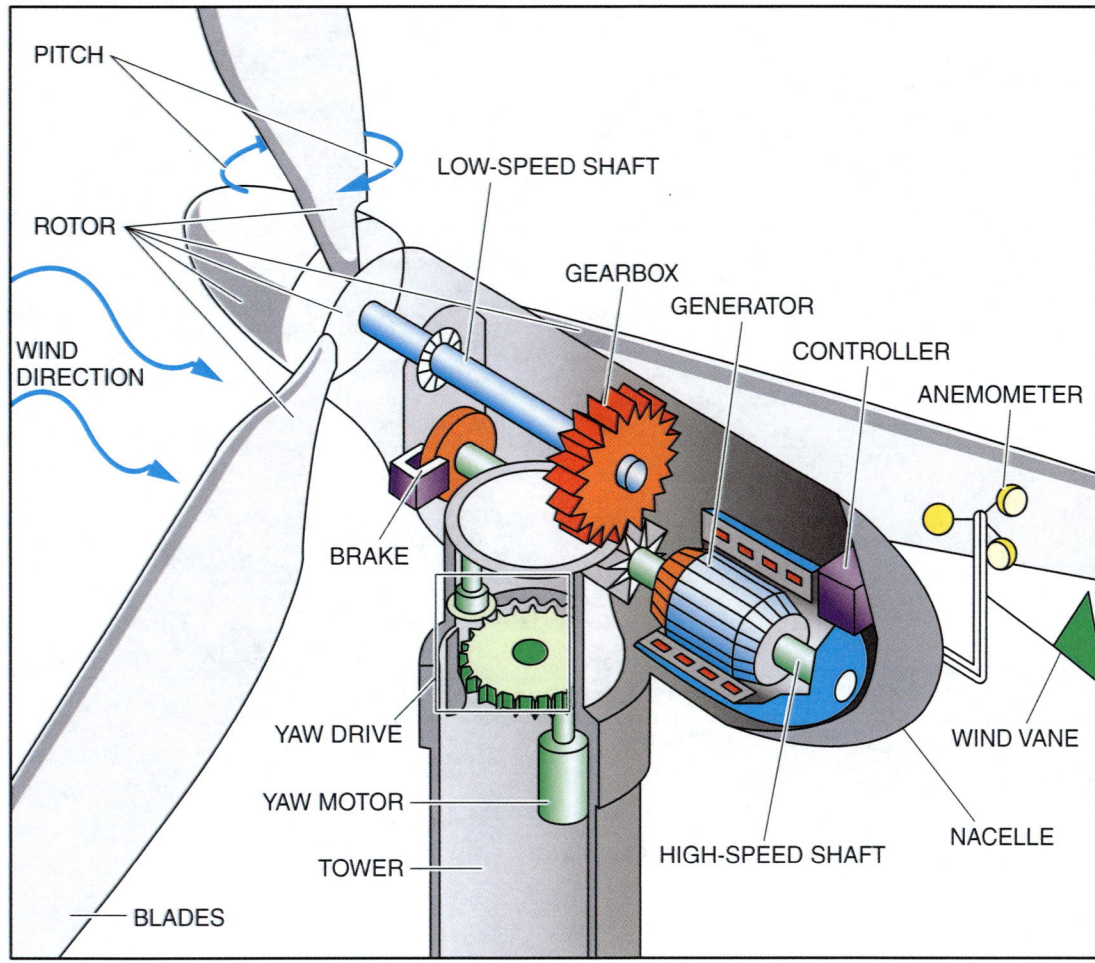

Figure 20 Basic turbine system components.

- *Yaw motor(s)* – Provide the power to pivot the nacelle assembly.
- *Yaw drive* – The gearing that turns the turbine assembly on its vertical axis.
- *Brake* – Equipped to slow the rotor and bring it to a stop for maintenance or in the event of an emergency such as a failure of yaw and/or pitch controls.
- *Tower* – Supports the nacelle and provides a means of access from the ground.
- *Blades* – Intercept the wind and turn wind energy into mechanical rotation.
- *Rotor* – Provides a mounting point for the blades and connects to the main low-speed shaft.

7.1.0 Wind Turbine Towers

The tower supports the primary turbine assembly, provides a conduit for power transmission to ground level, and usually provides maintenance personnel with some means of accessing the turbine. Tower designs and construction differ widely depending on the size and height of the turbine.

Height is a significant factor in wind availability and consistency. Taller towers place the turbine well above obstructions and allow access to faster and less turbulent winds. As the height increases, however, the complexity of the tower also increases.

7.1.1 Types of Towers

Towers fall into two broad categories: guyed and freestanding. Guyed towers (*Figure 21*) are primarily used with smaller turbines as they are more economical to build on this scale. They are, however, impractical for commercial and utility-scale turbines. Since the guy wires or cables are responsible for maintaining the tower's top position, the

Figure 21 Guyed tower.

Figure 22 Lattice tower.

weight-supporting portion of the structure is often consistent in size from base to turbine. The foundation is also minimal in scope and mass.

Freestanding towers are further broken down into two categories: lattice towers and tubular towers. Lattice towers (*Figure 22*) are sometimes referred to as truss towers. They may also use guy wires in smaller applications.

Depending on their size and height, lattice towers can be shipped in sections for site assembly or built on site. Heavy-duty hinges at the base allow the construction to be done on the ground, and then the tower can be pulled to its upright position. Opponents of lattice towers (and often of wind energy in general) cite their unattractive appearance as a reason to avoid their use. Although more expensive to construct, the wind industry in general prefers the tubular style over the lattice tower for several reasons, including appearance, security, improved personnel protection, and safer turbine access. The tallest towers may be equipped with elevators for easy access.

Tubular towers (*Figure 23*) are primarily used for medium- and large-scale turbines. They are also referred to as monopole towers. Although more expensive to construct, their advantages outweigh the cost factors.

Figure 23 Tubular tower.

Small tubular towers are built of steel pipe or tubing, concrete, fiberglass, and even wood in some cases. Like some lattice towers, simple pole towers can be hinged at their base to allow for easy erection. Multiple guy wires are then attached to maintain their vertical position. As the size and height increase, guy wires are eliminated and the tower is typically tapered like a candle. The larger base improves stability. Tubular steel towers can be fabricated in sections and shipped to the job site for erection (*Figure 24*).

Tubular towers can also be made of concrete, or a combination of both concrete and steel. Since the tower base needs mass for stability, concrete may be better suited for the lower portions of a tower, while steel has the advantage in the slender upper sections.

Large tubular towers offer the advantage of providing safe, protected access to the turbine for service (*Figure 25*). This type of access is preferable to that of a lattice tower, especially in cold, windy, or wet climates.

58101-11_F24.EPS

Figure 24 Tubular tower assembly.

58101-11_F25.EPS

Figure 25 Turbine access through the tower.

Depending on the height of the tower, there may be one or more decks. One deck is located just below the nacelle, acting as the final accessible location before the portion of the turbine that rotates for yaw positioning. This deck is known as the yaw deck.

7.1.2 Offshore Installations

Offshore wind farms offer many advantages, but they also present unique design challenges. Shallow water installations may use land-based towers that have been modified for increased corrosion control and with marine-class electrical systems, and then anchored to the seabed.

Deep sea installations offer better wind conditions, are out of sight from the shore, and are out of the path of coastal traffic. However, moving farther away from shore increases the complexity of the tower design. This is due both to the increased water depth and the fact that the largest systems are installed in these locations. An underwater guy system may suffice at some depths and tower sizes, but many deep-sea installations require the floating technology developed over the years for oil rigs and ocean drilling platforms (*Figure 26*).

7.2.0 Nacelles

The nacelle (*nuh-sell*) houses the wind turbine's primary working components and moving parts. On larger models, it also protects the technician from the elements during service and repair. The nacelle must be aerodynamic to avoid disrupting airflow as the wind passes through the rotor, creating undesirable backpressure. It must also have the strength to withstand all types of weather.

Figure 27 shows the internal layout of a nacelle from one manufacturer. Nacelle layouts can vary widely from turbine to turbine.

7.3.0 Electric Power Components

The major turbine electric power components include the generator, pad-mounted transformer, converter, and low-voltage distribution module.

7.3.1 Generator

The generator of a traditional HAWT unit is located in the nacelle and is driven by the high-speed shaft from the gearbox. The required input rpm is generally in the 1,000 to 1,500 rpm range. Some generators can provide power directly from the low-speed shaft of the rotor, eliminating the gearbox. These types are becoming

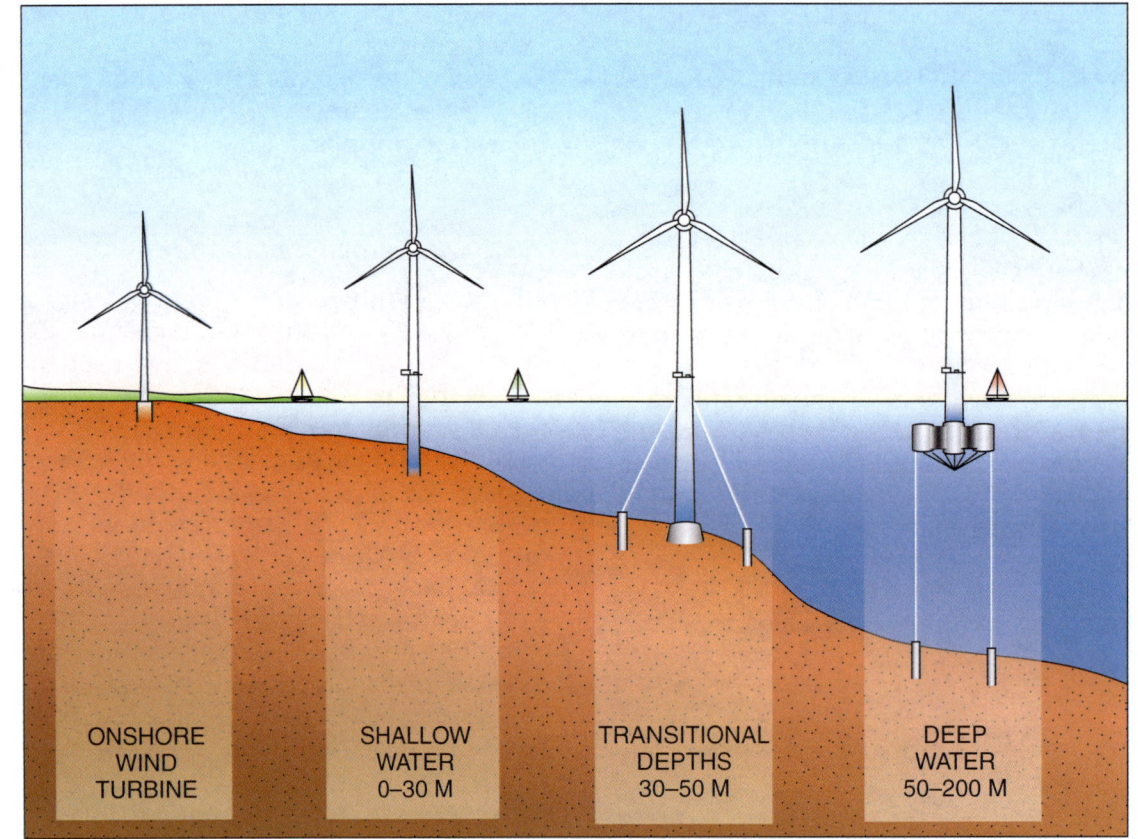

Figure 26 Offshore tower installation.

more popular and are referred to as direct-drive systems. Removing the gearbox eliminates the majority of failures and high-cost repairs in a turbine system.

Although older and smaller wind turbines may generate DC voltage to provide charging current for battery banks, it is more practical for larger grid-connected turbines to generate AC voltage.

This eliminates the need to change the generated power to an AC voltage before it leaves the tower site.

Figure 28 shows a typical commercial wind turbine-generator. Turbine-generators typically produce 575 VAC power, which is sent to a ground-mounted transformer where the voltage is increased to the 5,000V range for transmission

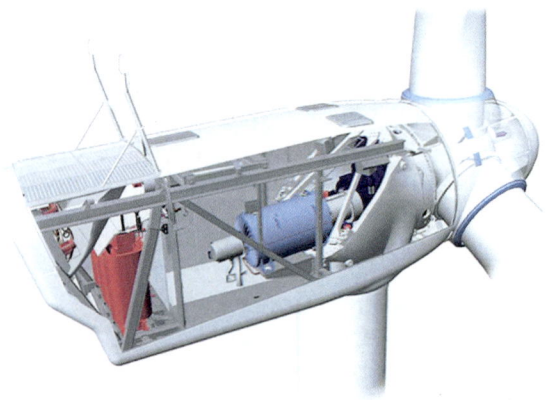

Figure 27 Vestas V90 nacelle.

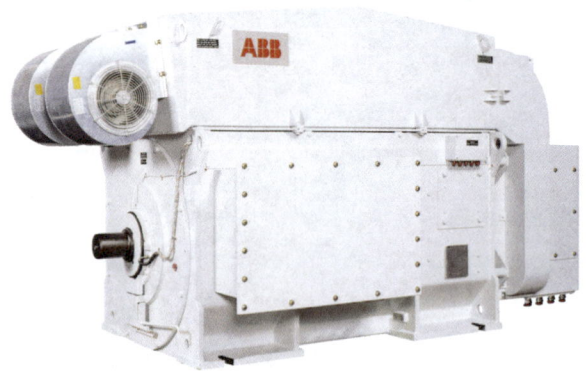

Figure 28 Wind turbine generator.

Gearbox-Free Power Generation

Several companies have developed turbine systems that eliminate the gearbox by using generators that can develop an equal amount of power at normal rotor speeds. However, the generator must be much larger to accomplish this. In spite of the larger size, the weight of the generator has been reduced by using permanent magnets in the generator rotor in place of heavy copper-coil electromagnets. This eliminates a high-maintenance component and reduces the nacelle weight by 15–20 percent. While the initial motivation was to eliminate the gearbox in offshore systems, continued success with this design will surely lead to their use in onshore units as well.

over significant distances. Some turbine systems do generate DC power, but it must eventually leave the turbine as an AC power signal. Other generator output AC voltages, such as 690 VAC, are also used. Constant design changes will result in other generated voltages.

The grid operator determines the specifications for the grid interface. As more wind farms are built and tied to the grid, operators must develop and enforce strict standards for the power provided. Utility grids require wind farms to operate like any other power substation, although the two differ significantly. In periods of light or zero wind, the turbine rotor may stop or rotate too slowly for the generator to function. In these cases, the rotor may be decoupled and allowed to turn without any load until its speed returns to the required level. As turbines drop off and return to the power grid, switching systems are required to avoid circuit damage.

Generators of this size must be reliable to avoid expensive downtime. Serviceability is also very important, since most of the work is done in place. Major overhauls or repairs may require the generator, if not the entire nacelle, to be removed and lowered to the ground by crane.

7.3.2 Pad-Mounted Transformer

Power produced by a nacelle-mounted generator is sent down the tower to a pad-mounted step-up transformer. This transformer increases the voltage to about 5kV for transmission. Since transformers are quite heavy, they are mounted at ground level to minimize the weight of the nacelle.

7.3.3 Converter

The power leaving the AC generator varies in both voltage and frequency. This is mostly due to the varying turbine speeds. AC converters (*Figure*

29) are used to condition the power and produce a consistent voltage and frequency. The converter first converts the AC power to DC and adjusts the power signal as necessary. Then, it converts it back to AC at the frequency required for delivery to the substation.

In 2010, the DOE awarded grants to several companies for the development of high-efficiency converters. These converters will provide grid-ready power using a new transformer technology. This will eliminate the need for a separate transformer.

58101-11_F29.EPS

Figure 29 Power converter.

Multiple Generators

One manufacturer has taken a different approach to the problem of large generator systems and the difficulty associated with their service and replacement. The Clipper Liberty uses four smaller generators powered by four separate high-speed shafts from a single gearbox. The generators can be lowered to the ground individually for service or replacement using tower-mounted hoists in place of large mobile or built-up cranes.

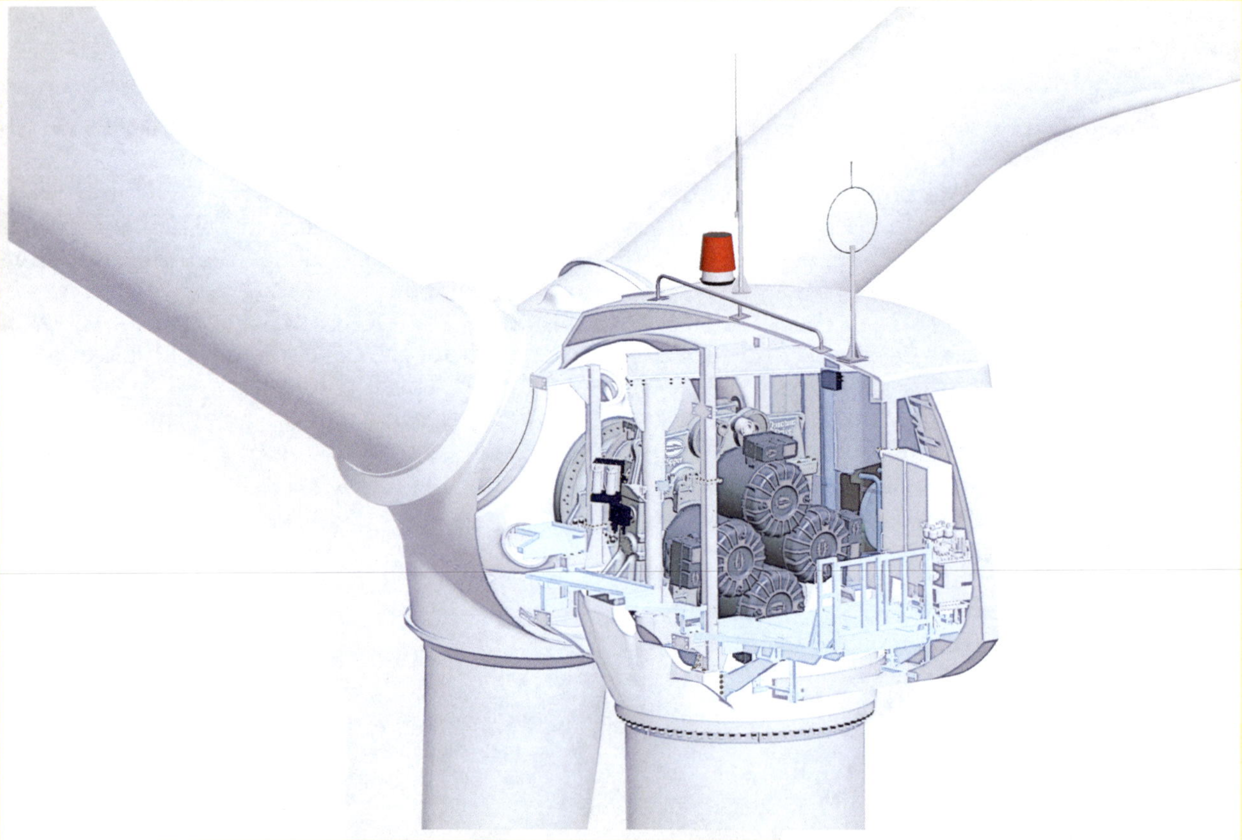

58101-11_SA04.EPS

7.3.4 Low-Voltage Distribution Module (LVDM)

The low-voltage distribution module, or LVDM, provides a location for distribution and protection of the turbine's low-voltage system. In this case, the low voltage is that produced by the generator, since it is boosted to a higher voltage for transmission to the grid or load. The LVDM houses and protects switchgear, surge protection devices, and circuit breakers related to generator power and devices in the turbine system that may use that power.

7.4.0 Drive System Components

This category of turbine components includes gearboxes, braking systems, hydraulic systems, and the yaw drive.

7.4.1 Gearboxes

The gearbox of a turbine is connected to the rotor through the low-speed shaft. It transfers the power from the rotor to the generator through the high-speed shaft. In this case, high speed is a relative term as the high-speed shaft often rotates below 2,000 rpm. The force and pressure placed on gearboxes is extreme. This is due to the size and weight of the rotor, coupled with the wind forces that are applied to it. Proper lubrication is essential. A failure of the lubrication system can be both mechanically devastating and costly.

The gearbox pictured in *Figure 30* is undergoing testing at the NREL's dynamometer testing facility. The gearbox is mounted in a nacelle on two large pins inserted into heavy-duty vibration isolators. Some manufacturers use the top of the

Figure 30 Wind turbine gearbox.

58101-11_F30.EPS

gearbox as a mounting location for the hydraulic fluid cooler coil.

Gearboxes are typically multi-stage, as the needed ratio of rotor speed to the speed required by the generator to begin delivering power ranges from 1:25 to 1:50. These figures indicate that the high-speed shaft of the gearbox must turn 25 to 50 times faster than the rotor speed. This is relatively unusual for a gearbox, as most applications take high-speed inputs and reduce the speed of the output shaft connected to the load.

Due to the stresses on gearboxes, they are responsible for more maintenance and downtime than any other major system component.

7.4.2 Braking and Hydraulic Systems

Wind turbines must include some method of rotation speed control. Without it, high winds could allow the rotor to rotate fast enough to self-destruct from extreme centrifugal force or bearing failure. Since the generator load tends to slow the rotor, a sudden loss of the generator will also cause a rapid increase in rotation, even at normal wind speeds. In a matter of seconds, a rotor may exceed speeds that braking systems are unable to counter before disaster occurs. Rotor speeds must be tightly controlled and include several levels of backup.

Rotor braking is both aerodynamic and mechanical. Using mechanical braking systems for normal speed control would be very costly in terms of component wear. Aerodynamic braking uses the wind itself to control the rotor speed. Types of aerodynamic braking include normal aerodynamic braking and tip braking.

Normal aerodynamic braking is initiated by the on-board controller or SCADA system. When the operating parameters indicate that the rotor speed is too fast or is increasing too rapidly, the controller uses pitch and yaw (aerodynamic) adjustments. These adjustments optimize performance and maintain the rotor at the ideal speed.

When normal aerodynamic control fails and the rotor speed continues to rise, fail-safe systems begin to engage. Simple centrifugal switches located in the rotor can be enabled by high rotational speeds, regardless of electronic controller inputs. Often referred to as centrifugal release units, these switches open hydraulic valves to position blade tip brakes (*Figure 31*). Tip brakes spoil airflow by turning the blade tips 90 degrees to the blade profile.

Mechanical braking is required for medium and large turbines as a backup to aerodynamic braking. It also ensures that the rotor does not move during service or maintenance. Mechanical braking relies on friction, often using disc brakes placed on the gearbox high-speed shaft. A high-speed shaft mounting position means that the stopping force applied to the rotor is transferred through the gearbox, rather than applied directly to the rotor shaft. These brakes are simi-

Pearson's Green Initiative

As part of a company-wide environmental initiative, Pearson, the world's leading learning company, plans to install a 95-kilowatt wind turbine to serve its Owatonna, MN location. The Owatonna Print Services facility consumes 4 million KWH of power annually and will be able to make use of 100% of the generated wind power. The wind turbine is expected to offset roughly 119 metric tons of CO_2 annually—equivalent to 5,540 barrels of oil—throughout its 20-year life span.

This project follows a previous renewable energy project at Pearson's Old Tappan, NJ data center, where solar panels have already been installed. The panels offset 4,000 tons of CO_2 annually, equivalent to the effects provided by planting 100,000 trees.

Pearson's green initiative also includes the acquisition of more efficient fleet vehicles, expanded use of online training and meeting tools, and upgrading of its HD video conferencing technology. These combined efforts help to reduce energy consumption and carbon emissions by reducing fleet fuel consumption and eliminating unnecessary travel.

GOING GREEN

Figure 31 Tip brakes deployed.

GOING GREEN

Big Horn Wind Farm

The Big Horn wind farm uses 133 GE turbines, each rising 119 meters (389 feet) from the ground to the top of the blade tips. The wind farm uses only a small percentage of the 15,000 acres dedicated to the project. The rest of the area remains in use for wheat farming and ranching, just as before.

The project invested further in the environment by working to conserve 455 acres south of the farm. In the nearby town of Bickleton, known as the bluebird capital of the world, funding from the project replaced 250 bluebird boxes, although research indicates that the turbines do not present a danger to them. Iberdrola Renewables, the owner and operator, continues to support the local environment by developing and funding community education programs to protect native bird species.

lar to auto brakes in that they use a rotor fixed to the shaft and a serviceable caliper-actuated brake pad. However, they operate at much higher temperatures due to the load imposed by slowing the huge mass of a turbine rotor. Temperatures can reach 700°C during braking.

Mechanical braking systems are often hydraulically actuated, as are blade tip brakes. As a fail-safe measure, hydraulic pressure must be present for the rotor to operate. This ensures that pressure is available for brake operation. Electrical braking systems are also used.

Many utility-scale turbines also use hydraulics to make pitch changes in the blades or drive the yaw system in place of an electric motor. Hydraulic pressure may be supplied by a standalone hydraulic power unit with an integral electric pump. It can also be provided by a pump driven from the gearbox, but this is less common. The hydraulic pressure is controlled by solenoid valves. Many systems use two valves in parallel to ensure that the failure of a single valve will not prevent operation. Some manufacturers go so far as to avoid using two solenoid valves of the same model or manufacturer on the same system. This avoids the potential of failure due to flaws in a given model or production run.

8.0.0 The Wind Farm

Wind turbines are often used in large groups on wind farms, also referred to as wind parks or wind industrial parks. One example is the Big Horn wind farm (*Figure 32*). It is located in the state of Washington and was built in 2007. This wind farm generates roughly 200 MW using 133 GE 1.5 MW turbine systems. The turbines are arranged in lines or strings.

Figure 32 Big Horn wind farm.

As shown in *Figure 33*, the power generated by each turbine is routed down through the tower to a pad-mounted transformer. Before it leaves the turbine tower, the power is generally conditioned for proper frequency and changed to AC voltage if needed. For utility-scale turbines, power from the turbine itself is generally in the range of 600 VAC.

The pad-mounted transformer boosts the AC voltage to reduce the wire size required to carry it long distances. The turbine-generator voltage is typically increased to roughly 34.5 kV by the transformer. The substation accepts power from the transformers on the site as a whole, and then boosts the voltage even higher to match the power grid. Grid voltages vary quite a bit, from 110 kV to as much as 765 kV.

The best wind farm sites are where wind resources are the greatest; average annual wind velocities of at least 11 mph are generally needed. Other crucial aspects to be considered prior to development include the following:

- The maximum capacity based on site size and grid interconnection issues
- Required setbacks from existing structures or protected areas
- The location of important sight lines and their relationship to the turbines
- The location of dwellings that could be impacted by blade flicker or shadows
- An analysis of the power buyer and the associated market

Each location must be examined carefully before construction. The turbines themselves create turbulence that may affect other turbines nearby. The designer must adhere to the manufacturer's specifications related to turbine spacing. Special software can be used to create a model of the entire facility.

8.1.0 Wind Farm Maintenance

Wind farm maintenance costs are lower than those at traditional power facilities. Periodic maintenance typically occurs twice per year, resulting in roughly 16 hours of turbine downtime. Only a few turbines in the farm are usually down at any one time. The entire farm may be taken offline for annual or semi-annual maintenance of the power substation. This is best done during periods when low winds are expected.

Unscheduled repairs are costly. Wind turbine warranty periods are often negotiable at the point of purchase from two to five years of service. The manufacturer often maintains the turbine during the warranty period. Once the warranty has ex-

① Rotating generator converts wind energy to electricity.
② Transformer increases voltage for transmission to substation.
③ Substation increases voltage for transmission over long distance.
④ Transmission to the grid.

58101-11_F33.EPS

Figure 33 Turbine power delivery to the utility grid.

Wind Energy Technician Needs

A weak maintenance program is a recipe for disaster with turbines. A high demand for trained technicians exists, as operating firms must ensure that maintenance is done on time and repairs are addressed quickly.

Gearboxes are the source of the most costly and frequent service and repairs. A simple bearing failure, which may have been prevented with proper maintenance, can easily cause $100,000.00 in damages.

Siemens manufacturers and maintains turbines on a worldwide scale. The leadership of their global service division cites the availability of trained technicians as their chief challenge. As of 2010, Siemens had 1,500 employees in their wind service organization. They expect to triple that number over the next four years.

pired, service work is often hired out to independent providers or the manufacturer. Wind farm owners may also use their own staff for maintenance and repair.

Record keeping is an essential part of maintenance for every wind turbine. This ensures that all maintenance is done correctly and in a timely manner. The use of checklists and maintenance reports help to ensure that inspection items and steps have not been skipped. Once completed, the documents provide a paper trail that can be analyzed for trends. *Figure 34* provides one example of a checklist used during a turbine inspection. Note that the checklist also lists the frequency of inspection for each item.

Beyond the maintenance for the turbines, a number of other tasks add to the workload. Required vehicles must be maintained. Roads must be kept passable, and the landscape must be groomed at some sites. Good communication is essential, using both radios and cell phones. All maintenance and repair activities must be tracked and recorded. This includes work orders, checklists, and accident or incident reports. These tasks are common to all operation and maintenance groups, although specific procedures often differ from site to site.

9.0.0 THE WIND ENERGY TECHNICIAN

A career in wind energy presents unique challenges, but also provides job security and a good income. As the growth of both onshore and offshore wind farms continues, the need for wind turbine technicians will grow as well. The natural laws of supply and demand dictate that wages and benefits will also rise over time. The training offered through this NCCER program provides the foundation needed for success in this industry.

A wind energy technician must work well both independently and as a part of a team. Due to the hazards involved with this work, team members must trust each other to work safely.

Workers in this field need good mechanical and electrical skills, as well as above-average math skills. Computer skills and knowledge of meteorology are also helpful. Prospective employers will make background checks. Most employers require a clean driving record.

This program prepares trainees for entry- and mid-level positions as wind turbine technicians. AWEA is preparing a list of the skills required for a wind turbine technician. It includes the following:

- Turbine fundamentals and an understanding of how various maintenance and service processes are completed
- Maintenance operations including the SCADA system, rotor blade inspection, oil analysis, and sensors/instrumentation
- Crane and rigging operations
- Gearbox service and inspection
- Hydraulic system principles, service, and maintenance
- Yaw system routine maintenance
- Electrical systems, from theory through motors and generators
- Basic computer skills
- Report writing and documentation
- Safety, including OSHA 30-hour training, electrical safety, emergency response and rescue, fire safety, confined space work, MSDS/hazardous waste protocols, personal protective equipment function and use, and risk management

The U.S. Department of Labor (DOL) has added the job of Wind Turbine Technician to the list of trades where proficiency can be gained through an apprenticeship. The DOL has also prepared a Work Process Schedule for the trade, which outlines the required categories of training and the number of hours required in each category. A minimum number of hours of instruction and on-the-job learning are required for completion. The DOL Work Process Schedule documents the specific training requirements under the broad headings of First Aid, Safety, Equipment and Practices,

INSPECTION AND MAINTENANCE SCHEDULE

Symbol meaning:

✿ = Tightening bolts
\+ = General inspection
× = Change
* = Lubrication
■ = Test
** = Inspect every two years
××× = Change every three years

EVERY VISIT	500H MAINTENANCE	1 YEAR MAINTENANCE	5 YEAR MAINTENANCE	10 YEAR MAINTENANCE		CHECKED BY
Tower						
+					Cleaning	
	✿	✿			Bolts (inside and outside)	
		+			Painting	
		+			Step	
		+			Welding	
		+			Door	
	+	+			Elevator	
+					Climbing equipment (both tower and elevator)	
	+	+			Lighting / emergency lighting	
		+		×	Fire-extinguisher (powder)	
		+			First aid box	
	✿	+			Ladder	
		+			Lifting winch	
Tower top floor						
	+	+			Cable work	
		+			Twist recorder	
		+			Nacelle position	
+					Climbing equipment	
Nacelle						
	+	+			Connection laptop / check parameters	
+					Cleaning / inspection	
	+	+			Cable passage	
	✿	✿			Bolt connections	
	+	+			Lighting and emergency lighting	
	■	■			The cable twist recorder	
		+		×	Fire-extinguisher (carbon dioxide)	
Yaw system						

58101-11_F34.EPS

Figure 34 Wind turbine maintenance checklist.

Electrical, Hydraulics, Mechanical, Braking, Blade Inspection and Maintenance, and Computers. The NCCER Wind Turbine Technician program will meet or exceed the DOL requirements for this occupation over a three-year period. To learn more about apprenticeship programs, visit the DOL at www.dol.gov. The U.S. Bureau of Labor Statistics has recently developed a category for the career field to facilitate tracking of employment and wage data. This information is available to the public at www.bls.gov. Once on the job, technicians often continue their education through manufacturer training.

Wind turbine technicians are required to work at great heights without fear (*Figure 35*). They may work in extreme weather or temperatures. Superior physical condition is required for climbing, as is the ability to work in the confined space of a nacelle. The freedom to travel is also important, as many technicians move from site to site for maintenance. Travel may also be required on short notice when problems arise. Technicians who are willing to travel may receive higher wages and other benefits. Technicians employed directly by manufacturers often travel frequently.

Technicians with good leadership skills may quickly move up the career ladder. Advanced positions include lead technician, craft instructor, and site manager.

10.0.0 STANDARDIZED TRAINING BY NCCER

The National Center for Construction Education and Research (NCCER) is a not-for-profit education organization established by the nation's leading construction companies. NCCER was created to provide the industry with standardized construction education materials, and a system for tracking and recognizing students' training accomplishments—NCCER's Registry. Refer to the *Appendix* for examples of NCCER credentials.

NCCER also offers accreditation, instructor certification, and skills assessments. NCCER is committed to developing and maintaining a training process that is internationally recognized, standardized, portable, and competency-based.

Working in partnership with industry and academia, NCCER has developed a system for program accreditation that is similar to those found in institutions of higher learning. NCCER's accreditation process ensures that students receive quality training based on uniform standards and criteria. These standards are outlined in NCCER's Accreditation Guidelines and must be adhered to by NCCER Accredited Training Sponsors.

58101-11_F35.EPS

Figure 35 Wind energy technician at work.

More than 500 training sponsors and/or assessment centers across the U.S. and twelve other countries are proud to be NCCER Accredited Training Sponsors and Accredited Assessment Centers. Millions of craft professionals and construction managers have received quality construction education through NCCER's network of Accredited Training Sponsors and the thousands of Training Units associated with the Sponsors. Every year the number of NCCER Accredited Training Sponsors increases significantly.

A craft instructor is a journeyman craft professional or career and technical educator trained and certified to teach NCCER's training materials. This network of certified instructors ensures that NCCER training programs will meet the standards of instruction set by the industry. There are more than 4,500 master trainers and 47,000 craft instructors within the NCCER instructor network. More information is available at www.nccer.org.

SUMMARY

The wind turbine industry is already a key player in the war against fossil fuel dependence in both the U.S. and abroad. A long and well-documented history of wind power has provided experience to an industry that hopes to achieve a goal of providing 20 percent of U.S. power from wind energy by 2030.

The science of wind and the energy it contains is well tested and understood, but consistency and perfect wind performance will likely always remain elusive. Performance has proven, at least for the time being, that the three-bladed, horizontal-axis rotor is best for medium- and large-scale wind turbine applications. The performance of a horizontal-axis wind turbine is further optimized by controlling the yaw of the rotor as it relates to wind direction, and the pitch of the individual rotor blades.

Wind energy technicians are needed to maintain the ever-growing number of turbine units in service. Physical conditioning is an important characteristic of the wind energy technician, due to the strength and stamina required to climb high towers and work in tight quarters. The field can be exciting and rewarding, generally appealing to those who are adventurous, intelligent, safety-minded, and willing and able to travel. For the technician with the right qualities, a lifelong and prosperous career awaits.

1. By the end of 2009, what percentage of U.S. electrical power was contributed by wind energy?

   a. Less than 1 percent
   b. 2.5 percent
   c. 3.5 percent
   d. 5.2 percent

2. One of the distinct advantages to the use of wind power is that _____.

   a. wind turbines are very inexpensive to purchase and install
   b. wind energy provides a means of energy independence from fossil fuels and those who control fossil fuel resources.
   c. wind energy can easily replace all fossil fuels as a source of electricity by 2030
   d. a sufficient amount of wind for power production is available everywhere

3. Remote or undeveloped areas that are beyond the electrical grid cannot make use of wind-generated power.

   a. True
   b. False

4. Prior to 0 BC, wind power was used to drive water pumps by the _____.

   a. Chinese
   b. Germans
   c. English
   d. Swiss

5. The first power-producing wind turbine in the U.S. was placed in service in _____.

   a. 1866
   b. 1888
   c. 1906
   d. 1918

6. Charles Brush' first 19th century wind turbine could produce _____ of electrical power.

   a. 2 kW
   b. 9 kW
   c. 12 kW
   d. 15 kW

7. Worldwide interest in wind power technology increased dramatically in the early 1970s due to _____.

   a. the first major oil crisis
   b. the end of the Viet Nam war era
   c. U.S. renewal of diplomatic relations with China
   d. the severe loss of life in Africa due to a lack of electrical power

8. By 2003, what country produced roughly 20 percent of its total electric power needs using wind energy?

   a. China
   b. Europe
   c. Germany
   d. Denmark

9. Through consistent research and development, 70 percent of the world's wind power as of 2003 was found in _____.

   a. Denmark
   b. the U.S.
   c. Europe
   d. China

10. As of 2010, what state led the U.S. in total installed wind energy capacity?

    a. Texas
    b. Iowa
    c. California
    d. North Dakota

11. How many megawatts of power will the Cape Wind project, the first offshore wind farm in the U.S., generate with its 130 installed turbines?

    a. 260
    b. 500
    c. 650
    d. 750

12. Research in wind energy systems is conducted at its primary facility in Colorado by the _____.

    a. DOE
    b. AWEA
    c. NREL
    d. NOAA

13. Kinetic energy is a measure of the energy within an object or mass that has been set in motion.

    a. True
    b. False

14. The two primary factors that determine the energy carried by the wind are its mass and its _____.

    a. velocity
    b. density
    c. pressure
    d. temperature

15. Doubling wind speed from 10 mph to 20 mph multiplies the available wind energy by a factor of _____.

    a. 2
    b. 4
    c. 6
    d. 8

16. The most accurate means of determining the available wind energy for a site is through a calculation based on the annual average of wind speed.

    a. True
    b. False

17. The terrain with the highest wind shear exponent would be _____.

    a. ice
    b. a suburban area
    c. a wooded area or forest
    d. snow-covered low crops

18. When the wind shear exponent is not clearly identified in a discussion, it is generally accepted to be the value associated with _____.

    a. an icy surface
    b. urban areas
    c. heavily wooded areas
    d. short or mown grass areas

19. Using the equation $A = \pi r^2$ and a value for π of 3.1416, the swept area of a horizontal-axis rotor with a total diameter of 44 meters would be _____.

    a. 138.23 cubic meters
    b. 1,520.53 square meters
    c. 1,520.53 cubic meters
    d. 6,082.13 square meters

20. When the diameter of a horizontal-axis rotor is doubled, how does it affect the swept area?

    a. It doubles the area.
    b. It triples the area.
    c. It quadruples the area.
    d. It quintuples the area.

21. The theoretical limit of how much wind energy a rotor could capture is known as the _____.

    a. Betz limit
    b. kinetic limit
    c. Hutter limit
    d. potential limit

22. Due to the enormous size of the largest turbines, they would likely be located _____.

    a. offshore
    b. in other countries
    c. near areas of high electrical load
    d. in rugged, remote mountain passes

23. What axis of control does yaw refer to?

    a. lateral
    b. vertical
    c. horizontal
    d. longitudinal

24. Which acronym stands for the system used to monitor and control turbine operation from a remote location?

    a. TCMS
    b. WEAS
    c. SCADA
    d. WTCDA

25. Before it is conditioned or transformed, the voltage typically delivered by a large turbine generator is _____.

    a. 12V DC
    b. 96V DC
    c. 230V AC
    d. 575V AC

26. The rpm required by most turbine generators to begin producing electricity is _____.

    a. 500 to 750 rpm
    b. 1,000 to 1,500 rpm
    c. 1,500 to 1,750 rpm
    d. 2,500 to 3,000 rpm

27. Within what range is the ratio of the gearbox high-speed shaft rpm to the rotor shaft rpm?

    a. 1:5 to 1:10
    b. 1:10 to 1:15
    c. 1:15 to 1:25
    d. 1:25 to 1:50

28. Mechanical braking is the primary means of controlling rotor speed.

    a. True
    b. False

29. Wind energy technicians must _____.

    a. be computer literate
    b. be licensed master electricians
    c. weigh no more than 86 kg (190 pounds)
    d. possess a bachelor's degree at a minimum

30. The demand for wind energy technicians continues to rise.

    a. True
    b. False

Trade Terms Quiz

Fill in the blank with the correct term that you learned from your study of this module.

1. A device used to measure wind speed is known as a(n) _____.

2. An apparatus that converts mechanical energy into electrical energy is called a(n) _____.

3. The streamlined enclosure containing the major working components of the turbine system is called the _____.

4. The action of controlling excessive wind turbine rotor speed aerodynamically by turning the rotor blades away from a direct wind facing is known as _____.

5. The 59.3 percent theoretical amount of wind power that a rotor could potentially capture from the wind is referred to as the _____.

6. A wind turbine that spins on an axis which is horizontal, or nearly so, is called a(n) _____.

7. The means of quantifying the power available in wind per unit of area, generally expressed as watts per square meter (w/m^2) is referred to as _____.

8. A(n) _____ is a device that uses centrifugal force to open or close hydraulic valves in response to rotor speed.

9. The _____ of a mass is the energy it stores as a result of its being placed in motion.

10. In the wind turbine industry, the _____ is the top-most surface that wind is actually passing over and must be considered in the process of determining the best height for a turbine.

11. Turbine rotor blades pass through the _____ as they rotate.

12. _____ is the term used to describe the wind speed variations that occur at different heights above the Earth's surface.

13. A(n) _____ uses a rotor that spins on a vertical or near-vertical axis and requires no yaw control.

14. Rotor blade ends that are designed to rotate independent of the rest of the blade, allowing it to spoil the blade's aerodynamic characteristics and reduce rotor speed are called _____.

15. The _____ is the final accessible platform located just below the nacelle assembly at the top of a tubular tower.

16. A computerized system used to supervise, control, monitor, and collect historical data from an individual wind turbine or a collection of turbine systems using real-time information and commands is called a(n) _____.

17. To determine the historic frequency that winds blow from a given direction at a given location, a circular graph known as a(n) _____ may be used.

18. Wind turbines use _____ to manage the direction of rotor facing through rotation of the nacelle on its vertical axis.

19. Load-bearing surfaces for the turbine nacelle that provide a solid but slick surface for its rotation during yaw adjustments are called _____.

20. A line extended along the length of an object forms its _____ axis.

21. The positioning of a turbine blade on its longitudinal axis to improve or spoil the airflow across it is called _____.

22. Power generated by the turbine is sent to the _____ for distribution and management.

Trade Terms

Anemometer	Furling	Low-voltage distri-	Supervisory control	Vertical-axis wind
Betz limit	Horizontal–axis	bution module	and data acquisi-	turbine (VAWT)
Centrifugal release	wind turbine	(LVDM)	tion (SCADA)	Wind rose
units	(HAWT)	Nacelle	system	Wind shear
Dynamo	Kinetic energy	Pitch control	Swept area	Yaw control
Effective ground	Longitudinal	Power density	Tip brakes	Yaw deck
level				Yaw pucks

Trade Terms Introduced in This Module

Anemometer: A device used to measure wind speed that often incorporates wind direction as well.

Betz limit: The theoretical limit of 59.3 percent of the available wind power that a rotor can capture. The theory is named for its German developer, Albert Betz.

Centrifugal release units: Devices that use centrifugal force to open or close hydraulic valves in response to rotor speed.

Dynamo: An apparatus that converts mechanical energy into electrical energy, typically in the form of direct current.

Effective ground level: The actual surface that air movement is passing across, as opposed to actual ground level. For example, the effective ground level for a wind blowing over a dense forest would be the tops of the trees rather than the ground itself.

Furling: One method of preventing excessive wind turbine rotor speed through yaw control by turning the rotor blades away from a direct wind facing.

Horizontal-axis wind turbine (HAWT): A wind turbine that spins on an axis which is horizontal or nearly so, much like the early windmills of the western and midwestern U.S. Also referred to as a conventional turbine or propeller-style, they are directional by design, i.e. the rotor must face into the wind for maximum performance.

Kinetic energy: The energy contained in a mass or body caused by its motion.

Longitudinal: Running lengthwise, or extending along the length of an object. A line drawn the length of an object would indicate its longitudinal axis.

Low-voltage distribution module (LVDM): The assembly of switchgear and circuit breakers that controls the flow of turbine-generated power. The voltage is low relative to the voltage that leaves the pad-mounted transformer on its way to the substation.

Nacelle: A streamlined housing or enclosure that contains the major working components of the turbine system at the top of the tower.

Pitch control: Management and turning of a turbine blade's position along its longitudinal axis.

Power density: The means of quantifying the power available in wind per unit of area, generally expressed as watts per square meter (w/m^2); in English units, it is expressed as watts per square foot (w/ft^2).

Supervisory control and data acquisition (SCADA) system: A computerized system used to supervise, control, monitor, and collect historical data from an individual wind turbine or a collection of turbine systems using real-time information and commands.

Swept area: The area that turbine rotor blades pass through.

Tip brakes: Rotor blade tips designed to rotate independent of the rest of the blade, allowing it to spoil the aerodynamic characteristics and reduce rotor speed.

Vertical-axis wind turbine (VAWT): A wind turbine with a rotor that spins on a vertical or near-vertical axis. VAWTs are generally omnidirectional and allow the drive train and generators systems to be mounted at ground level.

Wind rose: A circular graph that depicts the frequency at which winds blow from a given direction at a given location, generally reported as a percentage of time. Wind roses may also contain other information using color, such as how often the wind blows from a dirction at a given speed.

Wind shear: Term used to describe the wind speed variations that occur at different heights above the Earth.

Yaw control: Management of a wind turbine's facing direction by rotation of the turbine assembly on its vertical axis.

Yaw deck: The final accessible deck located at the top of a tubular tower and just below the nacelle assembly.

Yaw pucks: Load-bearing surfaces for the turbine nacelle that provide a solid but slick surface for rotation of the turbine during yaw adjustments.

NCCER CREDENTIALS

To access the Automated National Registry (ANR), visit

http://www.nccer-anr.org

and enter the number on the front of this card.

> If found, drop in mailbox. Postage paid by
> NCCER
> 3600 NW 43rd Street Building G
> Gainesville FL 32606

Showcase your credentials
Post your resume...Find jobs
http://careers.nccer.org

NATIONAL CENTER
FOR CONSTRUCTION
EDUCATION AND RESEARCH

Sample Student

2781481

58101-11_A04.EPS

November 10, 2010

Sample **Student**
XYZ Technical School
123 Test St
Gainesville, FL 32606

Dear Sample,

On behalf of the National Center for Construction Education and Research, I congratulate you for successfully completing NCCER's Contren® Learning Series program. I also congratulate you for choosing construction as a career.

You are now a valuable member of one of our nation's largest industries. The skills you have acquired will not only enhance your career opportunities, but will help build America.

Enclosed are your credentials from the National Registry. These industry-recognized credentials give you flexibility in planning your career and ensure your achievements follow you wherever you go.

To access your training accomplishments through the Automated National Registry, follow these instructions:
1.   Go to www.nccer-anr.org.
2.   Click the "Individuals" button.
3.   Enter the NCCER card number, located on front of your wallet card or transcript, and your PIN.
     Note: The default PIN is the last four digits of your SSN. You may change your PIN after you login.
4.   First-time users will be directed to answer a few security questions upon initial login.
5.   Contact the registry department with any questions.

NCCER applauds your dedication and wishes you the best in your future endeavors.

Sincerely,

Donald E. Whyte
President, NCCER

Enc.

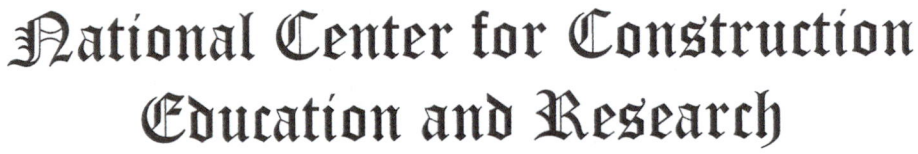

National Center for Construction Education and Research

This is to certify that

Sample Student

has fulfilled the requirements for

Introduction to Wind Energy

in NCCER's standardized training curriculum

this Tenth day of December, 2010

Donald E. Whyte

Donald E. Whyte

President

NATIONAL CENTER FOR CONSTRUCTION EDUCATION AND RESEARCH

BUILDING TOMORROW'S WORKFORCE

3600 NW 43rd St, Bldg G ∘ Gainesville, FL 32606
P 352.334.0911 ∘ F 352.334.0932 ∘ www.nccer.org

December 10, 2011

Official Transcript

Sample Student
XYZ Technical School
123 Test St
Gainesville. FL 32606

Current Employer/School:

Card #: 8176459

Course / Description		Instructor	Training Location	Date Compl.
00101-09	Basic Safety	Don E Whyte		1/1/2009
00102-09	Intro to Construction Math	Don E Whyte		1/1/2009
00103-09	Introduction to Hand Tools	Don E Whyte		1/1/2009
00104-09	Introduction to Power Tools	Don E Whyte		1/1/2009
00105-09	Introduction to Construction Drawings	Don E Whyte		1/1/2009
00106-09	Basic Rigging	Don E Whyte		1/1/2009
00107-09	Basic Communication Skills	Don E Whyte		1/1/2009
00108-09	Basic Employability Skills	Don E Whyte		1/1/2009
00108-09	Intro to Materials Handling	Don E Whyte		1/1/2009
49101-10	Introduction to the Power Industry	Don E Whyte		10/15/2010
58101-11	Introduction to Wind Energy	Don E Whyte		10/15/2011

NO ENTRIES BELOW THIS LINE

Donald E. Whyte
President, NCCER

Page 1

Additional Resources

This module presents thorough resources for task training. The following resource material is suggested for further study.

Wind Power. Paul Gipe. White River Junction, VT: Chelsea Green Publishing Company.

American Wind Energy Association (AWEA). http://awea.org/.

U.S. Department of Energy, Wind Powering America Program. www.windpoweringamerica.gov.

Figure Credits

NCCER CURRICULA — USER UPDATE

NCCER makes every effort to keep its textbooks up-to-date and free of technical errors. We appreciate your help in this process. If you find an error, a typographical mistake, or an inaccuracy in NCCER's Curricula, please fill out this form (or a photocopy), or complete the online form at **www.nccer.org/olf**. Be sure to include the exact module number, page number, a detailed description, and your recommended correction. Your input will be brought to the attention of the Authoring Team. Thank you for your assistance.

Instructors – If you have an idea for improving this textbook, or have found that additional materials were necessary to teach this module effectively, please let us know so that we may present your suggestions to the Authoring Team.

NCCER Product Development and Revision

13614 Progress Blvd., Alachua, FL 32615

Email: curriculum@nccer.org
Online: www.nccer.org/olf

❏ Trainee Guide   ❏ AIG   ❏ Exam   ❏ PowerPoints   Other _____

Craft / Level: _____   Copyright Date: _____

Module Number / Title: _____

Section Number(s): _____

Description: _____

_____

_____

_____

_____

Recommended Correction: _____

_____

_____

_____

_____

Your Name: _____

Address: _____

_____

Email: _____   Phone: _____

58102-11

Introduction to Wind Turbine Safety

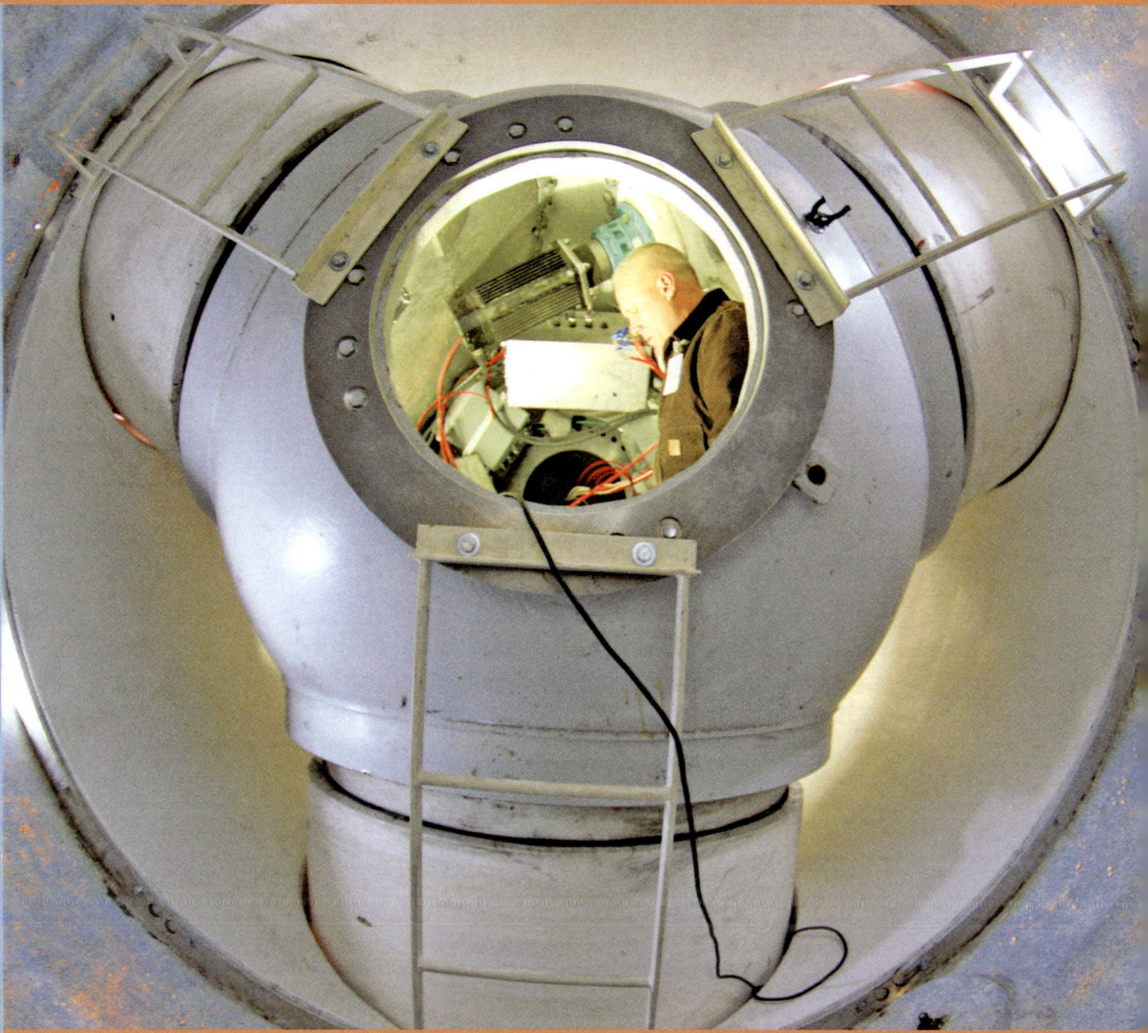

Module Two

Trainees with successful module completions may be eligible for credentialing through NCCER's National Registry. To learn more, go to **www.nccer.org** or contact us at **1.888.622.3720**. Our website has information on the latest product releases and training, as well as online versions of our *Cornerstone* newsletter and Pearson's Contren® product catalog.

Your feedback is welcome. You may email your comments to **curriculum@nccer.org,** send general comments and inquiries to **info@nccer.org**, or use the User Update form at the back of this module.

Objectives

When you have completed this module, you will be able to do the following:

1. Describe the purpose and value of the Occupational Safety and Health Administration (OSHA) 10-hour General Industry guidelines.
2. Describe the purpose of Job Safety Analysis (JSA) meetings as it relates to the wind turbine environment.
3. Prepare a JSA.
4. Describe confined spaces as they relate to the wind turbine.
5. Identify electrical system safety guidelines for wind turbine service.
6. Identify safe rigging practices in the wind turbine environment.
7. Describe wind energy site safe driving techniques.
8. Identify various aerial work platforms and their operating characteristics.

Performance Task

Under the supervision of the instructor, you should be able to do the following:

1. Prepare a JSA.

Trade Terms

Arc
Arc blast
Arc flash
Egress
Equipotential plane
Flash hazard analysis

Flash protection boundary
Jib
Limited approach
   boundary
Minimum approach
   distance (MAD)

Permit-required confined
   space
Prohibited approach
   boundary
Qualified worker
Reeving

Restricted approach
   boundary
Step potential
Touch potential

Industry Recognized Credentials

If you're training through an NCCER-accredited sponsor you may be eligible for credentials from NCCER's Registry. The module ID number for this module is 58102-11. Note that this module may have been used in other NCCER curricula and may apply to other level completions. Contact NCCER's Registry at 888.622.3720 or go to nccer.org for more information.

Contents

Topics to be presented in this module include:

1.0.0 Introduction 1
2.00 OSHA and the Wind Turbine Environment 1
3.0.0 Job Safety Analysis (JSA) 2
   3.1.0 Task Analysis 2
   3.2.0 Analysis Assistance 2
4.0.0 Confined Spaces 3
   4.1.0 Permit-Required Confined Spaces 4
   4.2.0 Worker Responsibilities in Confined Spaces 4
      4.2.1 Entrants 4
      4.2.2 Attendants 4
      4.2.3 Supervisors 7
      4.2.4 Rescue Workers 7
   4.3.0 Wind Turbine Confined Spaces 7
      4.3.1 Nacelle 8
      4.3.2 Rotor Hub 9
      4.3.3 Turbine Tower 10
5.0.0 Electrical Safety 10
   5.1.0 Shock Hazard 10
   5.2.0 Step and Touch Potentials 11
   5.3.0 Arc Flash and Arc Blast 13
   5.4.0 Hazard Boundaries 13
      5.4.1 Flash Protection Boundary 15
      5.4.2 Shock Protection Boundaries 16
      5.4.3 Minimum Approach Distance 17
   5.5.0 De-Energized Equipment 18
   5.6.0 Lockout/Tagout 18
      5.6.1 Lockout/Tagout Procedure 20
      5.6.2 Restoration of Energy 21
      5.6.3 Emergency Lockout/Tagout Removal 21
      5.6.4 *OSHA Standard 1910.269* 22
6.0.0 Rigging 23
   6.1.0 Hoists 23
   6.2.0 Lift Planning 24
   6.3.0 Critical Lifts 25
   6.4.0 Lift Plan Implementation 26
7.0.0 Safe Driving 27
   7.1.0 Driving Company Vehicles 27
   7.2.0 Off-Road and Inclement Weather Driving 27
      7.2.1 Steep Inclines 28
      7.2.2 Crossing Water 29
      7.2.3 Mud 29
      7.2.4 Ice and Snow 30
      7.2.5 Ground Clearance 30

8.0.0 Aerial Work Platforms 30

  8.1.0 Scissor Lifts 31

    8.1.1 Operating Precautions 32

    8.1.2 Controls and Indicators 32

    8.1.3 Operation 33

    8.1.4 Maintenance 33

  8.2.0 Boom Lifts 33

    8.2.1 Application 34

    8.2.2 Operation 34

    8.2.3 Maintenance 35

Appendix Job Safety Form 40

Figures and Tables

Figure 1   Technician in a wind turbine hub 1
Figure 2   Task analysis worksheet.. 3
Figure 3   Confined space of a rotor hub.. 3
Figure 4   Confined space entry permit 5–6
Figure 5   Confined space attendant 7
Figure 6   Wind turbine rescue training.. 8
Figure 7   Wind turbine access ladder 8
Figure 8   Confined space within a nacelle.. 8
Figure 9   Nacelle hatch 9
Figure 10 Connecting emergency descent lanyard 9
Figure 11 Departing the nacelle 9
Figure 12 HAWT hub area 9
Figure 13 Rotor hub access 9
Figure 14 Step and touch potential 12
Figure 15 Arc flash 13
Figure 16 Approach limits 14
Figure 17 Electrical hazard warning sign 15
Figure 18 Arc flash 16
Figure 19 Worker using appropriate PPE inside
           a restricted approach boundary 18
Figure 20 Lockout/tagout devices 19
Figure 21 Typical safety tags 19
Figure 22 Lockout devices 19
Figure 23 Multiple lockout/tagout device 20
Figure 24 Wind turbine rotor lift in progress 23
Figure 25 Hook-mounted electric hoist 23
Figure 26 Electric hoist on a trolley 24
Figure 27 Turbine hoist on a pivoting mast 24
Figure 28 Tag lines steady the load 25
Figure 29 Turbine blade in transit 26
Figure 30 Mounting the hub and rotor assembly... 26
Figure 31 Company-owned service vehicle 27
Figure 32 Rugged terrain 28
Figure 33 Climbing a steep hill 28
Figure 34 Crossing deep water 29
Figure 35 Truck winch 29
Figure 36 Reduced ground clearance on a hillcrest 30
Figure 37 Aerial lifts 31
Figure 38 Scissor lift with outriggers.. 32
Figure 39 Boom lifts 34
Figure 40 Boom lift control panel 35

Table 1   Effects of Current on the Human Body... 11
Table 2   Approach Boundaries for Shock Protection
           [Data from *NFPA 70E*® *Table 130.2(C)*].. 17
Table 3   Energized Line Work Minimum Approach Distances..................... 17

1.0.0 INTRODUCTION

As a trainee working towards a career in wind turbine generator service and maintenance, you may be somewhat surprised to be studying safety, before you have been exposed to task training. However, as in all trades, safety must be emphasized every day of your career.

The wind turbine environment is rather unique and offers a combination of hazards found in few other trades. Since the turbine exists as a means of generating electrical power, there are obviously electrical hazards to consider. Climbing and working at great heights also represents a significant safety hazard. The wind turbine nacelle and hub (*Figure 1*) qualify as confined spaces, representing additional hazards. For some tasks, you may be required to work in close proximity to rotating machinery driven by three wind-driven 14,000-pound blades. The vast majority of wind turbine services you will perform are done in relatively remote locations, far from instant assistance or rescue. This unique environment offers exciting and challenging career possibilities, but only to those technicians who think and function with safety in mind at all times.

This module focuses on safety and safe working habits required for success in the wind turbine environment, building upon basic safety principles.

2.0.0 OSHA AND THE WIND TURBINE ENVIRONMENT

The Occupational Safety and Health Administration (OSHA) is a part of the U.S. Department of Labor. OSHA's mission is to ensure safe and healthy working environments for all workers.

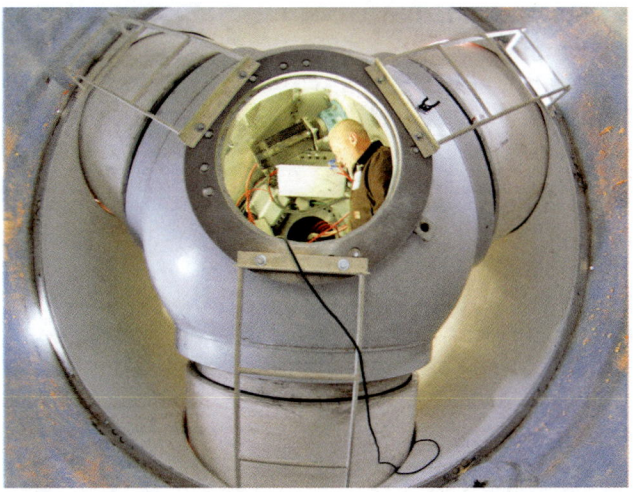

Figure 1 Technician in a wind turbine hub.

58102-11_F01.EPS

This is accomplished through the creation and enforcement of comprehensive safety standards. By providing the necessary training and guidance, both employers and employees can meet the standards. OSHA understands that a successful safety program requires planning, training, and commitment by all parties. A safe workplace is a right for U.S workers in every occupation, not a privilege.

Trainees wishing to pursue a career as a wind turbine maintenance technician should complete all the requirements of OSHA's 10-hour General Industry Safety Training. Although employers and workers are not required to complete the training by the *Code of Federal Regulations (CFR)*, many employers do require that employees attend and successfully complete the program. A number of states have enacted legislation that requires workers be provided with the 10-hour training program before working on publicly funded projects. The American Wind Energy Association (AWEA) recommends all wind industry workers complete a qualifying OSHA 10-hour General Industry program. OSHA's 10-hour program guidelines do include some electives and optional subject matter. The options chosen for instruction are usually based on the type of work being done. AWEA has identified subjects they feel best suit the wind turbine technician.

The 10-hour General Industry program provides a solid foundation for occupational safety and health training for most industries. A separate 10-hour program specifically for the construction industry is also available. Trainers authorized to present the 10-hour training programs present course completion cards for their students, providing documentation of program completion. Employers and trainers can add additional training to the basic 10-hour program, and some flexibility is allowed in the subjects they present. However, a few specific guidelines must be followed. As a rule, the training must be presented in person, although a limited number of online programs have been approved. Recent changes to OSHA's guidelines include the requirement that training time is limited to 7½ hours per day.

The safety training provided in this module, in conjunction with the Power Industry Fundamentals *Basic Safety* module and the Wind Turbine Maintenance Technician *Climbing Wind Towers* module covers the subject matter required under OSHA's 10-hour General Industry program. Trainees should be aware however, that safety training is not a subject to be studied only once. To ensure that safe work habits become true personal habits, safety training courses, such as the OSHA 10-hour program, may be repeated a number of times.

3.0.0 JOB SAFETY ANALYSIS (JSA)

A Job Safety Analysis (JSA) is a process that focuses on job tasks to recognize hazards before they cause accidents. This practice is also referred to as Job Hazard Analysis. The process focuses on the connection between the worker, the project or task, the tools, and the environment. After potential safety hazards are identified and brought to light, workers can take steps to eliminate or reduce the risk to themselves and their co-workers.

JSA is one tool used to help employers and workers manage hazards that are specific to tasks and work environments. JSA documentation is generally completed during the course of a meeting between involved workers and supervisors, but management may initiate the forms and thought process prior to the meeting. Tasks to be completed are first identified, and then they are broken down into smaller steps so potential hazards can be more easily identified. One example of a form used during a wind site JSA meeting is provided in the *Appendix*. Although the primary function of the JSA is to reduce or eliminate potential hazards and risks, it can also serve as a planning tool for the coming task.

The sample JSA provides a cover page that documents the team members, and offers a place for their signature to acknowledge both their presence at the meeting and their understanding of the safety issues related to the task. The next few pages provide the meeting leader with broad subject headings to discuss about the task. The final page is a comprehensive checklist of specific issues to consider. Checklists such as this one ensure that areas of concern are not overlooked. They assist all personnel involved to think clearly and bring up any potential hazard that comes to mind as the items are read aloud. Workers must be outspoken critical thinkers during the process, without ignoring diplomacy when addressing others.

JSA forms differ from site-to-site to fit the needs of the employer, the team, and the specific site. JSA forms should be considered living documents, in that they can and should be modified any time a deficiency is found or a new potential hazard is identified.

The analysis process is best considered a discovery activity, where supervisors and workers alike consider five important questions:

- What can go wrong?
- What are the consequences or results when it does go wrong?
- How could it happen?

- What other factors contribute to the possibility of it happening?
- How likely is it that it will occur?

As the activity progresses, answers to this series of questions should be documented in an organized and consistent manner. As a hazard scenario is developed, try to answer the questions in the following sequence:

- Where is it happening (in what environment or workstation)?
- Who or what is exposed to the potential for injury or damage?
- What happens that sets the hazard in motion?
- What will be the result or consequence if it happens?
- What other factors contribute to the sequence of events?

3.1.0 Task Analysis

Task safety analysis is very much like job safety analysis, except that it typically focuses on a smaller set of actions or steps. The analyzed task could be as simple as reaching into a box repeatedly to grab a part for cleaning. In this case, examples of items to consider would be the weight of the component and any obstructions that a worker might encounter while reaching into the bin. The complete task is dissected. A simple task analysis worksheet is shown in *Figure 2*. Identify potential hazards for a task and identify measures to avoid them by using the questions and answer patterns from the previous section.

A detailed task analysis for a given task may be repeated as the result of one or more accidents or injuries. Companies can review safety performance and incident reports to determine what tasks are related to the most serious or highest number of injuries. These tasks would be likely candidates for a fresh task safety analysis. Pictures and video taken while a task is being accomplished can help all personnel, including off-site consultants, to analyze the job from a different perspective.

3.2.0 Analysis Assistance

In some cases, outside assistance is requested to assist in task or job safety analysis. Private consultants, fire specialists, and insurance company representatives are just a few of the sources for assistance. Specialists often see things that others close to the task itself may not see. They are also consistently exposed to the many incidents that take place in a variety of trades, providing

Date: _____

Analyst: _____

Job Location: _____

Task Description:

Hazards (What can go wrong? What could result from it?):

Hazard Controls:

58102-11_F02.EPS

Figure 2 Task analysis worksheet.

them with a broad base of experience. However, it is essential that all personnel involved in a task remain involved even when a consultant is used. When workers remain a part of the process, they tend to understand, accept, and support changes more readily.

OSHA can also be consulted to assist in job and task safety analysis. Consultation is provided primarily to small employers who want help creating a safe and healthy workplace. The service was initially developed to assist employers who are engaged in relatively hazardous processes or products related to hazardous materials. When consultation is requested and provided, it is done at no cost to the employer. By agreement, no penalties are proposed or citations issued for hazards that are identified during the process. The employer, however, must agree to correct all noted deficiencies within a specified period.

4.0.0 CONFINED SPACES

Confined spaces can be broadly defined as any space where the entry, exit, or movement of workers is restricted in some way (*Figure 3*). A confined space is typically not designed as a work area to be

continuously occupied. It may also be a space that is poorly ventilated, which may reduce the availability of oxygen or increase the level of atmospheric contaminants. A confined space may not seem particularly problematic at first, but workers often contaminate or make the space more hazardous through the work performed. Welding is

58102-11_F03.EPS

Figure 3 Confined space of a rotor hub.

one example of work that can increase the hazard level of a confined space due to the smoke and fumes. Activities that increase the potential danger of a confined space may also change the entry requirements and preparations.

A written confined-space entry program can protect you by identifying the possible hazards. Following an inspection by a competent person, the confined space is classified based on the hazards present. The two classifications are non-permit-required and permit-required. A nonpermit-required confined space is a space that is free of any mechanical, electrical, physical, or atmospheric hazard that can cause death or injury. Once an area is classified nonpermit-required, workers can enter using the appropriate PPE for the task.

4.1.0 Permit-Required Confined Spaces

OSHA has a specific category of confined space known as permit-required confined space. Guidelines regarding permit-required spaces and their identification are documented in *OSHA Standard 1910.146*. Permit-required confined spaces can be identified by one or more of the following characteristics:

- The atmosphere in the space is hazardous or could become hazardous.
- The space contains a material or substance that could engulf or bury the worker.
- The space has an internal configuration that could cause an entrant to be trapped or asphyxiated by inwardly converging walls, or by a floor which slopes downward and tapers to a smaller cross-sectional area (like a funnel).
- Floors in the area slope downward and taper down to a smaller area where a worker could be trapped or denied breathable air.
- The area possesses some other recognized hazard, such as exposed live power wiring, unguarded machinery, or a space where a worker could potentially be overcome due to heat stress or extreme cold.

The text of the standard does allow for broad interpretation. However, it is always better to err on the side of safety. Broad interpretation of the guideline is also available to OSHA, which determines the level of compliance by an organization or worker.

Permit-required confined spaces require written authorization for entry. One example of a confined space entry permit is provided in *Figure 4*.

Workers may require specific training related to the space or hazard before entry is allowed.

4.2.0 Worker Responsibilities in Confined Spaces

Everyone involved in confined space work has certain responsibilities to ensure the task is carried out safely. This includes entrants working inside the space, attendants who monitor the situation just outside the space, supervisors, and rescue personnel.

4.2.1 Entrants

Entrants must understand the task to be done and the dangers associated with the work and environment. Entrants should:

- Make sure they have a valid and signed entry permit when required.
- Be aware of any specific hazards, such as the potential for extreme temperatures or high winds that may change while work is in progress. Most wind sites have specific standards that allow or prohibit access to the turbine area under certain weather conditions.
- Make use of all required PPE, including face and eye protection, gloves, fall protection and arrestance gear, and emergency descent equipment.
- Stay in contact with the attendant to ensure that any hazardous or important conditions are being monitored from outside the space as well as inside.
- Alert the attendant immediately if any discomfort or physical warning signs of exposure occur.
- Know how to escape should the need arise.
- Exit the space immediately upon hearing an evacuation alarm.

4.2.2 Attendants

Attendants stay outside of the confined space (*Figure 5*) and communicate with the entrant(s). In the wind turbine environment, when a worker must enter the rotor hub, an attendant must be in the nacelle and a third worker remains below. This places two workers in separate confined spaces. The nacelle entrant serves as the rotor hub entrant's attendant, while the third worker attends to both from outside the environment.

Attachment 16 – 2
Confined-Space Entry Permit
Master Card / Safe Work Ticket No. _____

1. Work Description: _____
  Equip. Name / Number & Location or Area _____
  Purpose of Entry _____
  Valid Start Date _____ Duration Time _____ to _____

2. Hazardous Materials:
  What did the equipment last contain? _____
  Will the work generate a hazardous atmosphere? ☐ Yes ☐ No If yes, specify hazards and controls.
  _____

3. Rescue Requirements:
  ☐ External, by attendant   ☐ Complex Rescue, by rescue team at point of entry
  ☐ Non-IDLH and/or Simple Rescue, by rescue team on-site ☐ IDLH, by rescue team at point of entry
  Has the rescue team been notified of the entry? ☐ Yes ☐ N/A Time of notification _____
  How will the rescue team be summonsed for an emergency? ☐ Radio Channel: _____ ☐ Other: _____

4. Gas Test Requirements:
  LEL/0_2 - Instrument Mfg./No. _____/_____   Bump Check Time/Gas Tester - _____/_____
  Toxicity - Instrument Mfg./No. _____/_____   Bump Check Time/Gas Tester - _____/_____
  Frequency of Testing: ☐ Continuous ☐ Other - Specify - _____
  • Continuous monitoring results must be recorded every three hours

Acceptable Levels	Results								
Oxygen: 19.5% - 23.5%									
Combustible Gas: %LEL - <10%									
Other _____ < PEL* _____									
Other _____ < PEL* _____									
Other _____ < PEL* _____									

  • Entry in excess of the PEL will require appropriate PPE.

5. Ventilation / Exhaust Equipment:
  ☐ None required, natural ventilation adequate ☐ Forced air ventilation ☐ Exhaust ventilation
  Equipment Type: ☐ Air powered horn ☐ Electric blower Volume Required - _____ cfm

6. Personal Protection:
  ☐ Gloves (type) _____   ☐ Respirator (type) _____
  ☐ Goggles or face shield   ☐ Self Contained Breathing Equipment
  ☐ Lifelines Attached to Harness   ☐ Other, specify: _____
  ☐ Chemical Resistant Suit, Specify Type _____

7. Fire Protection: ☐ None required ☐ Portable Fire Extinguisher – type and size:_____
  ☐ Fire Watch   ☐ Other, specify: _____

58102-11_F04A.EPS

Figure 4 Confined space entry permit (1 of 2).

8. Condition of Area and Equipment:

Required Yes	N/A	THESE KEY POINTS MUST BE CHECKED
		a. Equipment locked and tagged out?
		b. Piping is disconnected, capped or plugged and/or blinded.
		c. Equipment emptied, washed, purged & ventilated?
		d. Low voltage or GFCI protected equipment provided?
		e. Explosion proof electrical equipment provided?
		f. Provisions are made to barricade or post signs at entry points when attendant is not on duty.

Other Requirements:

9. Special Instructions: ☐ None   ☐ Check with issuer before starting work

10. Approval	Permit			Permit Acceptance	
	Supt. / Area Supv.	Date	Time	Maint. Supv. / Engineer / Contractor Supv.	Date
Issued by					
Endorsed by					
Endorsed by					

11. Individual Review / Entrant Roster: I have been instructed in the proper Work Permit, Confined Space Entry, Lockout/Tagout Procedures, associated physical and atmospheric hazards and have reviewed the gas testing results.

Entrants	Date	Time In / Out	Time In / Out	Time In / Out	Time In / Out	Time In / Out

I have been informed of the duties and responsibilities for an attendant, the associated physical and atmospheric hazards, and have reviewed the gas testing results.

Attendants	Date	Time On / Off	Time On / Off	Time On / Off

12. Job Completion:
☐ Yes ☐ N/A Has the rescue team been notified?
☐ Yes ☐ No   Is the work on equipment complete & the confined space ready to return to service?
☐ Yes ☐ No   Has the worksite been cleaned and made safe?
Workers answering above questions: _____

13. Post Job Review: Were any hazards encountered or created during entry operations?
☐ Yes ☐ No If yes, describe: _____
Possible solutions: _____

Forward to job file within 7 days of job completion.

58102-11_F04B.EPS

Figure 4 Confined space entry permit (2 of 2).

Figure 5 Confined space attendant.

58102-11_F05.EPS

The attendant's primary responsibility is the protection of the entrant. The attendant should have constant contact with the entrant through telephone, radio, or other dependable means. To properly carry out his or her responsibilities, attendants should:

- Set up a position near the exit of the confined space.
- Maintain a consistent count of the personnel inside.
- Visualize the entrant's work and movements to be prepared for any unusual events.
- Maintain contact with entrants and parties off-site who may need to provide valuable information.
- Order an evacuation of the space when conditions turn unfavorable.
- Know how to call in rescue assistance or set off an alarm system.
- Refuse entry to the space by any unauthorized or unnecessary personnel.
- If rescue, especially rescue-from-height, is required, follow all organizational guidelines for execution of the rescue attempt.

4.2.3 Supervisors

The entry supervisor is the person responsible for safe confined-space entry operations. This means he or she is responsible for the safety of all workers involved in the task. Entry supervisors should do the following:

- Authorize and oversee entry after critical information has been analyzed.

- Be well trained in all entry procedures, including the need for permits when required.
- Know the hazards with which entrants must contend.
- Make sure that all lockout/tagout (LOTO) procedures are performed.
- Make sure that all equipment required for entry, including emergency descent gear, is in place and workers know how to use it with confidence.
- Know the rescue plan and the authorized rescuers who will be called upon for assistance.
- Take responsibility for prohibiting or terminating the entry when unacceptable conditions are present.
- Evaluate task progress and constantly consider any potential unforeseen hazards.

4.2.4 Rescue Workers

In the wind turbine environment, the need to rescue an incapacitated worker from height is far more likely than in most other workplaces. Entrants and attendants alike should be qualified in rescue-from-height (*Figure 6*).

Rescues fall into one of two distinct categories: non-entry and entry rescues. For wind turbine service technicians, non-entry rescue is less likely. A worker who is conscious and mobile may be rescued from a position at height by a crane or similar equipment, but ideally, the worker would use automatic- or manual-descent self-rescue equipment instead. But if the worker is physically unable to perform self-rescue, entry into the confined space is likely to be necessary. Rescue workers should do the following:

- Be familiar with the hazards of the confined space.
- Be intimately familiar with the site's rescue plan and comply.
- Seek proficiency in rescue through consistent practice and training.
- Be trained in first aid and CPR.

Entry rescues can be very dangerous. With all service industries and trades considered, figures indicate that more rescue workers are killed in confined spaces than workers who are the subject of the rescue.

4.3.0 Wind Turbine Confined Spaces

The nacelle, hub, and tower areas of a typical horizontal axis wind turbine (HAWT) are considered confined spaces. Worker movement inside is restricted, and both entry and exit from these areas can be challenging. The hub may be considered a

Figure 6 Wind turbine rescue training.

permit-required confined space by some organizations or wind-site operators.

4.3.1 Nacelle

The nacelle of utility-scale turbines mounted on tubular, or monopole, towers is accessed by climbing the internal ladder (*Figure 7*). Nacelle heights of 75 to 100 meters (250 to 330 feet) are quite common. Accessing the nacelle area requires considerable physical effort, as does exiting the nacelle using the same ladder. The restricted entry and exit routes alone qualify the nacelle as a confined space. However, all nacelles have very little room to move and work, which additionally qualifies the area as a confined space. *Figure 8* shows the confined space within a nacelle.

Due primarily to the hazards associated with rotating machinery, access to the nacelle is generally denied when the turbine is rotating. Service switches that stop the turbine rotor are located in the base of the tower. By first using aerodynamic braking to slow the rotor by changing pitch and yaw, then engaging the mechanical brakes, the rotor is brought to a stop. Service braking systems that then lock the nacelle and rotor in place further ensure a safe work area. However, some turbines may require the technician to access the nacelle in order to apply the mechanical brakes. This places the technician in the nacelle with the rotor pin wheeling, although the generator is disengaged and aerodynamic braking may have been applied at the base. This exposure to rotating machinery in a confined space is a significant hazard. Wind turbine service requires a keen awareness of how to avoid the unique hazards of specific turbine types and brands.

Even though turbine systems may be equipped with a variety of braking and control systems, it is essential that technicians always be aware of the possibility of their failure, which could allow the rotor to turn. A plan for exiting the nacelle under such conditions should be studied and practiced. Some turbine systems may have locking pins that physically prevent the rotor shaft from turning while service is being performed.

In an emergency, workers may be unable to descend from the nacelle normally using the access ladder. Such a situation could be the result of a problem in the nacelle itself, or may be due to the sounding of an emergency horn, indicating lightning or another natural hazard is imminent. Emergency **egress** equipment should either be

Figure 7 Wind turbine access ladder.

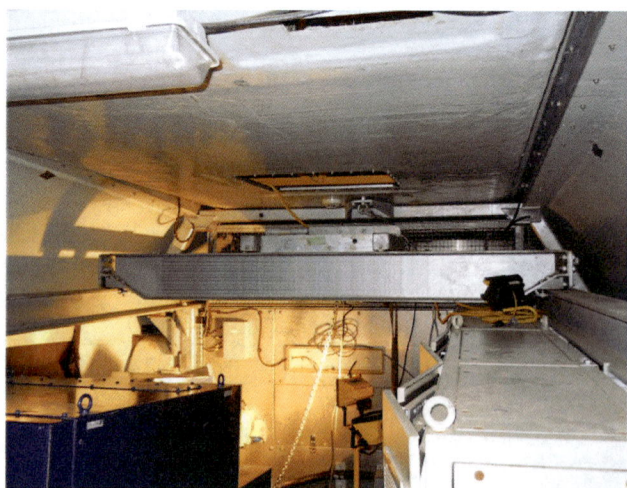

Figure 8 Confined space within a nacelle.

permanently available in the nacelle or a part of each workers standard climbing gear.

The primary point of rapid nacelle escape is the equipment hatch (*Figure 9*), typically used to raise or lower equipment to the nacelle. Self-rescue, controlled-descent devices are then used to quickly connect to an appropriate anchor point (*Figure 10*) and descend at a safe and controlled speed. The black cylinder-shaped device attached to the workers lanyard in *Figure 11* is an automatic descent control device. The worker controls his descent speed and stop at any time with virtually no effort. When multiple workers need to escape quickly, openings in the top of the nacelle offer a second means of exit.

4.3.2 Rotor Hub

The rotor hub of a wind turbine (*Figure 12*) houses a number of sensors, control mechanisms, and turbine blade pitch drive components. This small, confined space offers little room for movement, but it must be accessed for specific repairs or maintenance. The policies of most wind farms prohibit workers from accessing the nacelle during certain conditions, such as high winds or extremely low temperatures. Entering the hub requires a more enhanced set of weather conditions. Additional handholds and a cage around the hub hatch help (*Figure 13*), but high winds can easily separate the climber from the front of the rotor.

A second qualified climber should be in the nacelle and available for assistance for one worker to enter the hub. The worker in the nacelle can rescue

Figure 10 Connecting emergency descent lanyard.

Figure 11 Departing the nacelle.

the worker in the hub if he or she becomes incapacitated, and manipulate turbine controls to assist the service process during normal operations. A worker in the hub can escape using appropriate manual or automatic descent equipment, but sudden rotor rotation would physically prevent any exit from the hub.

Figure 9 Nacelle hatch .

Figure 12 HAWT hub area.

58102-11_F13.EPS

Figure 13 Rotor hub access.

4.3.3 Turbine Tower

The tubular tower of a HAWT does qualify as a confined space under OSHA's broad definition, since there is only one way out. It is very important to note that the guidelines and policies for climbing towers vary widely between service providers, manufacturers, and sites. Regardless of the service to be performed, wind tower climbing typically places two qualified climbers on site. It is generally required, subject to corporate policy, for one climber to remain at the base of the tower in a supporting role. This is for both safety and practical support of the task at hand. Some policies may even allow the supporting worker to support climbers in adjacent towers. When a worker or visitor without the proper climbing credentials is allowed to climb the tower, qualified climbers are required to take positions both above and below the unqualified climber.

5.0.0 ELECTRICAL SAFETY

Regardless of any specialty in wind turbine services a worker may pursue, working around wind turbines includes working very close to potential electrical hazards. This is especially true in the nacelle area. Although higher voltages are found at the nearby transformer, the tight confines of the nacelle places workers in very close proximity to electrical circuits. The electrical hazard is further elevated due to the many other physical hazards. Burns, electric shocks, or explosions can easily cause a worker to lose physical control of the body. If the rotor is turning, further injury can then occur from falling against rotating shafts. Even when the rotor is stationary, the nacelle area contains many sharp edges and protruding cor-

ners that can further harm an already-injured worker.

Even when electrical equipment is working properly, it can create hazardous situations for untrained workers. When a fault occurs, the results can be catastrophic. It is important that you have an understanding of the principles of electricity and its potential hazards so that you can protect yourself.

Electrical hazards are caused by current flowing through unintended paths. These paths can be any conductive object, such as metal rods, fences, or your body. Anyone working on or near electrical equipment may encounter one of the following electrical hazards:

- Shock hazard
- Step potential and touch potential
- Arc flash and arc blast

5.1.0 Shock Hazard

In the wind turbine system, the path of current flow is made up of the generator, converters and inverters, transformers, bus bars, and many other electrical devices that you will be servicing. Current travels along its intended path, so the circuit is safe unless contact is made with an energized part. When you contact ground (or grounded objects) and an energized part simultaneously, you complete a circuit and voltage and current flows through you. When a fault occurs, unintended paths of current flow may develop, causing normally de-energized parts to become energized. This condition is known as a ground fault.

Technicians should also be aware that electrical shock and injury is possible even in a de-energized circuit. Capacitors found in electronic assemblies and switchgear can hold an electrical charge for some time after power has been removed. Most are equipped with a resistor that drains the remaining charge, but it is not instantaneous. Large capacitors represent a significant and dangerous shock hazard.

The level of injury from electrical contact depends on the amount of current flowing through the person, which is a function of voltage and body resistance, as well as the parts of the body receiving current flow. Current flow through the chest cavity (heart) and the head (brain) is the most lethal. *Table 1* shows the effects of current on the human body.

A 9V battery has a typical current flow of 5 to 15mA. As shown in *Table 1*, a minor shock of 5mA can come from something as small as a 9V battery. The human body can easily resist this mag-

Table 1 Effects of Current on the Human Body

Current Value	Typical Effects
1mA	Perception level. Slight tingling sensation.
5mA	Slight shock. Involuntary reactions can result in other injuries.
6 to 30mA	Painful shock, loss of muscular control.
50 to 150mA	Extreme pain, respiratory arrest, severe muscular contractions. Death possible.
1,000mA to 4,300mA	Ventricular fibrillation, severe muscular contractions, nerve damage. Typically results in death.

58102-11_T01.EPS

nitude of electrical shock, but even a shock at this level can cause an involuntary movement away from the source. This can result in injuries as you jump back from the electrical shock source.

When the current is between 6mA and 30mA, the shock causes loss of muscular control. This may result in the worker falling from an elevated position or may cause the worker to fall into a more dangerous electrical source. Even though the shock itself may have no lasting effects, the immediate response to the shock may cause serious injuries.

As the current increases beyond 20mA, muscle contractions can prevent the victim from pulling away from the current source. At 50mA, respiratory paralysis may result in suffocation. Current levels above 150mA may cause the heart to beat in an abnormal rhythm. This condition is fatal unless the heart rhythm is corrected using a medical device called a defibrillator. Current levels of 4A or more may stop the heart altogether, resulting in death unless immediate first aid is provided. Remember, voltage doesn't kill; current kills.

Other effects of electrical shock include entry and exit wounds from high-voltage contact, and thermal burns from current flow of a few amps and greater. Thermal burns are often not apparent at first; however, the tissue in the current path may be destroyed, causing the skin to die from the inside over time.

Current follows all paths back to its source. Given the premise that current follows all paths to ground, you can prevent yourself from being shocked by being electrically isolated from ground. This means that no part of your body

touches a ground point. Another way to prevent electrical shock is to create an equipotential plane where you are working.

5.2.0 Step and Touch Potentials

The earth has resistance to current flow from one point to another. The amount of resistance depends on the type of soil in the area, the amount of moisture in the soil, and the presence of underground pipes and lines. In addition, you will be working with relatively high voltages, and because of the earth's resistance, it is possible for a voltage to develop between two points of the earth.

When potentials do develop in the earth, it is usually during a fault situation. These potentials are called step potential and touch potential (*Figure 14*).

Step potential is the voltage between the feet (usually about 1 meter) of a person standing near an energized, grounded object. It is equal to the difference in voltage between two points, each at different distances from the electrode. Near site substations, a covering of several inches of crushed gravel to provide a level of insulation between a person and earth may be provided. Although the voltage on the wind site itself is lower than that of the substation, no such layer of protection typically exists.

Touch potential is the voltage between the energized object being touched and ground. The ground path may be through any other body part of the person touching the object. For example, when you touch an object with one hand and a grounded object with the other, the current path is through the chest cavity so current flows through the heart; when you touch an object with one hand and you are standing on a grounded surface, the current path is from your hand through your body to your feet. The touch potential can be very high when the object is grounded far from the place where the person is in contact with it. As shown in the figure, there is also a transferred

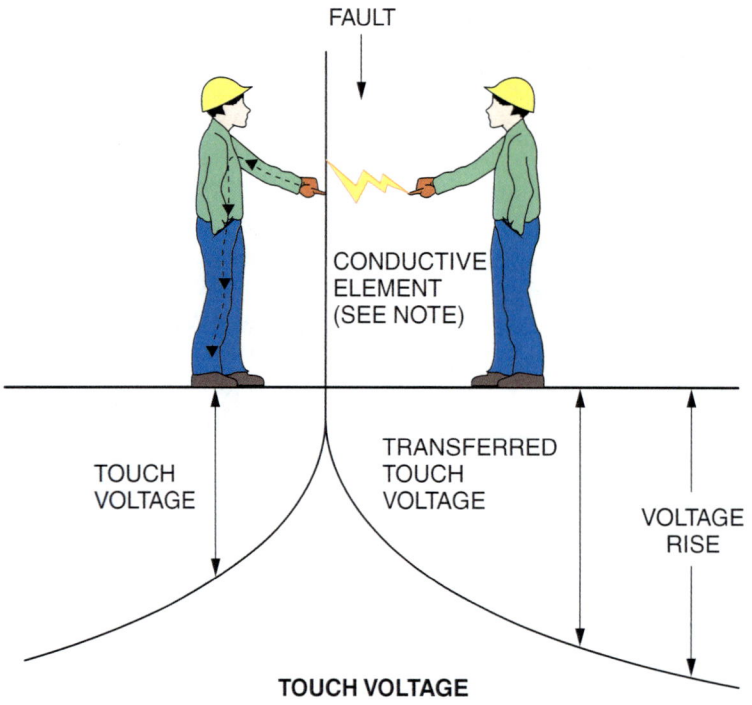

TOUCH VOLTAGE

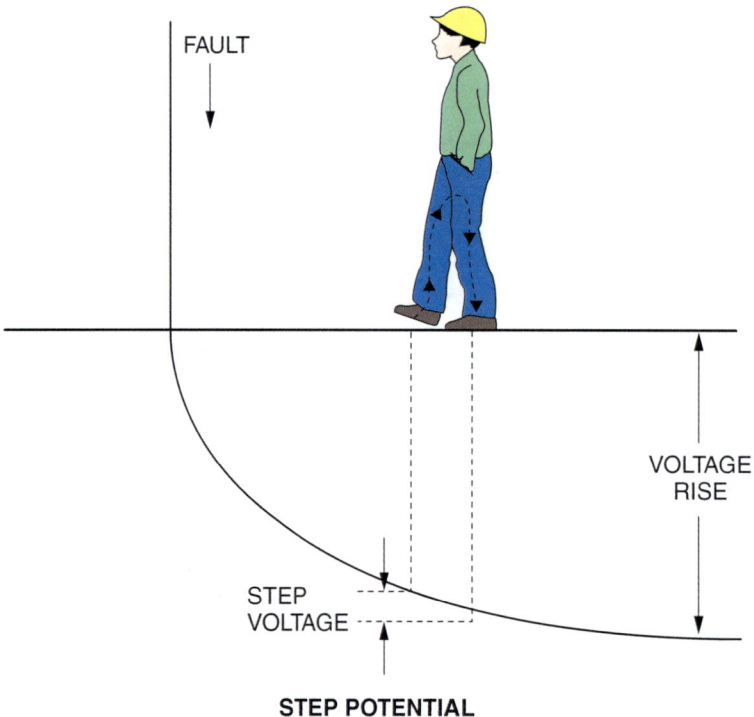

STEP POTENTIAL

NOTE:
A CONDUCTIVE ELEMENT MAY BE A TOOL, CABLE, VEHICLE, OR ANY OBJECT CAPABLE OF TRANSFERRING A FAULT.

58102-11_F14.EPS

Figure 14 Step and touch potential.

touch voltage where the difference in potential is great enough to create an arc between the grounded object and the person.

5.3.0 Arc Flash and Arc Blast

Air is not typically a good conductor of electricity. However, there are certain conditions in which current can jump through the air to reach another point and complete the path of current flow. This is called an arc. Arcs can happen anywhere there is voltage. The higher the voltage, the greater potential distance an arc can travel through air. This principle is demonstrated by a simple spark plug. High-voltage pulses are sent to a spark plug intermittently, causing an arc to jump across the gap each time.

In the winter when the air is dry, you may cause an arc when you touch a metal object and discharge static electricity. When this happens, the electrical charge can be seen passing through the air from one point to the other. That visible arc is called an arc flash. The static produces a small amount of current flow that can shock you when you discharge it. While the shock you receive from static electricity is uncomfortable, it is not dangerous because the voltage source is static. Once it is discharged, the source is gone.

When you are working around high voltage though, arcing can be very dangerous because there is a constant and powerful source behind it that allows a tremendous amount of current flow.

The current is called arc fault current and it causes an arc flash (*Figure 15*). Depending on the source, an arc flash can release an enormous amount of thermal energy to the point that an explosion occurs. This is called an arc blast and is comparable to a dynamite blast. It can destroy nearby equipment, melt metals, and cause serious injuries from flying shrapnel and molten metal. The flash intensity can damage the eyes before they have time to react and close, while pressure waves from the blast can damage hearing. A small controlled arc flash occurs when welding is done.

The intensity of the arc flash and blast are primarily dependent on two factors: the available short circuit current and the time required for the nearest upstream overcurrent protection device to actuate.

5.4.0 Hazard Boundaries

There is always the potential for an electrical shock when working around any level of voltage. It is safest to work around de-energized equipment. Since this is not always possible, it is necessary to establish hazard boundaries and understand how to use protective devices and equipment to reduce your risk when working around energized equipment.

OSHA issues regulations addressing safety issues in the workplace, including electrical safety issues. While OSHA is a federal agency and its regulations are law, OSHA also relies on national

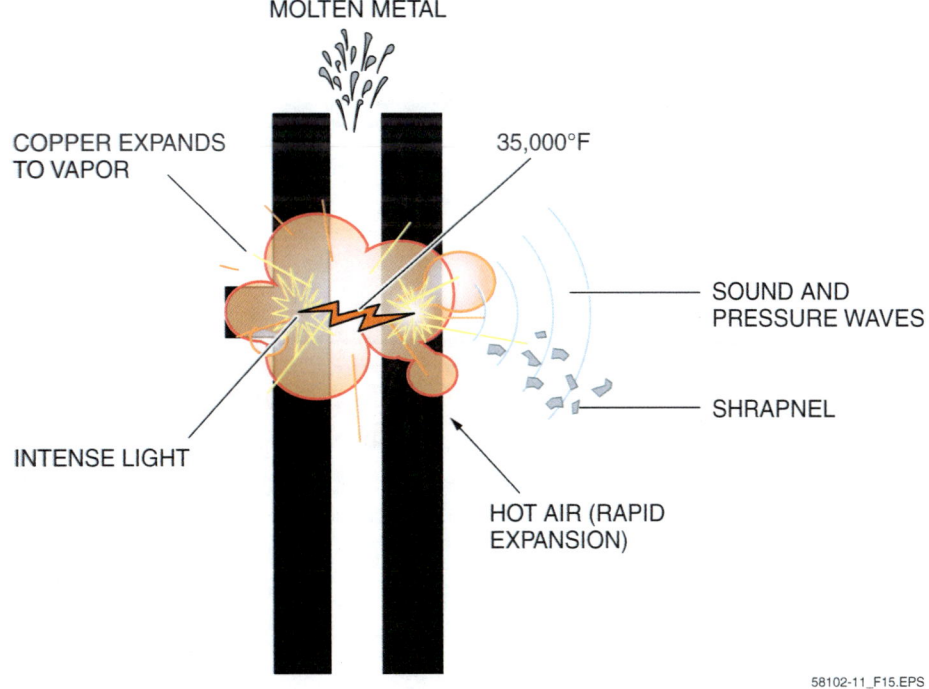

MOLTEN METAL

COPPER EXPANDS TO VAPOR

35,000°F

SOUND AND PRESSURE WAVES

SHRAPNEL

INTENSE LIGHT

HOT AIR (RAPID EXPANSION)

58102-11_F15.EPS

Figure 15 Arc flash.

consensus standards for certain requirements. For electrical safety, OSHA recognizes several standards from the National Fire Protection Association (NFPA). *NFPA 70, National Electrical Code® (NEC®),* provides requirements for electrical installations. *NFPA 70E®, Standard for Electrical Safety in the Workplace,* provides practical safe working requirements related to hazards arising from the use of electricity.

Special safety procedures are required when working on or near circuits with voltage levels of more than 50V line-to-line. Distance is the best protection against electrical hazards. That concept is so important it bears repeating. Workers who remain a safe distance from energized equipment increase their safety from electrical hazards.

Only a qualified worker may work unsupervised in areas with unguarded, uninsulated energized lines or equipment operating at 50V or more. An unqualified worker must remain well out of the danger zone at specified distances as follows:

- There is no distance specified for voltages less than 50V
- 10 feet (3.05 meters) for voltages from 50V to 750V
- 20 feet (6.1 meters) for voltages greater than 750V

Industry standards have established specific limits of approach to exposed energized parts. These limits are for personal protection and are called approach boundaries. Hazard boundaries include a flash protection boundary, shock protection boundary, and minimum approach distance (MAD). Only qualified workers are allowed within flash and shock protection boundaries and MADs. Unqualified personnel must be trained to stay away from potentially dangerous electrical equipment and processes. *Figure 16* shows a diagram comparing flash and shock protection boundaries. When working inside these boundaries, special PPE, tools, and other equipment must be used.

When working with electrical equipment, assume that the equipment is energized until you have personally verified otherwise. You should assume that all electrical equipment presents potential shock hazards, flash hazards, and blast hazards regardless of voltage sources until you have verified otherwise with the appropriate instruments.

Ideally, work on or near electrical equipment would always be performed with no electrical power applied (also known as an electrically safe work condition), but that is not always possible. Therefore, equipment installations are analyzed by specially trained workers to identify electrical hazards and to establish hazard boundaries.

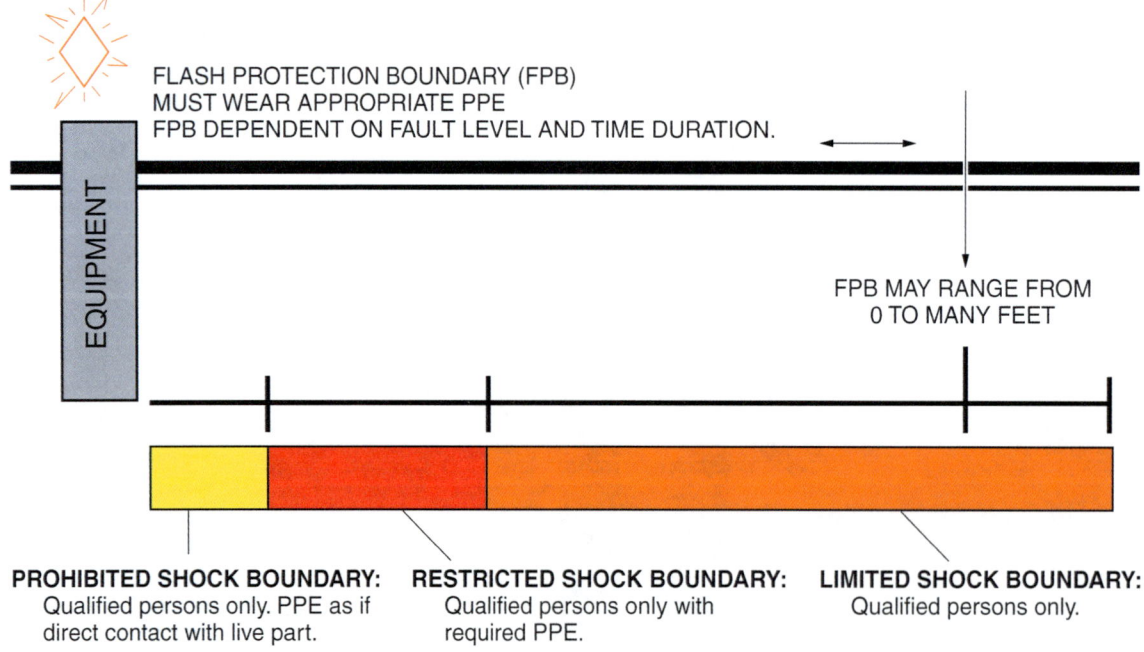

Figure 16 Approach limits.

Every possible electrical hazard within a work area must be analyzed and documented. This is called a flash hazard analysis. These areas must be clearly marked with appropriate signs (*Figure 17*) indicating the hazard as well as information about what PPE must be worn in the area. Specific PPE for a given situation are based on the information gathered from the analysis of a given hazard. That documented data includes all the electrical hazards (arc flash, blast, and shock). After all hazards have been documented, all personnel (qualified and unqualified) working in the area must be trained to recognize and avoid the identified hazards. Only qualified workers using all required PPE are allowed to enter and work inside the electrical hazard boundaries.

5.4.1 Flash Protection Boundary

When an arc flash hazard exists, an arc flash protection boundary must be established. This boundary is determined by how far away a person would need to be located to avoid receiving serious burns in the event of an arc flash. Anyone within the flash protection boundary is exposed to the possibility of both second-degree (blistering) and third-degree (tissue-destroying) burns should arcing occur.

Workers inside the arc flash protection boundary must wear flame-retardant clothing, including underwear. The heat released during arcing can be at temperatures high enough to melt fibers in clothing made from acetate, nylon, polyester, and rayon, which would increase the injuries to a worker. OSHA prohibits any clothing made from these fibers (fully or blends based on them) unless the fibers have been specially treated to resist burning and melting.

Depending on the situation, an arc flash protection boundary might be within, or outside of (exceed), a shock protection boundary. Many electrical safety programs establish both the flash protection boundary and the outer shock protection boundary at whatever distance is greater, as determined by the hazard analysis.

When an electrical fault causes an arc flash (*Figure 18*), the explosion produces both a fireball and a shock wave extending away from the arc flash location. Anyone within the flash protection boundary is exposed to searing heat as well as an extremely bright light that may cause pain and temporary loss of vision. The heat from arc flashes is often hot enough to melt metal fixtures, causing a hazard from flying molten metal.

The flash protection boundary is determined for thermal energy, but an arc blast accompanies the arc flash. The blast creates a shock wave that can blow equipment apart and blow people away from the area. Shrapnel, toxic gases, and copper vapor explode in all directions. The blast also creates sound waves that can damage hearing. The amount of current flowing through the fault affects the size of the arc flash.

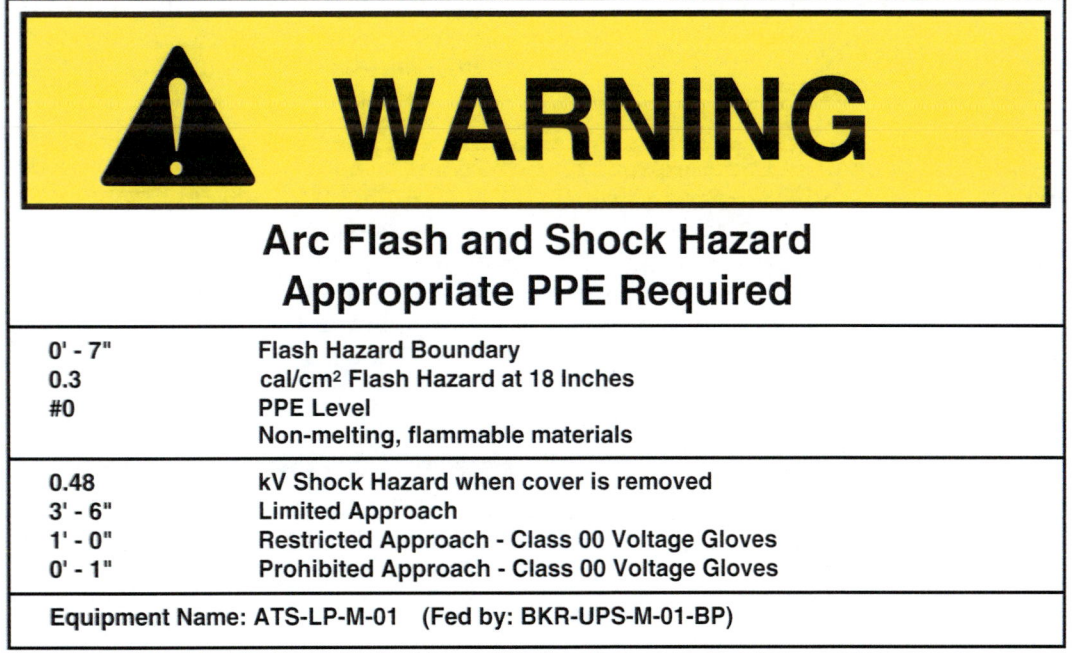

0' - 7"	Flash Hazard Boundary
0.3	cal/cm² Flash Hazard at 18 Inches
#0	PPE Level
	Non-melting, flammable materials
0.48	kV Shock Hazard when cover is removed
3' - 6"	Limited Approach
1' - 0"	Restricted Approach - Class 00 Voltage Gloves
0' - 1"	Prohibited Approach - Class 00 Voltage Gloves

Equipment Name: ATS-LP-M-01   (Fed by: BKR-UPS-M-01-BP)

58102-11_F17.EPS

Figure 17 Electrical hazard warning sign.

Figure 18 Arc flash.

58102-11_F18.EPS

5.4.2 Shock Protection Boundaries

There are three electrical shock protection boundaries or limits of approach. *NFPA 70E®* identifies the following electrical shock boundaries:

- Limited approach boundary
- Restricted approach boundary
- Prohibited approach boundary

Table 2 shows the shock protection approach boundaries to listed exposed energized parts. Column 1 shows the different voltage levels, measured phase-to-phase. Columns 2 and 3 cover the limited approach boundary. Column 2 shows the required distance from an exposed movable conductor (such as an overhead line), while Column 3 shows the distance from an exposed fixed circuit part or conductor. Column 4 covers the restricted approach boundary, and Column 5 addresses the prohibited approach boundary.

The exposed energized component can be a wire or a mechanical component inside the electrical equipment. All boundary distances are measured from that point. When establishing boundaries, exposed movable conductors are treated differently than exposed fixed circuit parts or conductors.

- *Limited approach boundary* – This is a shock protection boundary at a specified distance from an exposed energized part that can be crossed only by qualified persons. Unqualified persons may cross a limited approach boundary for purposes of on-the-job training and when escorted by a qualified person. All unqualified personnel in the area must be made aware of the hazards and warned not to cross the boundary.
- *Restricted approach boundary* – This is a shock protection boundary that, due to its proximity to exposed energized parts, requires the use of shock protection techniques and equipment when crossed. The restricted approach boundary may be crossed only by qualified persons using the required PPE (*Figure 19*) and authorized by an energized electrical work permit. Work within the restricted approach boundary requires that rubber-insulated tools and equipment be used within that boundary. It also requires the use of insulated tools for voltages of 1,000V and below, or live-line tools above 1,000V. The restricted approach boundaries include an added safety margin to compensate for inadvertent movement of the worker. Many shock and arc flash incidents occur because the tool or test-equipment lead becomes part of the circuit path to an exposed energized part. Some estimates are that 75 percent or more of arc incidents begin in this way.
- *Prohibited approach boundary* – This is a shock protection boundary at a specified distance from an exposed energized part. Work within this boundary is considered the same as making contact with the energized part. Any part of the body crossing the prohibited approach boundary must be suitably insulated and protected. There is no margin for inadvertent movement in the prohibited approach boundary.

> **WARNING!**
>
> Any workers exposed to electrical hazards must be qualified to manage the hazard and must use required PPE. If they are not exposed to an electrical hazard but have the potential to be, they must be informed of the hazard and instructed in how to avoid it.

Table 2 Approach Boundaries for Shock Protection [Data from *NFPA 70E*® *Table 130.2(C)*]

Nominal Voltage, Between Phases[1]	Limited Approach Boundary		Restricted Approach Boundary	Prohibited Approach Boundary
	Exposed Movable Conductors[2]	Exposed Immobile Circuit Part		
< 50V	N/A	N/A	N/A	N/A
50V – 300V	3.05m (10 ft)	1.07m (3 ft, 6 in)	Do Not Touch	Do Not Touch
301V – 750V	3.05m (10 ft)	1.07m (3 ft, 6 in)	305mm (1 ft)	25.4mm (1 in)
751V – 15kV	3.05m (10 ft)	1.53m (5 ft)	660.4mm (2 ft, 2 in)	177.8mm (7 in)
15.1kV – 36kV	3.05m (10 ft)	1.83m (6 ft)	787.4mm (2 ft, 7 in)	254mm (10 in)
36.1 kV – 46kV	3.05m (10 ft)	2.44m (8 ft)	838.2mm (2 ft, 9 in)	431.8mm (1 ft, 5 in)
46.1kV – 72.5kV	3.05m (10 ft)	2.44m (8 ft)	1m (3 ft, 3 in)	660mm (2 ft, 2 in)
72.6kV – 121kV	3.25m (10 ft, 8 in)	2.44m (8 ft)	1.29m (3 ft, 4 in)	838mm (2 ft, 9 in)
138kV – 145kV	3.36m (11 ft)	3.05m (10 ft)	1.15m (3 ft, 10 in)	1.02m (3 ft, 4 in)
161kV – 169kV	3.56m (11 ft, 8 in)	3.56m (11 ft, 8 in)	1.29m (4 ft, 3 in)	1.14m (3 ft, 9 in)
230kV – 242kV	3.97m (13 ft)	3.97m (13 ft)	1.71m (5 ft, 8 in)	1.57m (5 ft, 2 in)
345kV – 362kV	4.68m (15 ft, 4 in)	4.68m (15 ft, 4 in)	2.77m (9 ft, 2 in)	2.79m (8 ft, 8 in)
500kV – 550kV	5.8m (19 ft)	5.8m (19 ft)	3.61m (11 ft, 10 in)	3.54m (11 ft, 4 in)
765kV – 800kV	7.24m (23 ft, 9 in)	7.24m (23 ft 9 in)	4.84m (15 ft, 11 in)	4.7m (15 ft, 5 in)

[1] For single phase systems, determine the voltage range by multiplying the maximum phase-to-ground voltage by a factor of 1.732.

[2] Exposed movable conductors is a term usually assigned to overhead pole-to-pole utility lines. Technically, it is a condition where the distance between the conductor and the person is not under the person's complete control.

58102-11_T02.EPS

5.4.3 Minimum Approach Distance

OSHA has established MADs for working in the area of live, exposed electrical wiring and components. A MAD specifies the closest distance from any energized component that an unprotected worker may approach. It specifically refers to any unprotected body part of a worker or any part of a conductive object such as a tool carried by a worker. To penetrate the MAD, a worker must wear industry-specified PPE or use special live-line tools. PPE can include insulated gloves and sleeves, hoods, and suits. It can also include devices that are attached to equipment, such as rubber blankets, hose covers, and guards. The minimum distance is based on voltage levels found at the equipment. The MAD for qualified workers in areas with unguarded, uninsulated energized lines is shown in *Table 3*.

Table 3 Energized Line Work Minimum Approach Distances

Kilovolts (AC)	Phase-to-Ground	Phase-to-Phase
0.05 to 1.0	Avoid contact	Avoid contact
1.1 to 15.0	2'-1" (0.64m)	2'-2" (0.66m)
15.1 to 36.0	2'-4" (0.72m)	2'-7" (0.77m)
36.1 to 46.0	2'-7" (0.77m)	2'-10" (0.85m)
46.1 to 72.5	3'-0" (0.90m)	3'-6" (1.05m)
72.6 to 121	3'-2" (0.95m)	4'-3" (1.29m)
138 to 145	3'-7" (1.09m)	4'-11" (1.50m)
161 to 169	4'-0" (1.22m)	5'-8" (1.71m)
230 to 242	5'-3" (1.59m)	7'-6" (2.27m)
345 to 362	8'-6" (2.59m)	12'-6" (3.80m)
500 to 550	11'-3" (3.42m)	18'-1" (5.50m)
765 to 800	14'-11" (4.53m)	26'-0" (7.91m)

From *OSHA 29 CFR 1910.269 Electric Power Generation, Transmission, and Distribution*

58102-11_T03.EPS

Figure 19 Worker using appropriate PPE inside a restricted approach boundary.

> **NOTE**
>
> When you wear an insulating device, keep in mind that only the protected body part may violate the MAD. For example, when wearing insulating gloves, only your hands and the parts of your arms that are covered by the gloves can be inside the MAD. Uncovered parts of arms must remain outside the MAD.

5.5.0 De-Energized Equipment

Whenever possible, you will de-energize equipment before you begin working on it. Your employer will have written guidelines that you must follow to de-energize any equipment. It is important to follow all guidelines because some circuits receive power from multiple sources. To place equipment in a safe work condition, all energy sources must be removed. Once equipment is de-energized, LOTO devices must be attached to all sources to prevent someone from unknowingly energizing the circuit. Then you must verify that

no power is applied to the circuit using a voltmeter. Finally, temporary grounding devices must be placed at the area where the work is performed to protect workers from hazardous differences of electrical potentials should the circuit somehow become energized.

5.6.0 Lockout/Tagout

The understanding and proper application of lockout/tagout procedures is essential to the safety of the wind turbine service technician. Lockout and tagout procedures not only help prevent injury from electrocution, but also help prevent powered equipment from inadvertently starting and causing physical injury. In the tight confines of the turbine nacelle, there is little room to avoid an unexpected incident.

Lockout and tagout procedures safeguard workers against unexpected releases from various energy sources. An energy source can be electrical, mechanical, hydraulic, pneumatic, chemical, or even thermal. After the electrical power has been turned off, energy can still be stored in a device. For example, even when a hydraulic system has been turned off, hydraulic pressure may remain in all or part of the system. This energy can be released during service or repairs, creating an extremely hazardous situation. Additional dangers exist if the device or system contains chemicals, flammable liquids, high-temperature liquids, or gases.

When anyone is working on or around turbine systems, all equipment that could release energy must be shut down, drained, de-energized, or otherwise rendered harmless whenever possible. Switches, circuit breakers, valves or other components are switched off or closed, and then locks or tags (*Figure 20*) are applied so they cannot be re-energized while work is ongoing. There are generally three activities allowed, when necessary, that do not require LOTO:

- Visual inspection
- Equipment operation
- Measurements

When repair or component replacement is necessary, LOTO procedures apply for the affected panel. Some panels should not be accessed for any activity without de-energizing and locking out the energy source. The specific panels are dependent upon the turbine system you are servicing, but converters would be one example. The arc flash and blast potential inside the converter cabinet is likely too high, regardless of the PPE in use.

Generally, each lock has its own key, and the individual who applies the lock keeps the key. Plac-

Figure 20 Lockout/tagout devices.

58102-11_F20.EPS

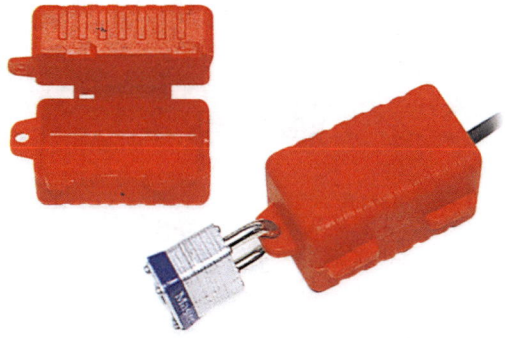

ELECTRICAL PLUG LOCKOUT

ing the key adjacent to the lock may encourage someone to remove it and restart the equipment without the installer's authorization. A variety of tags are used, depending upon the circumstances and the organization. Tags (*Figure 21*) typically have the word Danger on them, along with other printed information, writing space for comments, and a line for the installer's dated signature.

In a lockout, an energy-isolating device, such as a disconnect switch, is placed in the Off position and a lock is applied. Padlocks are popular and versatile, but other styles (*Figure 22*) are often specifically suited for the equipment being serviced.

CIRCUIT BREAKER LOCK

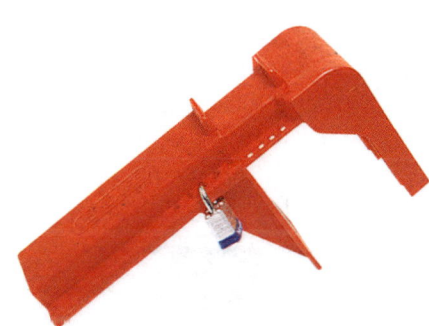

BALL VALVE LOCKOUT

Figure 21 Typical safety tags.

58102-11_F21.EPS

ELECTRICAL SWITCH LOCKOUT

58102-11_F22.EPS

Figure 22 Lockout devices.

Multiple lockout devices (*Figure 23*) are used when more than one person is accessing the equipment. When a worker needs to service a system and sees that another's lock has been applied, it is tempting to accept this as a safe situation. However, the lock may be removed by the first worker and the circuit re-energized while the second worker is still vulnerable. In these cases, the multiple lockout device allows several workers to apply locks, and each is ensured that the system cannot be restored until each lock is removed.

In a tagout, components that control power to equipment and machinery are set to a safe position and a written warning is attached. This method may be more appropriate for valves and other energy controlling devices that are difficult or impossible to lock, since tags alone cannot prevent a control device or switch from being repositioned.

It is important to emphasize that LOTO procedures are not solely related to electrical energy sources. Technicians cannot forget that other forms of energy may need to be rendered safe as well.

The exact procedures for LOTO related to wind turbine systems vary by organization. In addition, different brands and models of turbines vary dramatically in their design and construction. As a result, the location and type of control devices that must be rendered safe also vary. The procedures that follow provide one example of the steps required. Check with your instructor or site supervisor regarding the detailed LOTO procedure at your site and be sure you are familiar with the proper steps.

58102-11_F23.EPS

Figure 23 Multiple lockout/tagout device.

5.6.1 Lockout/Tagout Procedure

The following steps should be taken in sequence to complete a lockout or tagout procedure:

Step 1 Ensure that local guidelines and regulations for LOTO have not changed since you last used the procedure.

Step 2 Identify all authorized and affected workers involved with the pending LOTO. Notify those individuals that a LOTO is being invoked and explain why it is necessary.

Step 3 Shut down the equipment or system using the appropriate procedures for the situation.

Step 4 Lock out all energy sources, then physically test disconnects and other locked-out controls to ensure they cannot be changed to the On or activated position.

Step 5 As an added measure of protection, remove any key control wires from relays or contactors and isolate the wire end with electrical tape or another suitable material.

Step 6 Lock and tag other required switches. Each employee must affix a separate lock and tag when multiple service personnel are working on the same, or a related, system.

Step 7 Dissipate any stored electrical energy by attaching the equipment to an earth ground. Check as necessary to ensure hydraulic or pneumatic pressure has been relieved, and that the temperature of any surface or substance has reached a safe range.

Step 8 First, verify that any electrical or other testing device needed is functional using a known source. Then use testing devices and meters to confirm that all energy sources involved have been rendered safe for service.

Step 9 If it is necessary to leave the area temporarily, visually inspect and retest to confirm there has been no change in switch and control positions, and that all LOTO devices remain in place.

Wind Turbine Fatality

In August 2007, a wind turbine tower collapsed, killing one worker and injuring a second. No structural problem with the tower was found. The accident was the result of rotor overspeed, causing excessive stress in the tower structure.

58102-11_SA01.EPS

The tragedy occurred at a wind site in the United States. A service brake was applied to stop the turbine for service, and then a technician entered the turbine hub. Once inside, the technician positioned each of the three blades in its maximum wind resistance position. He then shut off power to them and installed lockout devices. Once the service was complete, the technician failed to remove the lockout device, exited the hub, and returned the turbine to operation while still in the nacelle. Since the pitch of the blades could not be controlled in the substantial winds, the rotor quickly exceeded its normal speed. As the blades flexed dramatically due to the force, one blade made contact with the tower and the entire turbine assembly self-destructed. The worker in the nacelle was killed, while the second worker who was descending the tower ladder was injured.

The OSHA investigation revealed that both workers had less than two months experience and were working unsupervised. They did not receive sufficient training and were not aware of the potential for disaster. It was further determined that the company's procedures did not fully comply with Oregon OSHA regulations. The turbine system interlocks have since been modified to prevent this specific sequence of events from happening again, and new procedures have been developed.

Trainees should take note that experience and training alone will not prevent disaster, but it will provide them with the skills and knowledge to do so.

5.6.2 Restoration of Energy

After work has been completed, use the following steps to restore energy sources.

Step 1 Completely reassemble and secure the equipment or system.

Step 2 Confirm through inventory that all equipment and tools are accounted for and removed from the work area.

Step 3 Restore any control wiring that may have been removed and isolated.

Step 4 Ensure that any required preparations for startup of the affected system or turbine have been made.

Step 5 Remove the locks and tags from any controls or disconnect switches. Each employee who has applied a lock or tag must remove their own safety devices.

Step 6 Notify all affected personnel that the LOTO period has ended and the equipment or systems will be re-energized.

Step 7 Operate or position any switches or valves as necessary to restore energy.

5.6.3 Emergency Lockout/Tagout Removal

There are times when the removal of a LOTO device by someone other than the installer is re-

P.R.O.P.E.R. Lockout/ Tagout Procedures

Using the acronym P.R.O.P.E.R. can help technicians ensure that the proper steps have been taken to prevent accidents and serious injury.

P — Process shut down
R — Recognize energy sources
O — Off
P — Place locks and tags
E — Release any stored energy
R — Return controls to neutral

quired. In these unique cases, only an authorized supervisor should remove the lock or tag.

The original installer must be informed of the action prior to removal if possible. This allows them the opportunity to provide any important information regarding the condition of the turbine or system being serviced. Written verification of the action, along with an explanation, should be prepared and filed with the job records.

5.6.4 OSHA Standard 1910.269

OSHA Standard 1910.269 Electric Power Generation, Transmission, and Distribution covers the operation and maintenance of the subject facilities. Since wind turbine generators generate electrical power, they are covered by this standard.

There are times when equipment service is required uptower, while the isolating controls and circuit breakers are located in the low-voltage distribution module (LVDM) or other panel downtower. In order to understand the impact of the standard on LOTO procedures for wind sites, several key points of the standard are presented here:

- *1910.269(d)(8)(v)* – If energy isolating devices are installed in a central location and are under the exclusive control of a system operator, the following requirements apply:
    - *1910.269(d)(8)(v)(A)*: The employer shall use a procedure that affords employees a level of protection equivalent to that provided by

the implementation of a personal lockout or tagout device.
    - *1910.269(d)(8)(v)(B)*: The system operator shall place and remove lockout and tagout devices in place of the authorized employee under paragraphs (d)(4), (d)(6)(iv), and (d)(7)(iv) of this section.
    - *1910.269(d)(8)(v)(C)*: Provisions shall be made to identify the authorized employee who is responsible for (that is, being protected by) the lockout or tagout device, to transfer responsibility for lockout and tagout devices, and to ensure that an authorized employee requesting removal or transfer of a lockout or tagout device is the one responsible for it before the device is removed or transferred.

The common interpretation of these entries, as it applies to the typical HAWT, identifies the central location as the downtower area and the cabinets within. A system operator is a worker who is authorized to apply locks and tags and is exercising exclusive control over the downtower area.

The standard calls for a procedure that is equivalent to personal installation of a lock or tag. One such method allows the system operator, while exercising control over the downtower area, to place locks and tags on behalf of the uptower worker. The system operator monitors the downtower area and maintains a view of, and access to, all electrical cabinets and breakers. The uptower worker, referred to as the autho-

---

On Site

Not for Lifting

Never use a come-along for vertical overhead lifting. Be careful not to confuse a come-along with a ratchet lever manual hoist. Ratchet lever hoists have both a friction-type holding brake and a ratchet-and-pawl load control brake. Come-alongs have only a spring-loaded ratchet that holds the pawl in place. If this simple mechanism fails, the load will fall. Come-alongs are used only for moving or pulling horizontally over the ground.

COME-ALONG
58102-11_SA02A.EPS

RATCHET LEVER HOIST
58102-11_SA02B.EPS

---

rized employee in *1910.269(d)(8)(v)(B)*, signs over responsibility for LOTO to the system operator. However, the system operator cannot remove locks or tags without the expressed consent of the uptower worker.

It is important to note this interpretation may not be the same as that of your future wind site maintenance organization or operating firm. Before initiating such a procedure, be sure that you are operating within the guidelines provided by the controlling organization.

6.0.0 RIGGING

The wind energy technician may be called upon to assist in the construction of turbines and the associated towers. In addition, much smaller rigging and lifting projects will occur during the normal working day for most wind energy technicians. Hoisting systems are a part of all commercial- and utility-scale HAWTs so that tools and equipment can be moved between the ground and the nacelle. The proper use of these hoisting systems requires some individual practice on the job with competent technicians, using the rigging skills learned in the Power Industry Fundamentals, *Basic Rigging* module.

Rigging and lifting projects for large HAWTs are some of the most complex projects of their kind, primarily due to the height and unusual configuration of the rotor (*Figure 24*). The wind energy environment is generally remote, without any pavement or other stable surface nearby. Only one turbine can be assembled once a crane is set up, since the turbines must maintain some separation in distance. Therefore, mobile cranes are a necessity. With tower heights of 100 meters common, mobile cranes used in wind turbine assembly must be relatively unique and very large. A significant amount of experience and training is required to supervise such a lifting project. Although advanced training for these positions is beyond the scope of this program, it is important to understand some of the requirements for these complex lifting tasks.

6.1.0 Hoists

Hoists provide a mechanical advantage for lifting loads to the nacelle of a HAWT. Chain hoists for wind turbine applications are typically electric. All chain hoists use a gear system to lift heavy loads. The load is hooked onto a chain and the gearing turns the sprocket. The hoist assembly can be suspended by a hook (*Figure 25*), or mounted on a trolley (I- or T-beam) as shown in *Figure 26*. Once the load has reached the nacelle,

58102-11_F24.EPS

Figure 24 Wind turbine rotor lift in progress.

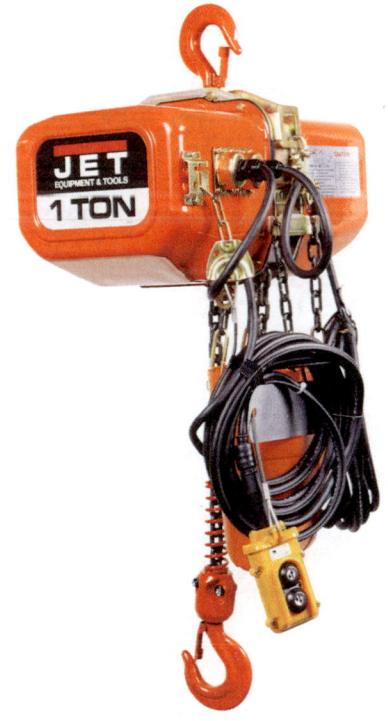

58102-11_F25.EPS

Figure 25 Hook-mounted electric hoist.

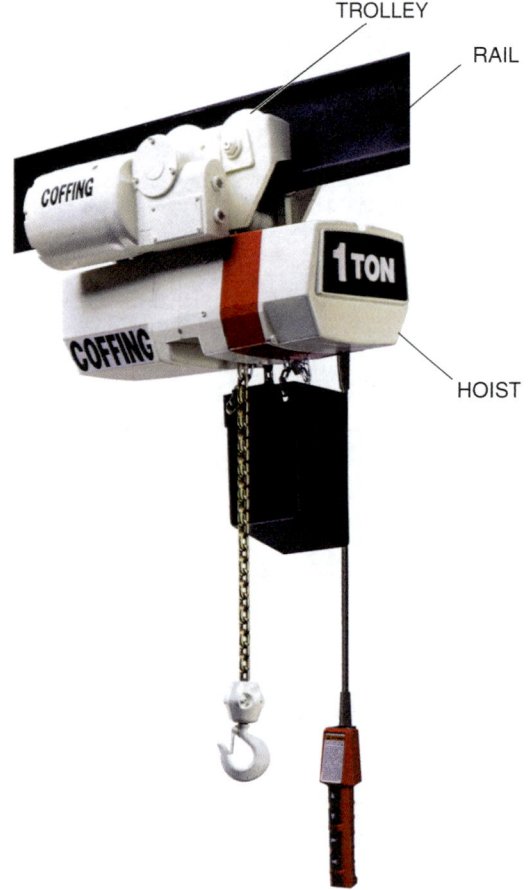

TROLLEY

RAIL

HOIST

Figure 26 Electric hoist on a trolley.

58102-11_F26.EPS

the trolley allows it to be moved to another position inside without physical strain. Some may also be mounted on a pivoting column or mast, such as the one shown in *Figure 27*.

To use an electric chain hoist, the operator in the nacelle positions the hoist over the equipment hatch in the nacelle and lowers the hook using a wired control pad. An attendant on the ground rigs the load for lifting. Since wind towers are very high, a long length of chain is required.

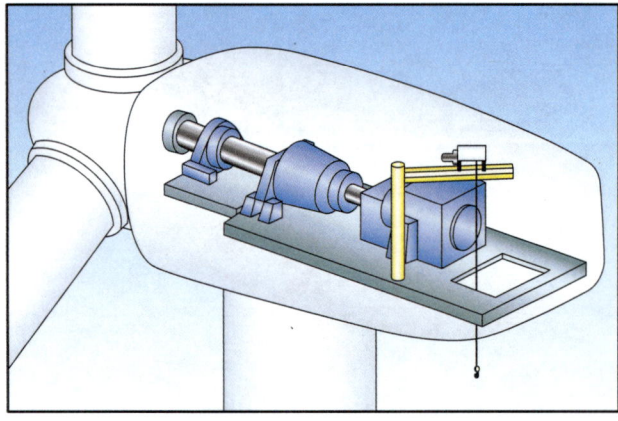

58102-11_F27.EPS

Figure 27 Turbine hoist on a pivoting mast.

Models used in wind turbines often have a heavy-duty bag attached to collect the excess chain. As a result, lifting in substantial winds may be a problem due to the potential for the load to swing uncontrollably. A tag line (*Figure 28*) may be necessary to allow workers on the ground to control the load as it ascends.

Remember the following important points regarding the use of hoists:

• Always use the appropriate personal protective equipment when working with and around hoists. While operating a hoist from the nacelle, connection of fall arrestance and fall protection lanyards is required.
• Make sure the load is properly balanced and correctly attached before initiating the lift.
• Keep gears, ropes, and chains clean.
• Lubricate gears periodically to keep the wheels from freezing up.

6.2.0 Lift Planning

Before conducting a crane lift, most construction sites and companies require that a lift plan be completed and signed by competent personnel. Since wind turbine and tower assembly at a given site is rather repetitive, the lift plan typically remains the same throughout the project. The primary difference at each turbine location is related to the terrain. Beyond that, a single lift plan for the entire project will likely suffice.

A lift plan contains information relative to the crane, load, and rigging. Any special instructions or restrictions would also be listed. It is important to remember that any change from the original plan, no matter how small, requires a review of the plan at a minimum. Significant changes may lead to a new plan. Since wind energy sites differ primarily by terrain, the lift plan must be reviewed for each specific turbine and tower site and changed accordingly.

A typical lift plan breaks down the following elements and information:

• Weight:
  – Equipment condition
  – Weight empty
  – Weight of overhead ball
  – Weight of block
  – Weight of lifting bar (when used)
  – Weight of slings and shackles
  – Weight of jib (if erected or stowed and not used)
  – Weight of overhaul ball on jib

- Weight of wire rope
- Weight allowance for any unaccounted material mounted on the equipment

- Jib:
  - Jib to be used
  - Length of jib
  - Angle of jib
  - Rated capacity of jib
  - Crane placement:
  - Smooth solid foundation within one percent of grade
  - Electrical hazards in the area
  - Obstacles or obstructions to lift or swing
  - Swing direction and degree
  - Area of operation (over-the-front, over-the-side, or over-the-rear)

- Rope:
  - Number of parts of rope – tag lines, etc.
  - Size of rope
  - Reeving capacity required

- Size and types of slings:
  - Sling selection
  - Shackle selection
  - Sizing and types

- Crane:
  - Type of crane
  - Crane capacity
  - Lifting arrangement required

- Other:
  - Special instructions or restrictions for crane, rigging, or lift
  - Diagrams for crane, load replacement, and rigging configuration
  - Signature of job supervisor, lift plan checker, and operator
  - Derigging plan

The personnel responsible for completing the plan should consult reference information to assist in a safe operation. The reference information can include the crane's specifications, site data

58102-11_F28.EPS

Figure 28 Tag lines steady the load.

relative to the condition of the ground and surrounding vegetation, and the specifications for each tower section, nacelle, hub, and rotor assembly.

There are OSHA and American Society of Mechanical Engineers (ASME) standards that cover the safe usage of mobile cranes. The related standards are the following:

- *OSHA 29, CFR 1910.180*
- *OSHA 29, CFR 1926, Subpart N*
- *ASME B30.5*

6.3.0 Critical Lifts

A designated person must classify each lift into one of three categories prior to beginning the lift planning process: ordinary, pre-engineered production, or critical. A lift must be designated as

---

Think About It

Consider a planned crane lift at a remote wind turbine site, with a crew of 16 workers involved. What do you think the overall cost would be if the lift planner forgets the necessary rope for two of the tag lines, stopping the project completely for half a day?

---

critical if collision, upset, or dropping could result in any one of the following:

- An unacceptable risk of personnel injury or significant adverse health impact (on- or off-site)
- A significant release of radioactive or other hazardous material
- Undetectable damage that would jeopardize future operations or the safety of a facility
- Damage that would result in unacceptable delay to the schedule or other significant program impact

Several characteristics of wind turbine assembly lifts cause them to be classified as critical:

- Due to their enormous size, transportation of the components to the site is very costly and challenging (*Figure 29*). Turbine blades and rotors especially are very expensive and can easily be damaged beyond use during the lift. The lift should be deemed critical since collision, upset, or dropping would cause a major delay in assembly and increase costs dramatically.
- Rotors can easily have a diameter of 100 meters. If dropped, it is impossible to determine how and where they will actually come to a rest. In addition, personnel must be on the tower or in the nacelle during the lift to secure components in place (*Figure 30*). Both issues represent a significant risk to personnel on the ground and uptower.
- Damage to rotor blades that may not be visually detectable could lead to serious structural failure once they are mounted and the turbine is placed into operation. The lift should be considered as critical due to the potential for undetectable damage that would place personnel and the entire turbine assembly at risk.

58102-11_F30.EPS

Figure 30 Mounting the hub and rotor assembly.

A lift should also be designated as critical if the load requires exceptional care in handling because of size, weight, close-tolerance installation, high susceptibility to damage, or other unusual factors. Wind turbine and tower assembly qualifies as critical for all of these reasons.

For critical lifts, the operating organization appoints a rigging superintendent or equally qualified person. The rigging superintendent must be present throughout the lift and must ensure that a lift plan is prepared that defines the operation and includes the following:

- Identification of the items to be moved, their weight, dimensions, and center of gravity
- Identification of all equipment to be used, along with their capacities
- Rigging sketches that include lifting points, methods of attachment, and other critical rigging information
- Operating procedures and special instructions to operators

Experienced operators who have been trained and qualified to operate the specific equipment to be used must be assigned to make the lift. Only designated, qualified signalers must give signals to the operator(s). However, operators are always trained to obey a Stop signal at any time, regardless of who gives this signal.

6.4.0 Lift Plan Implementation

For any lift plan to be successful, all critical data must be shared with all personnel involved at a pre-lift meeting. The signaler(s), riggers, operators, management, and site supervisors should all be involved in this meeting. At this time, the operator and signaler are assigned their tasks so that all others know who they are. The means

58102-11_F29.EPS

Figure 29 Turbine blade in transit.

for constant communication among the crew is also discussed. If tag lines are to be used, the responsibility for their management is assigned. Weather and other conditions that can change during the course of the job are discussed. Emergency procedures are reviewed, and then all personnel with responsibilities must sign the completed lift plan, acknowledging their understanding of it.

One of the most dangerous aspects of lifts for wind turbine assembly is the repetitive nature of the task. Repetitive tasks tend to cause workers to become complacent, especially after a number of assemblies have been completed without incident. With such great risk to personnel and the materials at stake, complacency is the last thing a team needs during a critical lift.

7.0.0 SAFE DRIVING

One thing you are not likely find at the average wind energy site is a stoplight. In spite of the lack of heavy traffic at the site itself, good driving skills for the terrain are essential. Many wind energy technicians spend a considerable amount of time behind the wheel, traveling between wind sites.

7.1.0 Driving Company Vehicles

Most wind turbine technicians are required to drive a company-owned vehicle (*Figure 31*). Trucks and toolboxes are needed daily to transport personnel, parts, and equipment to the individual turbines. Technicians are not be expected to utilize their own vehicle unless specifically contracted to do so.

Keep in mind that driving is a privilege, and without a valid driver's license, you may be of limited value to your new employer. In addition, your driving record needs to be as clean as possible. Employers often consider your driving record to be a direct reflection of your attitude towards driving, and possibly a direct reflection of your character in general. A poor driving record can prevent employment in any field where driving a company vehicle is a necessity.

Employers must insure the company's vehicles against damage. Be aware of the relationship between the cost of auto insurance and a clean driving record. Insurers who provide coverage for fleet vehicles must charge the client for the level of risk to which the insurance company is exposed. A poor driving record increases the potential risk, and, therefore the cost of insurance, for an employer. This additional cost, plus the added risk the employer assumes for the employee's driving behavior, may simply exceed an employee's value. A good driving

58102-11_F31.EPS

Figure 31 Company-owned service vehicle.

record can help you obtain employment, and maintaining a good record will help you keep it.

Regardless of what vehicle you are assigned as a driver, take the time to read and study the owner's manual. Be familiar with all aspects of its safe operation and periodic maintenance requirements. The manual provides all the necessary information for operation of the four-wheel drive system, if so equipped. Proper operation, maintenance, and cleaning of the company vehicle earns your employer's admiration and ensures that the vehicle will not leave you stranded. Always wear a seatbelt and use common sense behind the wheel of any vehicle.

7.2.0 Off-Road and Inclement Weather Driving

Although your career may require a great deal of highway travel, it may also require substantial travel through rugged terrain (*Figure 32*). Roads at wind turbine sites are seldom paved and are subject to a great deal of damage. Snow and ice, heavy rainfall, and heavy construction vehicles take their toll on the road. The first rule of off-road driving must be to slow down.

Figure 32 Rugged terrain.

58102-11_F32.EPS

Some service vehicles are equipped with four-wheel drive, also known as a 4 × 4, to allow for safer, more reliable transportation off paved roads. Simply driving a 4 × 4 can give the operator a false sense of security due to its available power and torque. Some off-road driving tips are provided here.

7.2.1 Steep Inclines

Steep hills (*Figure 33*) are a common obstacle at wind turbine sites, and can be highly dangerous. When traveling up an incline, be certain there is a clear exit or plateau at the top. Check in advance that there are no hazards at the top, since your vision ahead may be obscured as you crest the hill. Use second or even third gear to help avoid wheel spin and begin your ascent at a relatively low speed. An engine rpm of 1,500 to 2,000 allows you to climb the hill slowly and under control. Continue at the same pace all the way to the top. If there is not sufficient traction or power to maintain speed and control, stop and reconsider the route. Do not back up and charge the hill at a high speed, relying on momentum to help carry you to the top. This can easily result in a loss of control and lead to both vehicle damage and personal injury.

58102-11_F33.EPS

Figure 33 Climbing a steep hill.

During a descent, drive in first gear to allow the engine to assist in braking. Always approach the bottom of the hill at a low speed to keep the front bumper from making contact with the road ahead. Drive straight down the hill. Descending a hill at an angle can allow the vehicle to roll over sideways.

7.2.2 Crossing Water

Some sites may have a permanent creek or other water hazard to cross. Heavy rainfall can also turn otherwise dry land into a raging river (*Figure 34*). Always be certain that there is a firm, drivable surface beneath the water before attempting to make a crossing.

Crossing deep water should not be attempted. A 4 × 4 is not an amphibious vehicle, nor is it waterproof. Before entering unknown water, confirm the depth and underlying structure. Do not enter rushing water, walking or driving, if you do not know precisely what is ahead. If the rate of flow seems calm enough for walking, use a branch or other suitable object to confirm what is ahead of you before taking each step. It is essential that the greatest respect be given to deep and fast-moving waters. A life can be lost quickly by underestimating the speed or depth of moving water.

Allowing the vehicle to enter a water level that is above the bottom edge of the body can be extremely dangerous. For each foot of depth, 500 pounds of lateral force are applied to the vehicle by the water. In addition, water under the body causes the vehicle to float. One foot of water depth on the vehicle body effectively reduces the weight of the vehicle by 1,500 pounds. With water pushing both up and perpendicular to the vehicle, it can easily be washed away with no way for the driver to escape. As a rule of thumb, the depth limit of a vehicle is best considered as the bottom of the vehicle body. Even below this depth, water that is rushing too quickly should be avoided, especially if there is any incline in the direction of water flow.

If a crossing is reasonable, maintain a slow but steady speed. Do not stop while in the water. Once out of the water, check to ensure your brakes are working properly. Hold the brake pedal lightly for a few seconds while moving forward to dry them off.

7.2.3 Mud

Mud, deep or shallow, is encountered regularly. For smaller areas, build up some speed before hitting the spot and drive as straight as possible. The mud naturally slows forward motion, but try to keep the speed consistent. Use second or even third gear if needed. If the tires start to spin, ease off the gas. If you continue to lose traction, try turning the steering wheel a short way left and right in quick bursts as you accelerate. If you anticipate crossing very deep mud, it may be worth letting a little air out of the tires to improve traction and increase ground contact. Remember to reinflate to the correct pressure after passing the mud. It is always a good idea to carry a small 12-volt air compressor in off-road settings.

Large expanses of mud can be challenging. A bumper-mounted winch (*Figure 35*) can save the day when a vehicle becomes stuck. Look for ways around the obstruction first. Do not attempt to cross slippery, muddy areas that are near dropoffs or other terrain hazards, where vehicle control is often questionable. Avoid spinning the wheels when the vehicle is not moving forward to prevent the vehicle from digging holes.

58102-11_F34.EPS

Figure 34 Crossing deep water.

58102-11_F35.EPS

Figure 35 Truck winch.

7.2.4 Ice and Snow

Whether driving on pavement or off-road, ice and snow are surfaces that drivers typically travel on with the least amount of confidence. The primary point to remember is to drive slowly, allowing at least twice as much distance between yourself and the driver ahead as you would on normal surfaces. All actions, such as starts, turns, and stops should be done gently and at low speeds. Avoid sudden braking, acceleration, and turns. If the vehicle is equipped with cruise control, do not use it when driving on ice and snow. Remember that posted speed limits are for dry conditions. Traveling at the posted speed limit when ice or snow is present is not only unsafe, but may also result in a traffic citation.

Overpasses, bridges, and shaded areas are far more likely to be covered with ice. Patches of thin ice on roadways, sometimes referred to as black ice, can be difficult to see. Braking must be done very gently, well in advance of the desired stopping point.

Deep snow is a serious problem when driving off-road. The vehicle lacks traction and may be pushing snow ahead of it. The snow hides all obstructions, and even deep ditches or ravines can be concealed by a snowdrift. Drivers must be extremely careful driving off-road in snowy conditions, as dangerous or deadly terrain can look completely harmless in the snow. Use higher gears to help avoid wheel spin.

7.2.5 Ground Clearance

Be familiar with the ground clearance of your vehicle. Avoid obstructions in the middle of your path that could make contact with the drive train underneath. Also, be aware of your approach and departure angles when negotiating hills and obstacles. These angles are formed by a line drawn between the tires' point of ground contact and the bottom of the bumper or body.

The length of the wheelbase has a significant effect on ground clearance when cresting a small rise (*Figure 36*). The longer the wheelbase, the closer to the ground the drive train is when the vehicle center passes over the rise.

Always remember that both you and the vehicle represent major investments to your employer. Don't try to be a hero behind the wheel and jeopardize your safety as well as the safety of others. It is better to return another day than to take unnecessary risks reaching a turbine repair site.

58102-11_F36.EPS

Figure 36 Reduced ground clearance on a hillcrest.

8.0.0 AERIAL WORK PLATFORMS

Aerial work platforms, scissor lifts, and boom lifts are portable lifting devices that raise or lower workers to and from elevated job sites. All are manufactured in various models. Some are transported on a vehicle to the job site where they are unloaded. Others are trailer-mounted and towed to the job site by a vehicle. Some are permanently mounted onto a vehicle. Depending on the design, they can be used for indoor work, outdoor work, or both. Aerial work platforms and boom lifts have a single arm that extends a work platform/enclosure capable of holding one or two workers. Some models have a jointed (articulated) arm that allows the work platform to be positioned both horizontally and vertically. Scissor lifts raise a work enclosure vertically by means of crisscrossed supports. Most aerial work platforms, boom lifts, and scissor lifts are powered either hydraulically or by electric motors. *Figure 37* shows examples of aerial lifts.

OSHA Standard 1926, Subpart N defines and governs the use of aerial lifts. The following lists some guidelines for safely using lifts:

- Know the capacity and operating characteristics of the aerial platform. Do not overload the platform. Use a lift only on a solid, stable surface. Lock the wheels, especially on an incline. Avoid using a lift outdoors in stormy weather or in strong winds.
- Inspect the equipment before each use, as specified by the manufacturer, to make sure everything is in proper working order. Have any defects repaired before using the lift.
- Never modify or remove any part of the equipment unless authorized to do so by the manufacturer.

BOOM SUPPORTED WORK
PLATFORM (BOOM LIFT)

SELF-PROPELLED
ELEVATING WORK PLATFORM
(SCISSOR LIFT)

58102-11_F37.EPS

Figure 37 Aerial lifts.

- Check the job site for any hazards that may cause the lift to overturn.
- Check for hazards above, below, and all around the path of travel. Maintain a safe distance from electrical power lines and other electrical hazards.
- Prevent people from walking beneath the work area of the platform.
- Use personal fall arrest equipment (body harness, lanyard, and lifeline) if required for the type of lift being used.
- Lower the lift and lock it into place before moving the equipment. Lower the lift, shut off the engine, set the parking brake, and remove the key before leaving it unattended.
- Keep the lift free from oil, grease, wet paint, mud, or any slippery material.
- Do not lean over the guardrails of the platform, and never stand on the guardrails.
- Only use approved cages and platforms that conform to OSHA regulations. Field-improvised personnel lifts are dangerous and potentially deadly. Repairs to cages must be made by

qualified welders. A weld on a cage that fails when the cage is suspended can injure or kill those in or below the cage.

8.1.0 Scissor Lifts

Scissor lifts are powered personnel lifts. They use hydraulic rams to raise and lower the platform. Power is supplied to the hydraulic system by a battery, a gasoline engine, or a propane-powered (LPG) engine. The controls for the lift are located at an operator's station on the platform itself. Some models also include an operator's station at both the ground and the platform level. In addition to controlling the height of the platform, the operator is also able to move the unit to the front or rear and steer it as it moves.

Lift heights and capacities vary by model and manufacturer. Lift heights may be as high as 41 feet and capacities may reach 2,500 pounds. Both the height and capacity of the lift can be found on a decal conspicuously placed on the lift. Some manufacturers offer optional, manual outriggers

Figure 38 Scissor lift with outriggers.

58102-11_F38.EPS

for added stability. *Figure 38* shows an example of a large-capacity lift with outriggers.

8.1.1 Operating Precautions

No one is permitted to operate a scissor lift without training in the proper operating procedures for the particular model. This section of the module discusses general operating procedures, but is not intended to serve as an all-inclusive guide to scissor lift operation. Keep the following safety considerations in mind when operating a scissor lift:

- Read and observe the warnings and cautions placed on the unit.
- Keep the vehicle and work platform at least 15 feet from electrical power lines. Scissors lifts are not insulated to protect against arcing or electrical contact.
- Anything that is not working correctly must be repaired before using the lift.
- Do not remove any parts or safety devices.
- Make certain the immediate area is clear of all personnel and obstructions before raising or lowering the lift.
- Keep the guardrails and chains in place when operating the lift.

- Do not bridge between two scissor lifts or between a scissor lift and a building.
- Never crawl into the scissors at any point.
- Operate the vehicle only on firm, solid ground. Do not operate the vehicle near a drop-off.
- Do not impose side loads on the work platform.
- Never exceed the rated capacity of the lift.
- Do not operate the vehicle in an explosive atmosphere.
- Never leave the key in a vehicle.

8.1.2 Controls and Indicators

Each manufacturer uses different control systems. Always familiarize yourself with the particular controls of the unit you are using before you use it. The following controls are typically found on scissor lifts:

- *Power switch* – This may be key-operated and may include three positions: Off, On, and Start.
- *Steering* – This may be a toggle switch that is used to turn the vehicle to the right or left.
- *Platform* – This may be a toggle switch that moves the platform up and down.
- *Drive* – This switch may control both the speed and direction of the vehicle. Some models have dual speeds for both forward and reverse directions.
- *Emergency stop* – This switch may have two positions: Run and Emergency Stop. It is usually a protected toggle switch. In order for the power switch to work, the Emergency Stop switch must be in the Run position.

The following controls and indicators may also be found on a scissors lift:

- *Manual override* – This may be a push-button switch to lower the platform manually during an emergency. It is usually labeled with a decal and may be found at an opening on the hydraulic panel cover.
- *Emergency release for brake* – This is used if the vehicle stalls during operation. It may be a lever located behind the hydraulic tank and labeled with a decal. Once the vehicle has been restarted, the switch should be returned to the normal position.
- *Fuel gauges* – These indicate the levels of propane or gasoline. Propane-powered models also have a supply valve to open or shut the LPG supply.
- *Oil dipstick and hydraulic oil indicators* – These allow the operator to check the supply of oil in the crankcase of engine-powered models and the amount of hydraulic oil in the reservoir.

You should learn to check oil levels before operating the vehicle.

8.1.3 Operation

The following section discusses the general operating procedures for scissor lifts. It is not intended to serve as a substitute for the operator's manual for your particular unit.

Startup and shutdown procedures are different depending upon how the vehicle is powered. For battery-powered models, turn the key to the Power position (or the equivalent). Then, activate all functions before applying load to warm up the hydraulic oil. To shut down a battery-powered model, first lower the platform and then turn the key to the Off position. Remove and store the key.

Follow these steps to operate engine-powered models:

Step 1 Turn the key slowly to the Start position. Release it quickly when the engine starts. If the unit does not start, turn the key to the Off position and wait about five seconds.

> **CAUTION**
>
> To keep from damaging the starter, never hold the key in the Start position for more than 10 seconds.

> **NOTE**
>
> If the vehicle does not start, turn the switch to Off and check the gasoline or LPG supply gauge. Check the LPG supply valve to make sure it is open. Then check the condition of the batteries.

Step 2 Let the engine warm up for about two minutes. This allows the hydraulic oil to warm up.

To shut down an engine-powered model, lower the platform and turn the key to the Off position. Remove and store the key.

Before traveling with a scissor lift, check the immediate route. Avoid personnel, obstructions, other traffic, rocks, holes, debris, soft surfaces, and steep grades.

On most models, the switches that control direction and steering must be held manually. When pressure is released, these switches usually return to the center position. It may be necessary to use both hands when traveling and steering. Follow these steps to move the scissor lift to another location:

Step 1 Choose the speed range, if the vehicle is equipped with this control.

Step 2 Push and hold the Forward-Reverse switch with one hand in the desired direction of travel. At the same time, push the Steering switch to Right or Left with the other hand.

> **NOTE**
>
> On most models when the Forward-Reverse switch is released, the vehicle decelerates and stops. When the Steering switch is released, the wheels remain in the same position until moved again.

On most models, the Up and Down switches must be held in position to raise or lower the platform. The platform stops automatically when the switch is released.

> **CAUTION**
>
> Before operating the platform, be sure that the areas above and below the platform are clear of personnel, obstructions, and power lines, and that the vehicle is on firm, level ground. These precautions are necessary to prevent personal injury.

8.1.4 Maintenance

Always follow the manufacturer's recommendations for daily and periodic maintenance. General safety checks for loose or missing parts and damage must be conducted before operating the vehicle. Always check the oil before starting the engine and never operate a vehicle that is low on hydraulic oil.

At the end of the day, the batteries on electrically powered models should be recharged. Periodically check the batteries for proper electrolyte level, the tightness and condition of the cables, and any signs of corrosion or visible damage.

Check the hydraulic oil reservoir when the unit is shut down. The oil level should be up to the neck strainer, unless otherwise specified by the manufacturer. Always check the hydraulic connections for signs of leaks.

8.2.0 Boom Lifts

Boom lifts are work platforms mounted on the end of an arm that can be raised and lowered and may be able to extend beyond the mobile base. The boom section may consist of two or more telescoping sections. The machines usually can be driven by the operator from the platform. Boom lifts are generally restricted to supporting one or two workers and their tools at a time. *Figure 39* shows some typical boom lifts.

8.2.1 Application

Boom lifts come in two basic categories, those designed as off-slab units and those designed for use on paved/slab surfaces. The off-slab units have truck-type tires or even deep tread mud-type tires. Some units have oscillating axles and four-wheel drive to maneuver more easily on rough terrain. These units have a wide variety of uses in construction and outdoor maintenance.

The paved/slab units are designed for use on improved surfaces. Some are even intended for indoor use and are small enough to fit through interior doorways. The units designed for indoor use are battery powered to avoid any exhaust problems. These units are primarily used by maintenance personnel.

8.2.2 Operation

Boom lifts have two sets of controls, the first on the work platform and a second on the base of the unit. The work platform controls are the primary control station. The base-mounted controls are mostly used in emergency situations. If the worker is injured, the base controls can be used to lower the worker to the ground to provide the necessary assistance. As an additional safety feature, all boom lifts can be lowered even when there is no power to the unit.

Be sure to read and understand the vendor-provided operator's manual before attempting to use any boom lift. The following section is provided as a general guideline for operating a boom lift, but is not specific to any individual lift. *Figure 40* shows a typical boom lift control panel.

Notice that some control panel switches are guarded. This serves two purposes: it protects the switch handles and it allows the switch to move in a logical direction. For example, the switch to move up or down moves up and down. This control panel is for an engine-powered lift similar to the lifts shown in *Figure 39*. The engine could be either gasoline or LP fueled. The engine is started by setting the ignition switch to the On position and pressing the Start push button. If necessary, the Choke switch may be used during startup. Be sure to release the button as soon as the engine starts to avoid damaging the starter and engine flywheel. Once the engine is started and allowed to warm up briefly, the lift can be

ARTICULATING BOOM LIFT

TELESCOPIC BOOM LIFT

58102-11_F39.EPS

Figure 39 Boom lifts.

maneuvered into position. The engine provides power to move the unit along the ground as well as to operate the boom lifting, telescoping, and swing. The High Engine and High Drive

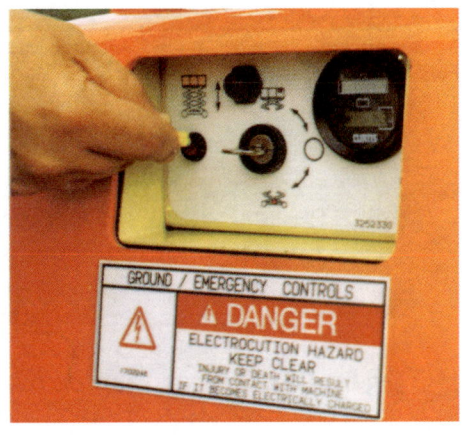 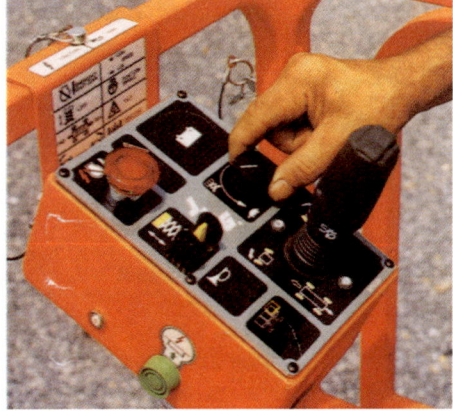

Figure 40 Boom lift control panel.

switches are set to the On position while moving the unit from place to place with the boom retracted and in the down position. The hydraulic pumps used to raise and maneuver the work platform are not powered while the High Engine and High Drive switches are in the On position. Once the unit is situated these switches are returned to the Off position.

Movement of the boom is controlled using levers or a joystick. Operators must be very careful to avoid overhead hazards while moving the unit with the boom raised.

8.2.3 Maintenance

Maintenance of boom lifts must be done in accordance with the manufacturer's recommendations. Lubrication of the boom is a large part of the maintenance requirements. Normal engine maintenance is required on the LP- or gasoline-powered units and recharging is required on battery-powered units. Servicing the hydraulic-power units of boom lifts requires checking fluid levels, cleaning or replacing filters regularly, and keeping the units clean to allow any leaks to be easily detected.

SUMMARY

Safety on the job is everyone's responsibility. In this regard, the wind turbine environment is no different from any other. To ensure the safest working conditions possible, OSHA has created many regulations to guide employers. In addition, OSHA provides assistance and consultation to employers who request it.

JSA provides employers and workers with a process to identify and avoid potential hazards. To ensure that no possible hazard is overlooked, individual tasks can be broken down to their smallest action and analyzed.

Nacelles, rotor hubs, and turbine towers all represent confined spaces, each presenting its own unique challenge. Workers must understand the hazards associated with confined spaces, including the unique risks that are associated with permit-required confined spaces.

Wind turbine systems are power-generating systems, and specific OSHA regulations and guidelines pertain to them. Since medium-voltage electrical systems represent a significant hazard, workers must be qualified to work with them. Technicians must respect all approach boundaries and take the required precautions when working around energy sources of any kind.

Wind turbine assembly includes unique and complex lifting and rigging tasks. A great deal of lift planning is necessary to ensure an accident-free and successful project. Wind turbine technicians use hoists on a regular basis and should become proficient in their safe operation.

Access to wind turbines for service often requires workers to travel long distances on the highway or journey across challenging terrain. Technicians are often assigned company vehicles to complete their daily work. It is important that employees take personal responsibility for vehicle care. Off-road driving requires some practice and specific knowledge about different terrains and the safest way to address the many possible hazards that will be encountered.

1. OSHA is a part of the U.S. _____.
   a. Department of Labor
   b. Department of Health and Safety
   c. Bureau of Workplace Management
   d. Environmental Protection Agency

2. The *Code of Federal Regulations (CFR)* requires that construction and industrial workers successfully complete an OSHA 10-hour General or Construction Safety Training.
   a. True
   b. False

3. The primary purpose of a JSA is to _____.
   a. reduce or eliminate potential hazards and risks
   b. plan the steps that are required to complete a task
   c. ensure that all that all materials needed are on hand
   d. determine if LOTO procedures need to be used

4. Task safety analysis differs from JSA _____.
   a. since it focuses on much larger projects as one single task
   b. because they are completed only by the workers who do the task
   c. because they are required only when an accident or injury has occurred
   d. since it pertains to a smaller set of actions or steps

5. OSHA will provide safety consultation and assistance with JSA to some companies upon request.
   a. True
   b. False

6. A confined space _____.
   a. must have only one possible exit
   b. requires written permission before entering
   c. must never contain more than one worker
   d. is not typically designed for continuous occupancy

7. One responsibility of the attendant for a permit-required confined space is to _____.
   a. acquire and hold the document authorizing entry
   b. authorize and oversee entry after critical information has been analyzed
   c. refuse entry to the space by any unauthorized or unnecessary personnel
   d. minimize all communication with the entrant to avoid any confusion

8. One of the components housed in the turbine rotor hub is the _____.
   a. generator
   b. gearbox
   c. yaw drive components
   d. pitch drive components

9. When a worker is electrically shocked with a current in the range of 150mA, _____.
   a. thermal burns are likely
   b. the heart typically stops beating altogether
   c. entry and exit wounds are generally present
   d. the heart may begin beating in an abnormal rhythm

10. Workers who are inside of a flash protection boundary must _____.
    a. wear flame-retardant clothing
    b. be protected by rubber insulating garments and tools
    c. be observed by a second qualified worker at all times
    d. first complete OSHA's 10-hour Electrical Safety program

11. It is estimated that 75 percent of arc incidents in the workplace are the result of _____.
    a. a lack of training
    b. a lack of system design standards
    c. direct contact of an energized surface by tools or equipment
    d. unqualified workers working on electrical systems beyond their skill

12. A MAD specifies the _____.

   a. steps to take before beginning work on systems over 50 volts

   b. specific garments which must be worn before work may begin

   c. skills and training required before working on a given electrical system

   d. closest distance from any energized component that an unprotected worker may approach

13. Locks used to lock out energy sources at a single site should all be keyed the same.

   a. True

   b. False

14. The primary difference in lift plans from one turbine to another at a given site is related to the _____.

   a. terrain

   b. size of the turbine

   c. crane

   d. rigging

15. When driving down a steep hill, _____.

   a. use lower gears

   b. use higher gears

   c. descend at a 45-degree angle to the hill

   d. use only the parking brake until you are at the bottom

Trade Terms Quiz

Fill in the blank with the correct term that you learned from your study of this module.

1. A study done to minimize a worker's potential risk of exposure to arc flash injury is known as a(n) _____.

2. A(n) _____ is defined as the distance from exposed energized electrical conductors or components within which work is considered the same as making physical contact with the energized surface.

3. When calculating the amount of rope needed for a rigging task, a great deal of rope can be taken up in the _____.

4. The component often added to the main lifting arm of a crane to lengthen its reach is called a(n) _____.

5. The distance from energized electrical conductors or components that a qualified person may approach before donning insulating PPE made of rubber is considered the _____.

6. The _____ is defined as the voltage between the feet (a distance of roughly 1m) of a person standing near an energized grounded object.

7. When an emergency occurs in the nacelle of a wind turbine, workers need a rapid means of _____.

8. When workers are within the _____, they are at a distance from exposed energized electrical conductors or components within which they could be shocked.

9. A visible and bright bridge that is seen between two surfaces when electrical power jumps from one to the other is called a(n) _____.

10. Forming a(n) _____ makes the electrical potential between two points equal, preventing an electrical hazard.

11. The voltage between an energized object being touched and ground is referred to as the _____.

12. An area that requires workers to receive written authorization before they can enter due to potential hazards is called a(n) _____.

13. A(n) _____ is a dangerous electrical situation caused by the release of an enormous amount of thermal energy.

14. The distance from exposed energized electrical conductors or components within which a person could receive a second-degree burn if an arc flash incident occurred is called the _____.

15. Electrically speaking, a(n) _____ has received safety training to recognize and avoid electrical hazards, and has the appropriate skills and knowledge of the electrical equipment and installation involved.

16. An explosion that can be compared to dynamite and is initiated by an arc flash is called a(n) _____.

17. The distance from exposed energized electrical conductors or components within which there is an increased risk of electrical shock, not just an electrical hazard, is called the _____.

Trade Terms

Arc	Flash hazard	Limited approach	Permit-required	Reeving
Arc blast	analysis	boundary	confined space	Restricted approach
Arc flash	Flash protection	Minimum approach	Prohibited approach	boundary
Egress	boundary	distance (MAD)	boundary	Step potential
Equipotential plane	Jib		Qualified worker	Touch potential

Tom Lieurance

Electronics Instructor, Renewable Energy Technology
Columbia Gorge Community College
The Dalles, Oregon

How did you get started in the renewable energy industry?
I always had an insatiable curiosity about how things work. Mechanical systems were pretty easy to figure out. Electronics and electricity, on the other hand, were a mystery to me. I went back to school at the tender age of 27 when the aluminum plant I was working at as a laborer closed and laid everybody off. I researched community colleges for one that had a good reputation in electronics engineering technology.

My first job after graduating school was in the communications industry. I was the electronics bench technician at a radio and communications shop. Occasionally I did field repairs on communications and electrical equipment outside of the shop. I found I was well prepared by the community college to execute the duties of my job. This lasted 11 years.

My second job was at Columbia Gorge Community College where I was able to teach my electronics skill to others so they would have more career choices than they had before. At CGCC we started the Renewable Energy Technology program in 2006 with the idea of placing students with the wind industry. We also deduced that, with a little more training in electronics, students would be able to work in many power generation fields, such as hydro, solar, wind, natural gas, and coal. All these fields had one thing in common: everything is controlled by electronics. These electronic control systems are expensive enough to keep fixing. An employee with skills in electricity and electronics could troubleshoot and repair any of these control systems. At Columbia Gorge Community College we based the program on the electronics engineering technology program and added electrical, hydraulics, and mechanics. We also stress one of the most important things: industrial (personal) safety.

Who inspired you to enter the industry? Why?
My dad, who fixed everything. I was also inspired by my own curiosity and a desire to find a career at a good wage.

What do you enjoy most about your job?
I enjoy my job because of the satisfaction of training students for new careers that will ensure their economic prosperity. Also, there are so many electronic toys for us to figure out how they work.

Do you think training and education are important in the renewable energy field? If so, why?
Training in renewable energy is a great track to success in a career. There are so many opportunities because the price of fossil fuels will only go up, which will create more opportunities for electrical generation and control. Everything is based on an electronic control system.

How has training impacted your life and your career?
Electronics have been very good to me. Everything I have done since working at the aluminum plant has paid better and the caliber of people has greatly improved. The most entertaining part is that even the boss doesn't know what I am doing. Stuff comes in, I fix it, and it works; they just stand back in amazement!

Would you suggest renewable energy technology as a career to others? If so, why?
Yes, I have seen my friends at the aluminum plant get laid off and rehired and eventually put out of a job when another plant closed, while I am still working at a job I really like. I have even trained some of them in renewable energy. The field of renewable energy is growing rapidly with the price of fossil fuels increasing. Also there are many fields one can enter with training in renewable energy.

How do you define craftsman?
A craftsman is a person trained in an occupation who knows how to do it and why it is done that way, and who takes pride in doing the job right. Also, a craftsman never stops learning.

JOB SAFETY FORM

JOB SAFETY ANALYSIS FOR THE WIND TURBINE ENVIRONMENT

Use in conjunction with JSA Preparation Checklist

Facility and / or Turbine #: _____64_____ Date: __5-30-23__

Job Task Performed: __inspection__ PIC of Job: __tucker__

1) **Assign the right team:** Consider knowledge and experience. Reconsider those who lack either attribute.

2) **Understand the steps / Define the hazards:** Think through the task and envision the hazards.

3) **Execute:** Influence, manage, lead, assign responsibility to recommended actions to reduce / eliminate hazards.

Team Members (Print Name)	Job Title	Employer	Signature

Emergency Phone		Police Phone	
Fire / Rescue Phone		Company Contact Phone	

Site Address or GPS Coordinates	P64183 4681

Communication	☐ Cell phone (Check service on site ☐ Y ☐ N) ☐ Radio (Channel:)
	☐ Phone (Note phone number, location, special dialing, other contacts, etc.)

If procedures should change or the JSA is revised, STOP the job!
You MUST get approval from person-in-charge (PIC) of job before continuing.

REMEMBER
- Is there anything associated with the work I am getting ready to do that could cause a fatality or serious injury to my coworkers or me?
- How can the work I'm about to do adversely affect the other workers in or around my work site?

Job Safety & Environment Analysis

Page 1 of 5

58102-11 A01.EPS

JOB SAFETY ANALYSIS FOR THE WIND TURBINE ENVIRONMENT

Add detail as needed to steps, hazards, and controls

Facility and / or Turbine #: _____   Date: _____

Job Task Performed: _____   PIC of Job: _____

_____

Basic Job Steps	Considered Hazards	Controls for Identified Hazards	Name of Responsible Party for Putting Controls in Place
Transportation to site	Collision / soft shoulders pulling vehicle into ditch / hitting wildlife	Drive slowly / obey all traffic rules as posted / 15 mph max speed limit on site / 5 mph in lay down yard / wear seat belts / no cell phone or radio usage while driving vehicles	Site supervisor / job lead
Parking at site	High winds damaging vehicle doors / collision by passing vehicles	Park with nose of vehicle into the wind / park as far off the traveled portion as possible	Site supervisor / job lead
Tower climb	Fall / dropping tools onto personnel below / injury / slipping on rungs and decking / back injuries while pulling tools up to spill deck from saddle deck with rope	Full fall harness and use of climb assist for climbing / fully inspect harness and Lad Saf for any damage and full functionality / tie off when exiting ladder system / close each section of ladder doors / announce "clear" to other personnel climbing from below / use approved and secured carrying sacks / use buckets with tools secured to prevent dropping / use service elevator for awkward / or bulky tool transport up and down tower. Ensure all personnel have empty pockets and are not transporting tools in pockets. Use proper body mechanics at all times.	Site supervisor / job lead
LOTO	Shock / Arc flash / electrocution / severe burns / hearing damage	Full arc flash gear including hearing protection.	Site supervisor / job lead and authorized personnel only

58102-11 A02.EPS

JOB SAFETY ANALYSIS FOR THE WIND TURBINE ENVIRONMENT

Add detail as needed to steps, hazards, and controls

Facility and / or Turbine #: _____   Date: _____

Job Task Performed: _____   PIC of Job: _____

_____

Basic Job Steps	Considered Hazards	Controls for Identified Hazards	Name of Responsible Party for Putting Controls in Place
Walking on working surfaces	Slippery / oily surface causing fall and potentially serious injury / poor footing for generator removal / uneven ground, wet / snow / ice	Watch step / communication with crew on footing, 100% tie off / fall protection / clean up and prevent oil from spilling onto surfaces with pigmats and rags. Watch where walking and ensure good footing for any walking across site or while performing other duties. Apply anti-slip tape across top shroud of generators 2 and 4 to prevent slipping and place on any surface that may be needed for safe footing	Site supervisor / job lead
Nacelle work	Fall while door and guard rail open while lowering and raising parts out nacelle door.	100% tie off with fall harness	Site supervisor / job lead
All phases of work	Eye injuries	Wearing proper safety glasses at all times while on site / safety face shields for any grinding work required	Site supervisor / job lead
All site work	Head injuries / facial injuries	Wear hard hats while on site. ABSOLUTELY NO PERSONNEL WILL BE DIRECTLY UNDER OTHER WORK BEING DONE ABOVE THEM.	Site supervisor / job lead
Chemical usage	Possible chemical reactions / eye splash, injury / flammable hazard	Wear all appropriate PPE / obtain MSDS and know properties and cautions of chemicals / ensure fire extinguisher charged and accessible. First aid kit readily available and stocked with eyewash solution.	Site supervisor / job lead

Job Safety & Environment Analysis

58102-11 A03.EPS

JOB SAFETY ANALYSIS FOR THE WIND TURBINE ENVIRONMENT

Add detail as needed to steps, hazards, and controls

Facility and / or Turbine #: _____ Date: _____

Job Task Performed: _____ PIC of Job: _____

_____

Basic Job Steps	Considered Hazards	Controls for Identified Hazards	Name of Responsible Party for Putting Controls in Place
Fire or emergency egress required	Injury or death	Ensure emergency descent equipment is present and in good condition and readily accessible. Know the phone number / contact information of emergency services systems in response area. Know location of site by physical address	Site supervisor / job lead
Rigging and hoisting material	Chain, lifting fixtures and slings breaking during lifts	Inspect all lifting and rigging hardware prior to start of work. Do not load shock the jib crane during operation. Use hand signals by designated person for stop, lift, etc.	Site supervisor / job lead
All phases of work	Spine injuries and muscle strains / sprains	Use proper lifting techniques and body mechanics. Use two person lift for heavy, awkward equipment (>50 lbs). Use available forklift / crane for lifting heavy equipment and materials. Stretches prior to beginning work especially in cold weather. Practice good housekeeping for oily surfaces to prevent slipping on walking, working surfaces. Use anti-slip tape on surfaces.	Site supervisor / job lead

Job Safety & Environment Analysis

58102-11 A04.EPS

Preparation Checklist

Procedures To Review
- [ ] specific job procedures
- [ ] permit to work*
- [ ] confined space entry*
- [ ] hot work
- [ ] cold cutting
- [ ] hot bolting
- [ ] energy isolation*
- [ ] lifting operations*
- [ ] ground disturbance*
- [ ] working at heights
- [ ] rescue plan for use of fall protection
- [ ] emergency action plan
- [ ] MSDS
- [ ] other:

Adequate Personnel
- [ ] knowledge / skills / experience
- [ ] training / orientation
- [ ] short service employees
- [ ] approved EHS contractor's list
- [ ] PIC of job identified
- [ ] communication among co-workers
- [ ] other:

Personal Protective Equip.
- [ ] hard hat / safety glasses / steel toe boots
- [ ] hearing protection
- [ ] fire resistant clothing / flash gear
- [ ] gloves
- [ ] face shield / goggles
- [ ] respiratory protection
- [ ] fall protection gear
- [ ] other:

Tools & Equipment
- [ ] proper tools & equipment available
- [ ] proper use of tools & equipment
- [ ] tools & equipment inspected for damage
- [ ] qualification of operator
- [ ] proper certification
- [ ] safety devices in bypass
- [ ] service / maintenance
- [ ] hot / cold surfaces
- [ ] equipment guarding
- [ ] other:

Energy Sources (Electrical)
- [ ] bare wiring
- [ ] shock hazards
- [ ] grounding on tools / equipment
- [ ] electric power tools
- [ ] electrical
- [ ] other:

Energy Sources (Mechanical)
- [ ] rotating equipment
- [ ] pinch points
- [ ] sharp objects

- [ ] moving / dropped objects
- [ ] other:

Energy Sources (Pneumatic)
- [ ] sandblasting
- [ ] painting
- [ ] volume bottles
- [ ] air hoses
- [ ] air tools
- [ ] air taggers
- [ ] cylinders (O$_2$ / acetylene / N$_2$)
- [ ] natural gas (vessels / tanks / piping)
- [ ] other:

Lifting
- [ ] lifting with crane
- [ ] lifting with hoist / come-a-long
- [ ] hoisting of tools w/ rope
- [ ] proper rigging practices
- [ ] manual lifting (body position)
- [ ] other:

Body Position / Movement
- [ ] prolonged awkward body position
- [ ] bending / twisting
- [ ] climbing / overextending
- [ ] crawling / crouching
- [ ] reaching / pulling / pushing
- [ ] carrying materials
- [ ] pinch points of body / hands
- [ ] other:

Weather Conditions
- [ ] rain / lightning
- [ ] excessive cold / heat
- [ ] wind / sea conditions
- [ ] other:

Physical Surroundings
- [ ] cluttered walkways & work area
- [ ] slippery walking surfaces
- [ ] slips / trips / falls
- [ ] housekeeping
- [ ] work over open water
- [ ] open holes
- [ ] emergency egress identified
- [ ] low hanging pipes / supports
- [ ] lighting levels
- [ ] noise levels
- [ ] ambient temp (heat / cold stress)
- [ ] other:

Safe Location Factors
- [ ] geographic area (wildlife refuge)
- [ ] terrain (rough / wet / rocky / muddy)
- [ ] excavation (shoring / water table)
- [ ] adequate access / egress roads
- [ ] overhead wires
- [ ] lasers / x-rays / microwaves / UV
- [ ] other:

Roadways
- [ ] flagmen (PPE & training)
- [ ] signs / cones / flares / lights / reflectors
- [ ] transition, high / low asphalt to dirt
- [ ] speed of traffic on highway
- [ ] other:

Hazardous Materials
- [ ] crude oil / natural gas / condensate

- [ ] flammable / combustible / explosive
- [ ] H$_2$S / SO$_2$ / CO
- [ ] acids / bases / corrosives
- [ ] radioactive materials
- [ ] dusts / gases / fumes
- [ ] poisons
- [ ] reactive chemicals
- [ ] other:

Potential Pollution Factors
- [ ] spill procedures
- [ ] controlling waste (rags, pads, etc.)
- [ ] trash containment
- [ ] breaking flanges
- [ ] drip pans
- [ ] proper absorption pads
- [ ] moving hoses & containers
- [ ] identify waste generation and
- [ ] procedures to use
- [ ] other:

Check waste generated:
1. For waste(s) checked, review procedures from Waste Management Manual.
2. Review jobsite after work completion to ensure jobsite is cleaned.

- [ ] absorbent materials
- [ ] acetylene
- [ ] acid
- [ ] aerosol cans
- [ ] antifreeze
- [ ] aviation fuel
- [ ] barite (excess)
- [ ] batteries
- [ ] bulk containers (sacks)
- [ ] caustic soda
- [ ] cement (excess)
- [ ] ceramic pkg material
- [ ] chlorine tablets
- [ ] completion fluids
- [ ] cooking oil
- [ ] crude oil / condensate
- [ ] diesel fuel
- [ ] drilling fluids / cuttings
- [ ] drums (empty)
- [ ] filters (used)
- [ ] fluorescent light bulbs
- [ ] food waste
- [ ] freon TF solvent
- [ ] glycol
- [ ] halon
- [ ] metal cuttings

- [ ] metal, scrap (NORM free)
- [ ] methanol
- [ ] methyl ethyl ketone
- [ ] natural gas
- [ ] nitrogen
- [ ] NORM, equip, pipe
- [ ] NORM, produced sand
- [ ] NORM tank bottoms
- [ ] oil (used)
- [ ] oxygen
- [ ] paint / paint waste
- [ ] paraffin
- [ ] pigs
- [ ] pipe dope
- [ ] produced water
- [ ] refuse, debris, contaminated
- [ ] sand, produced sand
- [ ] tank btm (NORM free)
- [ ] sandblasting materials
- [ ] sanitary wastewater
- [ ] thread protectors
- [ ] tires (used)
- [ ] trash
- [ ] Vertrel MCA solvent
- [ ] other:

Notes & Comments
_____
_____
_____
_____
_____
_____
_____
_____
_____

Job Safety & Environment Analysis

58102-11_A05.EPS

Trade Terms Introduced in This Module

Arc: A visible, luminous bridge formed between two surfaces when electrical power jumps between them.

Arc blast: An explosion similar to the detonation of dynamite that occurs during an arc flash incident.

Arc flash: A dangerous condition caused by the enormous release of thermal energy when arcing occurs.

Egress: An escape or emergence from one area into another.

Equipotential plane: An electrical condition when all points have the same potential.

Flash hazard analysis: A study investigating a worker's potential exposure to arc flash energy, conducted for the purpose of injury prevention, the determination of safe work practices, and the identification of appropriate PPE.

Flash protection boundary: An approach limit at a distance from exposed energized electrical conductors or components within which a person could receive a second-degree burn if an arc flash incident occurred.

Jib: The projecting arm of a crane from which the load is suspended. Generally, a jib is considered an extension of the main arm of the crane, rather than the main arm itself.

Limited approach boundary: An approach limit at a distance from exposed energized electrical conductors or components within which a shock hazard exists.

Minimum approach distance (MAD): The distance from energized electrical conductors or components that a qualified person may approach before donning insulating PPE made of rubber.

Permit-required confined space: A confined space that has been evaluated and found to have specific actual or potential hazards, such as a toxic atmosphere or other serious safety or health hazard. Workers need written authorization to enter.

Prohibited approach boundary: An approach limit at a distance from exposed energized electrical conductors or components within which work is considered the same as making physical contact with the energized surface.

Qualified worker: Electrically speaking, one who has the skills and knowledge related to the electrical equipment and installations, and has received safety training to recognize and avoid the hazards involved.

Reeving: The passing of rope through one or more pulleys or openings. A significant length of rope can be reeved through pulleys and lifting blocks, adding to the necessary length needed for the task.

Restricted approach boundary: An approach limit at a distance from exposed energized electrical conductors or components within which there is an increased risk of electrical shock.

Step potential: The voltage between the feet (usually about 1m in length) of a person standing near an energized grounded object.

Touch potential: The voltage between the energized object being touched and ground.

Additional Resources

This module presents thorough resources for task training. The following resource material is suggested for further study.

OSHA Standard 1926, Safety and Health Regulations for Construction, available at www.osha.gov.

Safety Orientation, 2004. Upper Saddle River, NJ: NCCER/Pearson Education, Inc.

Figure Credits

Courtesy of DOE/NREL, Module opener, Figures 1, 24, and 28–30

Paul Anderson, Figures 3, 12, and 13

Photo courtesy of Environmental Management, Confined Space Rescue Team Website: EMIOK.com, Figure 5

Brian Shultz, Figures 6, 7, and 9

Topaz Publications, Inc., Figures 8, 20, 23, 34, 36, and 39 (bottom photo)

Tech Safety Lines, Figures 10 and 11

Jim Mitchem, Figure 17

Salisbury Electrical Safety, Figures 18 and 19

U.S. Department of Labor, Occupational Safety & Health Administration, Table 3

Accuform Signs, Figure 21

North Safety Products, Figure 22

Iberdrola Renewables, SA01

JET brand of WMH Tool Group, Figure 25 and SA02 (ratchet-lever hoist)

Coffing Hoists, Figure 26

Konecranes, Inc., Figure 27

Lug-All Corporation, SA02 (come-along)

Courtesy of H&N Electric Co., Figure 31

© 2010 Photos.com, a division of Getty Images. All rights reserved., Figure 32

© iStockphoto.com/georgethefourth, Figure 33

Westin Automotive Products, Figure 35

JLG Industries, Inc., Figures 37, 39 (top photo), and 40

Genie, a Terex Brand, Figure 38

NCCER CURRICULA — USER UPDATE

NCCER makes every effort to keep its textbooks up-to-date and free of technical errors. We appreciate your help in this process. If you find an error, a typographical mistake, or an inaccuracy in NCCER's Curricula, please fill out this form (or a photocopy), or complete the online form at **www.nccer.org/olf**. Be sure to include the exact module number, page number, a detailed description, and your recommended correction. Your input will be brought to the attention of the Authoring Team. Thank you for your assistance.

Instructors – If you have an idea for improving this textbook, or have found that additional materials were necessary to teach this module effectively, please let us know so that we may present your suggestions to the Authoring Team.

NCCER Product Development and Revision

13614 Progress Blvd., Alachua, FL 32615

Email: curriculum@nccer.org
Online: www.nccer.org/olf

❏ Trainee Guide   ❏ AIG   ❏ Exam   ❏ PowerPoints   Other _____

Craft / Level: _____   Copyright Date: _____

Module Number / Title: _____

Section Number(s): _____

Description: _____

_____

_____

_____

Recommended Correction: _____

_____

_____

_____

Your Name: _____

Address: _____

Email: _____   Phone: _____

58103-11

Climbing Wind Towers

Module Three

Trainees with successful module completions may be eligible for credentialing through NCCER's National Registry. To learn more, go to **www.nccer.org** or contact us at **1.888.622.3720**. Our website has information on the latest product releases and training, as well as online versions of our *Cornerstone* newsletter and Pearson's Contren® product catalog.

Your feedback is welcome. You may email your comments to **curriculum@nccer.org,** send general comments and inquiries to **info@nccer.org**, or use the User Update form at the back of this module.

Objectives

When you have completed this module, you will be able to do the following:

1. Identify the purpose and use of required safety equipment for proper wind tower climbing.
2. Demonstrate the ability to inspect and don required personal protective equipment before use.
3. Identify various environmental hazards that may affect or prevent climbing.
4. Describe the elements of a proper pre-climb meeting and wind tower area inspection prior to climbing.
5. State the practices for safely ascending and descending wind towers.
6. Under the supervision of an instructor, properly and safely ascend and descend a 30-foot tower or comparable structure.

Performance Tasks

Under the supervision of the instructor, you should be able to do the following:

1. Demonstrate the ability to inspect and don required personal protective equipment prior to use.
2. Properly and safely ascend and descend a 30-foot tower or comparable structure.

Trade Terms

Anchor point
Body harness
Cable grabs
Carabiner
Connecting devices

Fall arrest
Fall restraint
Lanyards
Newtons

Occupational Safety and Health Administration (OSHA)
Personal fall arrest system (PFAS)

Point of daylight
Rappel
Suspension trauma
Swing zone

Industry Recognized Credentials

If you're training through an NCCER-accredited sponsor you may be eligible for credentials from NCCER's Registry. The module ID number for this module is 58103-11. Note that this module may have been used in other NCCER curricula and may apply to other level completions. Contact NCCER's Registry at 888.622.3720 or go to nccer.org for more information.

Contents

Topics to be presented in this module include:

1.0.0 Introduction 1
2.0.0 Climbing Safety Equipment 1
  2.1.0 Personal Fall Arrest Systems 1
    2.1.1 Anchor Points.. 2
    2.1.2 Belts and Harnesses 4
    2.1.3 Connecting Devices.. 5
  2.2.0 Lanyards 8
    2.2.1 Shock-Absorbing Lanyards.. 9
    2.2.2 Self-Retracting Lanyards 10
  2.3.0 Safe Climbing Assistance Devices 10
  2.4.0 Powered Climbing Assistance 12
  2.5.0 Other Safety Gear 13
  2.6.0 Equipment Inspection 14
    2.6.1 Belts and Harnesses 14
    2.6.2 Ropes and Lanyards 15
    2.6.3 Self-Retracting Lanyards 15
    2.6.4 Hooks, Carabiners, and Cable/Rope Grabs 16
3.0.0 Hazards of the Environment 16
  3.1.0 Noise and Communication 17
  3.2.0 Living Dangers 17
    3.2.1 Birds 17
    3.2.2 Insects 17
    3.2.3 Snakes 18
    3.2.4 Humans 18
  3.3.0 Weather 18
    3.3.1 Sunshine 18
    3.3.2 Rain, Snow, and Ice 18
    3.3.3 Temperature.. 19
    3.3.4 Lightning 19
    3.3.5 Wind 20
4.0.0 Climb Preparations 20
  4.1.0 Site and Tower Assessment 20
  4.2.0 The Pre-Climb Meeting 21
  4.3.0 Common Climbing Policies and Guidelines.............................. 23
5.0.0 Basic Climbing Skills 24
  5.1.0 Preliminary Considerations 24
  5.2.0 Ascent.. 25
  5.3.0 Maneuvering and Positioning 25
  5.4.0 Descent.. 26
6.0.0 Rescue at Height 26
  6.1.0 Suspension Trauma 26
  6.2.0 The Rescue Plan.. 27
  6.3.0 Basic Rescue Equipment...28

Figures and Tables

Figure 1   Wind turbine 1
Figure 2   Body harness labeling 2
Figure 3   Personal fall arrest system 3
Figure 4   Typical anchor point 3
Figure 5   Back D-ring 4
Figure 6   Full body harness 4
Figure 7   Donning the body harness 6
Figure 8   Suspension trauma strap use..7
Figure 9   Carabiner..7
Figure 10 Lanyard hook 7
Figure 11 Shock-absorbing lanyard 9
Figure 12 Nonshock-absorbing lanyard 9
Figure 13 Y-configured shock-absorbing lanyard 10
Figure 14 Self-retracting lanyard 10
Figure 15 Permanent safe climb lifeline system..........................11
Figure 16 Cable grab..11
Figure 17 Temporary safe climb rope system..............................12
Figure 18 Rope grab 12
Figure 19 Lad Saf® powered climb assist system........................12
Figure 20 Climber's hard hat 13
Figure 21 Stitching inspection...15
Figure 22 Lanyard webbing inspection 15
Figure 23 D-ring inspection 15
Figure 24 SRL casing inspection 16
Figure 25 Braking mechanism test 16
Figure 26 Hook load indicator 16
Figure 27 Osprey, or fish eagle 17
Figure 28 Paper wasps...17
Figure 29 Sunrise 18
Figure 30 Antarctic turbines 19
Figure 31 Lightning 19
Figure 32 Offshore wind turbines 21
Figure 33 Entering the turbine tower..23
Figure 34 Suspended technician 27
Figure 35 Automatic descent control unit.....................................28
Figure 36 Manual rescue system 29

1.0.0 INTRODUCTION

Many wind turbine maintenance opportunities are available to those who possess the skills of a competent climber. Of the many technicians who have the aptitude for mechanical or electrical work, few also have the ability to climb and work at extreme heights. Wind turbines (*Figure 1*) represent an important step toward independence from fossil fuels and greater use of renewable energy sources. Their numbers continue to grow at a rapid pace. Not only are talented and capable technicians needed as a result, but each of those technicians must also be a strong and competent climber. Some members of this group also possess the advanced skills to rescue a fallen colleague should the need arise.

In addition to the challenge of learning and understanding the safe working habits and rules involved with climbing, this module also provides you with an opportunity to test your mental and physical ability to withstand the stress of climbing and working in an elevated environment. The vast majority of this module focuses on personal safety, and it is essential that you learn this material. The smallest lapse in judgment or attention can prove disastrous. There is no substitute for proper training and preparation for a climbing assignment.

2.0.0 CLIMBING SAFETY EQUIPMENT

As one may guess, the list of safe working policies and regulations can seem endless. A number of organizations, both government and private, have attempted to create a widely accepted version of comprehensive climbing regulations or guidelines. However, there is still considerable disparity between the creators of such publications. The **Occupational Safety and Health Administration (OSHA)** provides a number of regulations that apply to the trade, primarily consisting of directives for fall protection. Remember that OSHA has been charged with the responsibility of ensuring that employers protect the worker from unnecessary hazards in the workplace. Its requirements should be considered the minimum level of action only. Beyond that, individual organizations and employers implement additional requirements and policies for internal use. Also, other entities, such as land or equipment owners, might possess the right to invoke their own set of guidelines when work is being conducted on their property.

Over 100,000 injuries or deaths occur each year from falls in the workplace. OSHA regulations require that specific precautions be taken to protect employees who work at heights. When workers are exposed to the potential for falling 6 feet or more, OSHA requires the use of one or more of these three primary fall prevention systems: a guardrail system, safety net system, or a personal fall protection system. Obviously, guardrail systems and safety nets are not likely to be found on wind towers. That places the focus squarely on the personal fall protection system.

The fall protection requirement can be satisfied through the use of either **fall restraint** equipment or **fall arrest** equipment. Fall restraint equipment helps prevent any significant fall from occurring, while fall arrest systems stop or take control of a fall in progress. The climber must take responsibility for using fall arrest and/or fall restraint equipment 100 percent of the time.

2.1.0 Personal Fall Arrest Systems

Personal fall arrest systems (PFAS) combine several pieces of equipment into a complete fall protection system. Many different types and styles exist, with each one designed for specific uses. In addition, the personal aspect of a PFAS must also be emphasized. One size does not fit all, and proper fit is essential to avoid injury to the worker when a fall is arrested. *OSHA Standard 1926.502, Subpart M, Section (d)* lists specific requirements, such as the test strength and use of D-rings and snap hooks and the specifications for **anchor points** on the structure.

The referenced OSHA standard requires several important characteristics of a PFAS:

* It must limit the maximum arresting force imparted to the body to 1,800 pounds with a full **body harness**.

Figure 1 Wind turbine.

58103-11_F01.EPS

- The free fall distance must be limited to 6 feet and be rigged to prevent contact with anything below.
- It must bring the body to a stop within an additional 3½ feet (1.1 m).
- The system must be strong enough to withstand twice the possible force of a body falling from a distance of 6 feet.

You will often see references to American National Standards Institute (ANSI) standards regarding the performance of PFAS products. ANSI publishes standards for PFAS equipment that provide manufacturers with guidelines for products. Its standards cover items such as design, testing, markings, and performance. Items that bear the ANSI certification markings let the user know that a product has been built and tested to ensure its integrity. Many organizations require that PFAS equipment used by their employees or agents meets ANSI standards and is marked accordingly, without exception.

It should be obvious that any PFAS is only as good as its weakest component. The systems should never be used for any other task, such as lifting tools or materials to the work area. The specifications and usage instructions for components such as the full body harness (*Figure 2*) identify the working load limit, which should be strictly followed. Climbers must use common sense and not overload themselves with equipment, even if the weight does fall within the limits of the PFAS specifications.

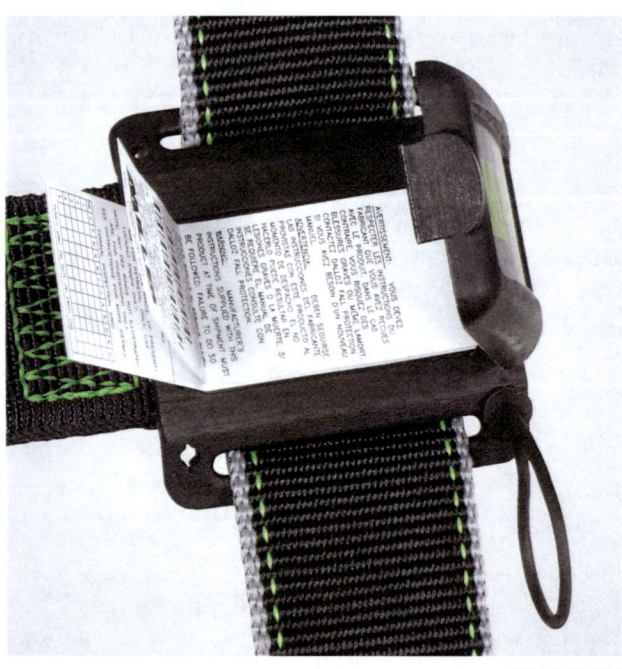

Figure 2 Body harness labeling.

A complete PFAS (*Figure 3*) is made up of three primary components. Remember that this a system, not a single piece of equipment. Anchor points are related to the structure, and the type and availability of anchor points help determine what other equipment should be chosen. The body harness comprises the system of belts, rings, or hooks worn by the climber. Connecting devices or connectors are used to maintain attachment between anchor points and the PFAS or positioning belts, and includes both lanyards and various pieces of hardware.

2.1.1 Anchor Points

OSHA requires that anchor points (*Figure 4*) be rated at or equal to 5,000 pounds breaking strength, and they cannot be part of the support used for tower platforms. This can be a challenging requirement for the climber, as some towers may not appear to have any structures that conform, and no global standard presently exists for clearly marking or installing anchor points during construction. Final selection of the anchor point during a climb is left to the professional judgment of the climber. When no 5,000-pound point anchor point is clearly identified, the next standard a climber must meet is to calculate the PFAS anchor point to provide a safety factor of two. This means that the chosen anchor point must be capable of supporting twice as much weight as the maximum amount that could be applied to it during a fall arrest situation. In other words, if a chosen anchor point is expected to support 2,500 pounds, then you must ensure that you only apply 1,250 pounds of force in a worst-case scenario of fall arrest. This is a challenging structural strength-

Personal Fall Arrest System

Three key components of the Personal Fall Arrest System (PFAS) must be in place and properly used to provide maximum worker protection.

Individually these components will not provide protection from a fall. However, when used properly and in conjunction with each other, they form a Personal Fall Arrest System that becomes vitally important for safety on the job site.

Anchorage/ Anchorage Connector

Anchorage: Commonly referred to as a tie-off point (Ex: I-beam)

Anchorage Connector: Used to join the connecting device to the anchorage (Ex: beam anchor)

Connecting Device

Connecting Device: The critical link which joins the body wear to the anchorage/anchorage connector (Ex: retractable lifeline *(shown)*, or shock-absorbing lanyard, see inset below)

Body Wear

Body Wear: The personal protective equipment worn by the worker (Ex: full-body harness)

Figure 3 Personal fall arrest system.

58103-11_F03.EPS

evaluation to make mid-climb, but it could become necessary in unusual situations. It is also difficult for a climber to have any real control over the potential load imposed on an anchor point in a fall.

Two climbers can connect their fall arrest lanyard to a single anchor point. However, the anchor point must then meet a 10,000-pound standard. Since the potential anchor load from a fall has doubled, so must the anchor point's rated capacity.

Note that the anchor point shown in *Figure 4* is one that can be installed either temporarily or permanently, and is only one example of the many different styles you may encounter. Properly installed anchor points are typically found on a professionally built tubular or lattice wind tower and should not be difficult to identify. However, there are no mandated tower construction standards that dictate anchor installation

58103-11_F04.EPS

Figure 4 Typical anchor point.

locations. Some builders or developers do maintain related standards on their own.

The ideal fall arrestance anchor point is located directly above the back D-ring (*Figure 5*), selected to minimize any swing zone hazards as well as the possible free fall distance. Swing zones are minimized when the anchor point is directly above the climber. Horizontal climbing situations can be far more hazardous in terms of the swing zone, as an anchor point directly above may not exist. Serious injury and damage can occur when a human body strikes an immoveable object while swinging as a pendulum. Although the PFAS may do its job by preventing the climber from falling a great distance, serious injury or death can still occur by striking an object in the swing zone.

Anchor points are often needed to secure the position of the climber, leaving the hands free to accomplish the task. Ideally, climbers should select anchor points, and connect as needed, to maintain a potential fall distance of two feet. Positioning connections are a factor in fall arrest, as the connection to a positioning strap is likely to modify the swing zone and/or distance of the fall during an accident. However, they cannot be considered the primary fall arrest anchor. Therefore, positioning anchor points are required by OSHA to be rated at a 3,000 pound strength instead of the 5,000 pound rating for primary fall arrest. Some rails or railings on wind towers may be useful as positioning anchors but may not suffice as an anchor point for fall arrestance. Remember the positioning lanyards, connected to D-rings on the harness other than the back or front chest D-ring,

are fall restraints rather than fall arresting connections. A positioning lanyard does not take the place of a fall arrestance lanyard or anchor point. If you make a connection with the intention of placing weight or stress on it in any way, consider it a positioning connection.

2.1.2 Belts and Harnesses

The full body harness (*Figure 6*) is unquestionably the center of the PFAS. It should be worn any time work is conducted more than 6 feet above ground. A wide variety of rings are part of a full body harness assembly. The back D-ring is the only one used to connect the harness to the anchor point for primary fall arrest purposes unless you are climbing a ladder. When climbing a ladder such as those installed in tubular wind towers, the front chest D-ring is the likely choice. D-rings located at the hips are used for positioning and fall restraints only. D-rings mounted to shoulders are often used for rescue situations. All of them can be used for fall restraint, but the back D-ring is the primary connection for fall arrestance, except for a ladder climbing system.

58103-11_F06.EPS

Figure 6 Full body harness.

58103-11_F05.EPS

Figure 5 Back D-ring.

A body harness must fit correctly to ensure proper protection. Do not place additional holes or openings in harness components under any circumstances. No field modifications to a body harness or lanyard should be attempted. Installation, maintenance, and inspection instructions are provided for every harness, and it is the responsibility of the climber to read and understand the details regarding his or her personal equipment. Harness straps are generally designed with some stretch to help absorb some of the potential force. This means good, taut installation on the body is essential so that any slack, plus stretch, does not allow the climber to fall out of the harness.

Figure 7 demonstrates the proper procedure for donning a common full body harness. The most important adjustments to be made include the chest straps, the groin straps, and the final position of the back D-ring. The related details that follow must be considered during the fitting and wearing of a full body harness:

> **NOTE**
>
> These guidelines are general in nature and the instructions provided for specific equipment by the manufacturer must always take precedence. It is the climber's responsibility to be intimately familiar with the duty of each and every ring and belt.

- The back D-ring location is vital to proper fall arrest. Position this ring between the shoulder blades. If it is too low, it tends to cause the body to hang in a more horizontal position during fall arrest, increasing pressure on the diaphragm and affecting breathing. If it is positioned too high or with too much slack, the D-ring may strike the climber's head at the base of the fall, and the shoulder straps may be pulled too tightly into the neck and restrict blood flow. A climber must remember that the impact at the base of the fall arrest can be dramatic, and that this force must be spread all around the body to prevent injury to any one portion.
- Chest straps generally form either an "H" pattern or an "X" pattern. Adjust "H" pattern straps to land between the bottom of the sternum and the navel. This helps to ensure that the horizontal portion of the "H" does not contact the throat during a fall, choking the climber. Some harness designs may not allow for this adjustment, with the final position of the "H" based solely on a properly sized harness.
- The position of the chest straps is also crucial for "X" pattern harnesses. Position the "X" at or just below the sternum.

- Groin straps are an integral and required part of a PFAS. Adjust the groin straps for a good, snug fit. Too much slack here will cause extreme discomfort in a fall, when the impact snatches them up tight and you are left suspended this way. Climbers may add some padding both here and in the shoulder straps to avoid some discomfort. When done properly and without any interference to the working parts of the harness, this practice is not considered a modification by definition.
- A suspension trauma strap (*Figure 8*) may be required as part of the PFAS gear. Additional information regarding suspension trauma is provided in a later section of this text. The suspension trauma strap is stored in a convenient pouch that is pre-connected to the harness. This is done by either one end of the strap being permanently sewn to the harness (by the manufacturer), one end of the strap attached to the harness with a carabiner, or by choking the pouch around a harness strap or hip D-ring. The strap can then be quickly removed and used without any possibility of the climber dropping it. Once connected, the strap allows the worker to stand up in the harness, relieving suspended weight and pressure from the hips and groin. This helps open the path for blood flow from the legs back to the heart, preventing blood from pooling in the lower extremities.
- A separate waist or tool belt, while not considered a necessary component of the fall arrest system, must be fitted properly. It is best used for body positioning with the D-rings precisely located at the hip sides, rather than in the front or rear. Do not adjust it in a way that could apply pressure to the kidneys or lower back. If the waist belt is not an integral part of the harness, it must not be worn on the outside of the harness – don it first, and then add the harness over it. This is also true of any added tool belts.
- Saddles, like waist belts, are also not considered an integral or required part of the PFAS. They are optional and often detachable. They are generally used by workers to allow a seated, suspended position when the task may require long periods in the same location.

2.1.3 *Connecting Devices*

Connecting devices connect the PFAS or positioning belts to anchor points. They include several different types of hardware, as well as the lanyards. Lanyards are discussed in detail in the next section.

6 Easy Steps That Could Save Your Life

How To Don A Harness

1 Hold harness by back D-ring. Shake harness to allow all straps to fall in place.

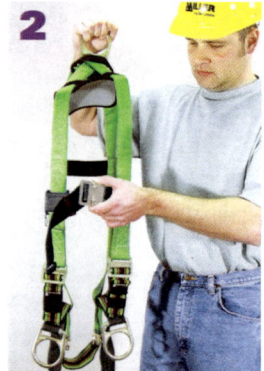

2 If chest, leg and/or waist straps are buckled, release straps and unbuckle at this time.

3 Slip straps over shoulders so **D-ring is located in middle of back between shoulder blades**.

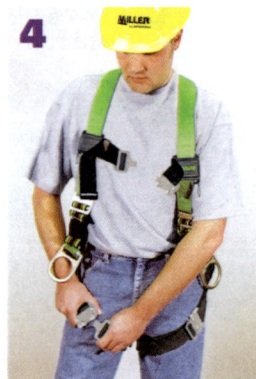

4 Pull leg strap between legs and connect to opposite end. Repeat with second leg strap. If belted harness, connect waist strap after leg straps.

5 **Connect chest strap and position in midchest area.** Tighten to keep shoulder straps taut.

Snug Fit

6 After all straps have been buckled, **tighten all buckles so that harness fits snug but allows full range of movement.** Pass excess strap through loop keepers.

58103-11_F07.EPS

Figure 7 Donning the body harness.

It is important to note that not all hardware qualifies as a component of a PFAS. Some hardware is to be used only for attaching tools and equipment to the climber or to structures, due to its limited specifications or testing. Hardware used as connecting devices as part of the PFAS must be drop-forged steel, and have a finish resistant to corrosion from salt spray per ANSI standards. Any type of hook or carabiner (*Figure 9*) must be equipped with safety gates or keepers to prevent the hooked object from being disconnected accidentally. In most cases, these safety gates are required to be two-step, also called double action. Designs for these features vary, but those designed so that both movements required to open the gate can be done with a single hand are generally better. Using both hands to manipulate a single connector can be a hazard in itself.

Figure 8 Suspension trauma strap use.

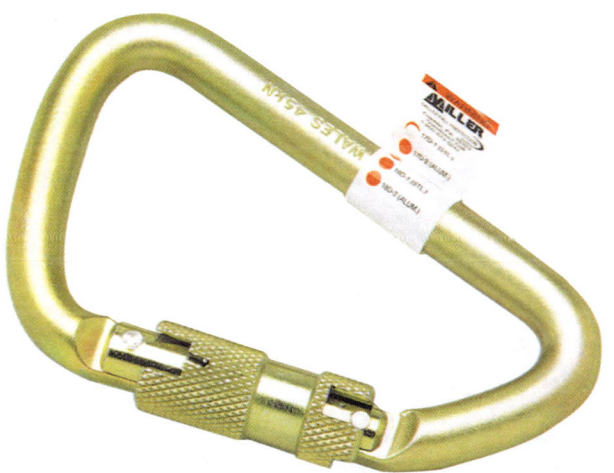

Figure 9 Carabiner.

A carabiner is like an elongated ring, with some style of safety gate that is shaped much like a chain link. The safety gate on some may be screw-type, requiring the gate to be screwed closed. Carabiners are used to attach lines and have no sharp edges. They are tested by impact to determine their strength and are rated in **Newtons** of force. For use in a fall arrest system, they must be rated at a minimum of 22 kilo-Newtons, or 22kN.

Hooks (*Figure 10*) are usually curved and have an opening to allow connection to a line that can then be securely closed. They are usually not as consistent in appearance as carabiners, and come in somewhat different shapes. Some are referred to as pelican style due to the shape of the hook and the closure. When used as part of a PFAS, the security closure should be automatic. Hooks are tested under direct load rather than by force applied, and thus are rated in pounds of load as the minimum breaking strength. Hooks are designed to connect to D-rings primarily – not to each other. In most cases, hooks are already connected to a lanyard or rope to ensure the integrity of the connection and are not purchased separately.

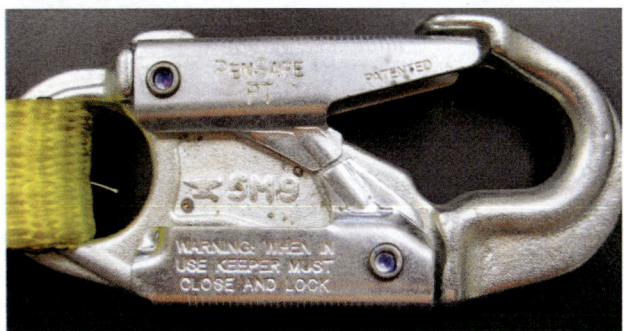

Figure 10 Lanyard hook.

2.2.0 Lanyards

Lanyards consist of their primary material of construction (rope, webbing, aircraft cable, etc.) with a connecting device attached to the ends. Lanyards for fall restraint or arrestance must never be field-fabricated. They are available in a great variety of lengths from as small as 1½ feet up to 30 feet, and should never be connected together to increase their length. Depending upon the use, padding may be added during fabrication or in the field to protect against sharp edges. Never tie a knot in a lanyard, as knots can severely reduce the load limit. Never wrap lanyards around a structure then choke by placing a pelican hook around the lanyard itself, unless they are of a design specifically allowing this use. A special large D-ring is usually attached to a lanyard for this purpose.

There are two main categories of lanyards: shock-absorbing (*Figure 11*) and nonshock-absorbing (*Figure 12*). Lanyards for fall arrest should always be shock-absorbing. Nonshock-absorbing lanyards are used for positioning and fall restraint. Lanyards used for positioning are not considered part of the fall arrest system; they are fall restraints attached to D-rings on the harness other than the back D-ring. Since fall restraint is all about preventing a fall from happening, lanyards used for positioning should not allow a fall or movement greater than 2 feet (0.61 m). That is a relatively short distance, and

Figure 11 Shock-absorbing lanyard.

58103-11_F11.EPS

it means that the positioning lanyard may have to be wrapped several times around the anchor point to get the length required. Again, do not choke the lanyard with a pelican hook in an effort to adjust the lanyard to the desired length. Some lanyards with special D-rings are designed for this purpose.

The line of support used for fall restraint, fall arrest, or positioning should always be toward the centerline of the body. Lanyards may have to be wrapped around an anchor point to allow the correct allowable working distance. When the threat of cuts from the structure exists, use a wire or aircraft cable lanyard or a lanyard with proper padding to guard against it.

2.2.1 Shock-Absorbing Lanyards

Shock-absorbing, or deceleration, lanyards have shock-absorbing properties built in and are designed for fall arrest. There are also some shock-absorbing devices designed for connection to nonshock-absorbing lanyards to provide the needed protection. These types of lanyards are capable of reducing the fall arrest force on the body by as much as 80 percent. They should not be used for fall restraint or positioning.

A section of these lanyards is made to extend under severe stress, such as that encountered in a fall. ANSI outlines the specifics and standards of a shock-absorbing lanyard's construction. The section or end that contains the energy-absorbing feature should be connected to the climber's back D-ring rather than the anchor point. Because of their duty, deceleration lanyards should be no more than 6 feet in usable length and allow no more than an additional 3½ feet (1.06 m) of extension when the shock-absorbing feature is activated. The climber must consider lanyard length to know what the total fall will be. Remember that the primary anchor point should be straight above the body whenever possible to minimize the potential fall distance and swing zone. The projected fall distance must be less than the distance to any structure below that could be struck during deceleration.

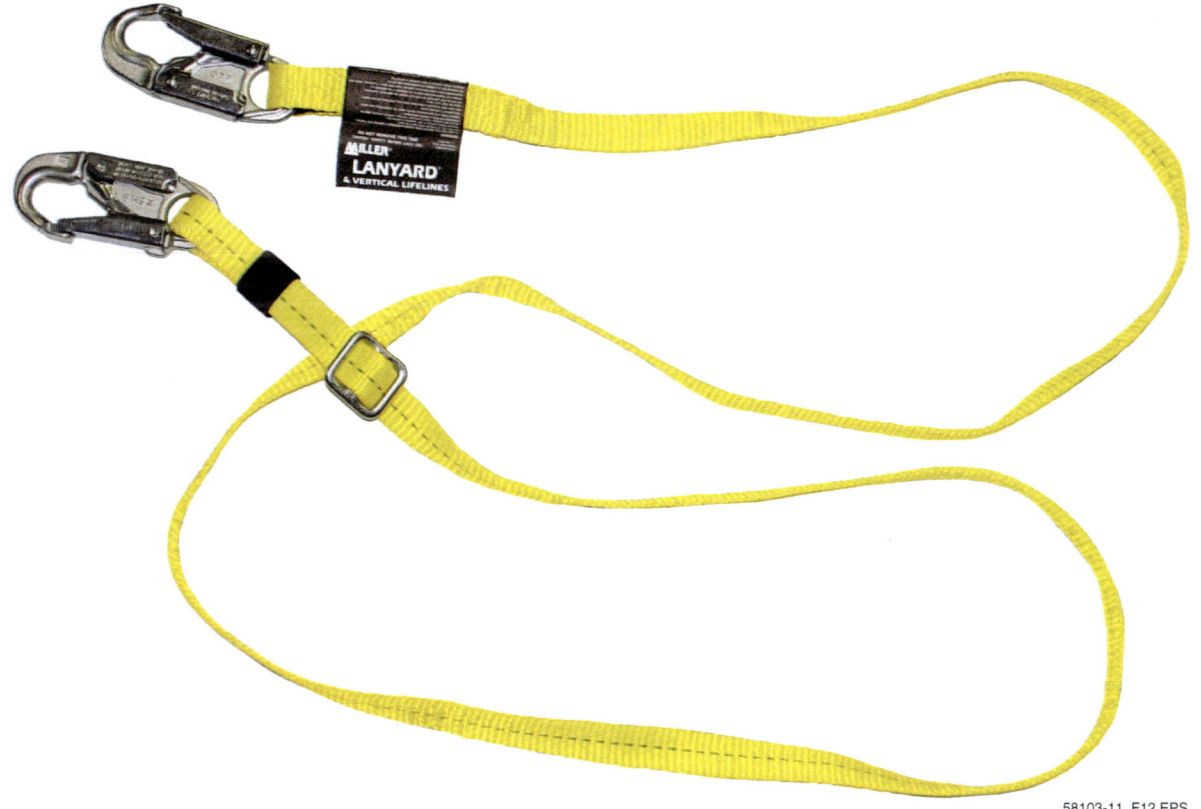

58103-11_F12.EPS

Figure 12 Nonshock-absorbing lanyard.

The D-ring used on the PFAS for fall arrest may only be connected to one live connection at a time. This can be challenging when trying to move from one point to another, especially horizontally. The climber must be able to reach back and disconnect a lanyard, while having one connected at all times. There are Y-configured lanyards (*Figure 13*) used for this purpose, where a single point of attachment at the D-ring is used to accommodate two lanyards.

2.2.2 Self-Retracting Lanyards

Self-retracting lanyards (SRLs) arrest falls as they occur by their reaction to tension. As is the case with deceleration lanyards, climbers must consider the operating principles as they decide on anchor points, then calculate fall distance and swing zones. Self-retracting lanyards (*Figure 14*) generally restrict free fall to 2 feet (0.61 m), then deceleration of the weight may take up to an additional 1½ feet. The total fall distance then would be 3½ feet (1.06 m), just as is the case with the deceleration lanyard by design. SRLs allow far greater mobility for the climber, both horizontally and vertically.

One very simple type of SRL with an effective length of up to roughly 10 feet (3.05 m) is often used. These shorter styles are often built without true braking components and are more like deceleration lanyards. Other units allow for much greater freedom of movement for the climber, and are equipped with advanced braking systems. This more sophisticated style of SRL can be found in lengths approaching 200 feet (61 m). The line moves in and out slowly as the climber moves about, unless the line begins paying out at high speed. Braking operation then begins to slow and stop the fall. It is important that too much slack is not allowed to develop in the line connection to the D-ring. If slack or loose line is present, then

58103-11_F14.EPS

Figure 14 Self-retracting lanyard.

the extra length adds to the distance of the fall before the braking mechanism begins to function.

Because of the freedom allowed by long SRLs, climbers can place themselves in dangerous situations unless constant vigilance is maintained. It is all too easy to wrap the lanyard in and out of structural members as you move, or to move too far from the anchor point vertically or horizontally. Never use the SRL in a manner that prevents it from doing its job. Dangerous use of an SRL may also prevent you from doing your job in the future.

2.3.0 Safe Climbing Assistance Devices

There are several different systems and devices that allow a climber to climb more confidently and safely while helping to eliminate dangerous situations. In some cases, a climber must connect to an anchor; climb; connect to another anchor; disconnect the first anchor; climb further, etc. Safe climbing devices and systems help eliminate the many disconnects and allow for a smoother,

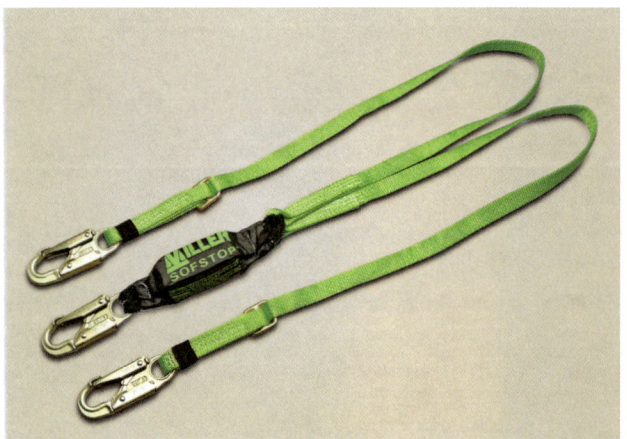

58103-11_F13.EPS

Figure 13 Y-configured shock-absorbing lanyard.

safer climb. They are generally considered permanent attachments to the structure, although this section also covers the use of temporary vertical lifelines.

Permanently installed systems vary in design. Some have a metal rail with a sliding puck to which the climber connects. The puck locks up when pulled in the opposite direction or when upward tension is not maintained. However, wind tower climbers more often encounter styles that use cable routed up the preferred climbing path (*Figure 15*). Most wind towers are equipped with such a permanent system alongside the ladder, providing an excellent fall arrest system. These systems are sometimes referred to as lad safe systems, popularized by a line of such products called Lad Saf® from one manufacturer. During the assembly of a tubular tower, the lifeline is often pre-attached in the upper section and left rolled up on the yaw deck. Once the upper section is set in place, a climber can ascend the now-completed ladder and drop the cable to the bottom for permanent attachment at the base.

Cable grabs (*Figure 16*), also called shuttles, pucks, or guided fall arresters, are mounted on the cable for the climber to use as a connection point. The cable grab locks down on the cable to help restrain a fall when movement is too rapid or aggressive. In normal use, the cable grab simply rides the cable up with the climber, sliding along without resistance. Its freedom of movement up-

58103-11_F16.EPS

Figure 16 Cable grab.

ward is important to ensure that it does not suddenly snag or stop, interfering with the climber's rhythm.

When using systems such as the one shown in *Figure 15*, the front D-ring, located on the chest just below the sternum, is used as the primary point of connection. This is one of the very few occasions when the back D-ring is not used as the fall arrest connection.

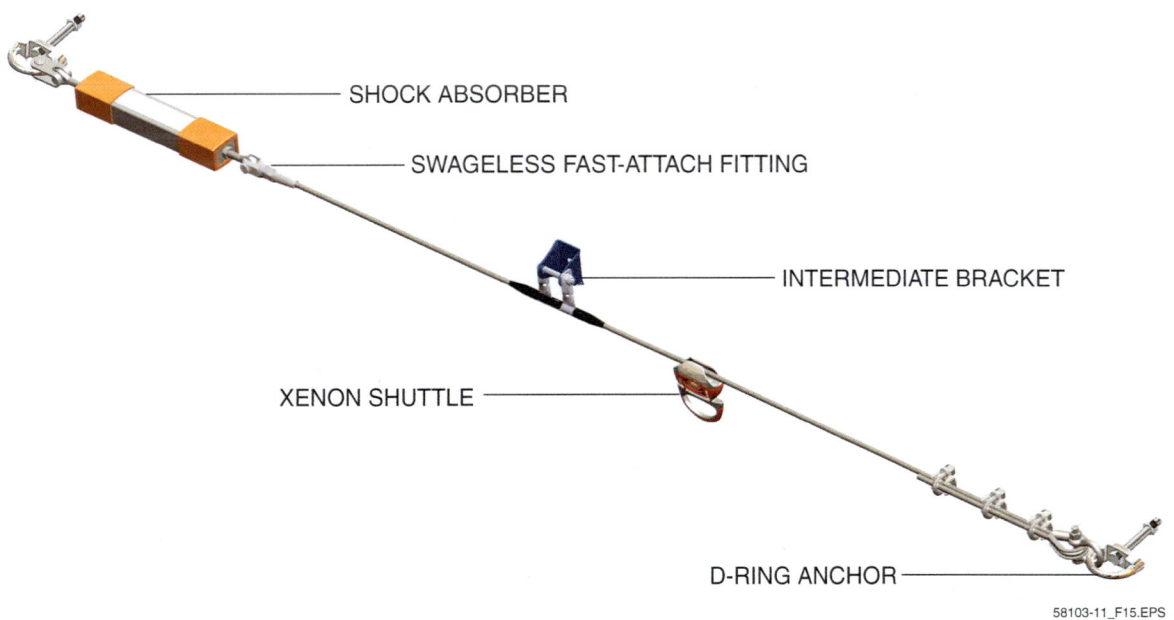

SHOCK ABSORBER

SWAGELESS FAST-ATTACH FITTING

INTERMEDIATE BRACKET

XENON SHUTTLE

D-RING ANCHOR

58103-11_F15.EPS

Figure 15 Permanent safe climb lifeline system.

Temporary vertical lines, using a safety rope, are also sometimes necessary (*Figure 17*). They are rigged at the time of need and are connected to anchor points qualifying for fall arrest service. Rope grabs (*Figure 18*) are much like cable grabs and provide the point of connection for these systems. A deceleration-type lanyard should be used and connected to the back D-ring for proper fall arrest protection. As the climber ascends, the connection to the rope trails below and slightly behind the climber, as far as the connecting lanyard allows. This adds to the potential fall distance of the climber before arrest begins. For this reason, and since this is a fall arrest application, the lanyard used should be no more than 6 feet (1.8 m) in length to avoid a fall that is too long and imparts a force to the rope beyond its ratings. Permanent rope lifelines are seldom used since rope is not as durable or long-lasting as cable, especially when exposed to weather.

2.4.0 Powered Climbing Assistance

Permanent safe climbing lifelines have integrated some advanced features over the years that are welcomed by wind tower climbers. A powered climb assist system (*Figure 19*), for example, attaches directly to the front D-ring of the climber during ladder ascent or descent, just like a typical lifeline connection. Thanks to an added electric motor, the system also pulls up on the harness and climber, relieving a significant portion of the weight. Climbing then requires far less effort,

especially when tools or repair parts add to the climber's load. The degree of assistance is adjustable in a common range of 45 to 120 pounds (20 to 55 kilograms). A safe anchor point for a lanyard is installed at the top to allow the climber to disconnect from the lifeline upon reaching the destination. These systems are normally used only on ladder systems in tubular towers where the components are protected from the weather.

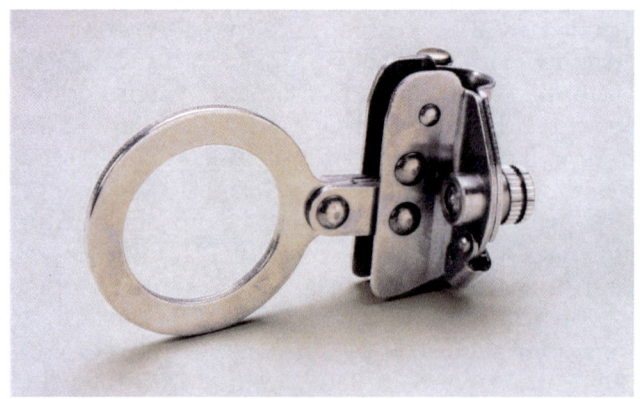

58103-11_F18.EPS

Figure 18 Rope grab.

58103-11_F17.EPS

Figure 17 Temporary safe climb rope system.

58103-11_F19.EPS

Figure 19 Lad Saf® powered climb assist system.

2.5.0 Other Safety Gear

In addition to the PFAS and the equipment associated with it, climbing workers require other common safety gear.

First aid kits must be readily available but are not carried by the climber in most cases. Wind tower work sites and/or vehicles should be equipped with minimal EMT gear, and workers should be properly trained in its use. Other types of equipment that must be used when appropriate include:

- *Head protection* – Head protection (*Figure 20*) is required by *OSHA Standard 1926.100(a)*. A number of different styles are acceptable and many are better suited to one or more applications or crew positions. Hard hats that are well-fitted and securable to the head with a chin strap are essential. Most organizations require that a chin strap be used. Many climbing hard hats were originally developed around the recreational rock climber and have since been adapted to meet construction standards. Check the chosen hard hat to ensure that it meets the required OSHA standards and has been ANSI-tested for the projected use. The hard hat pictured in *Figure 20* is approved for use under *ANSI Standard Z89.1-2009 Type I, Class C*. Other accessories, such as protective visors, are available. Before purchasing a hard hat, be sure that it meets the ANSI type and class ratings required by your employer.

- *Hand protection* – Wind towers and their ladders can be rough structures, with ladders often constructed of hot-dip galvanized steel. The galvanizing process sometimes results in sharp burrs that easily penetrate hands. Climbing presents an endless number of ways

to experience a hand injury. The choice of gloves is a personal one and should be based on the environment and task, but simple cotton gloves alone do not provide much protection. Climbers may also make a different choice for summer versus winter. The gloves should fit snugly and allow the freedom of movement and manual dexterity to do the job. A good tight fit also helps minimize any potential for the glove itself to be a safety hazard. Gloves

Figure 20 Climber's hard hat.

58103-11_F20.EPS

On Site

Wind Tower Elevators

As the wind turbine market and the number of turbines grows, so does the need for service. To date, most turbines have either a limited climbing assistance system or no assistance at all. However, companies are beginning to develop and market elevator systems that have a tremendous impact on turbine service.

The elevator unit shown here, from the Danish company Hailo, is designed to transport up to two people at a time. In addition, it can also be used as a service elevator to transport parts or materials with no passenger at all. The model depicted here has a maximum payload of 550 pounds (250 kg). Although it is best installed at the point of tower construction, it can also be retrofitted into existing turbine towers.

58103-11_SA03.EPS

with integrated knuckle protection may also be in order. Many high-quality gloves that allow maximum dexterity have come on the scene in recent years, some inspired by the auto racing industry and adapted to the trades with specific features incorporated.

- *Footwear* – ANSI and OSHA standards require footwear that contains steel or composite safety toes and reinforced soles on the job site. Since wind turbines may use permanent magnet generators, composite safety toes or shanks may be a better choice. Problems with steel toes are unlikely, but composite materials are still a good choice. They should have a well-defined heel, a non-slip sole, and provide good ankle support. One style of shoes may not be a proper choice for every climbing situation and every climber. Due to the consistent and unique stress that is placed on the foot by climbers, an excellent fit is crucial. Soft-soled shoes without reinforcement can quickly tire or injure a climber, since the stress is not spread across the foot. Also, keep in mind that bootlaces can become entangled with ropes, lanyards, and hardware. Laces should be tucked in, secured in some other positive way, or avoided entirely.
- *Eye protection* – This is as important for wind tower work as it is for any other trade work. An eye injury can become far more complex when the worker is in the air and is unable to see well enough to descend safely.
- *Hearing protection* – Earmuffs or earplugs should be used as needed.
- *Kneepads* – These are optional, but can greatly increase comfort depending upon the task.
- *Clothing* – Although no specific clothing is mandated, it is suggested that long pants and long-sleeve shirts with collars always be worn regardless of outdoor conditions. This helps prevent minor skin injuries and irritations that come from the tower surface, cables, ropes, and webbing. It also helps to prevent unnecessary exposure to the sun and minimizes access to the skin for insects. Regardless of your personal clothing choice, a good fit ensures that movement is not restricted and clothing cannot become entangled in the PFAS, lanyards, or ropes. Climbers should also carefully consider other personal grooming choices. Long hair that is not properly controlled, necklaces, and rings can become safety hazards of their own. Jewelry should not be worn, and hair should be contained to avoid it being caught up in the PFAS webbing, especially during a fall arrest.

2.6.0 Equipment Inspection

All personal safety equipment should be inspected by the user before each climb. Any equipment found to have defects must be taken out of service. It is also important to note that any such equipment that has been involved in a fall arrest situation must be taken out of service and replaced. Do not re-use harnesses, lanyards, or ropes that have played a role in a fall arrest situation. Depending upon the policies of your employer, the equipment should be clearly tagged as having been involved in a fall arrest and either be destroyed or returned to a certifying organization for testing prior to reuse.

It is important to note that the guidelines presented here are not intended to replace the recommendations for use, inspection, or maintenance provided by the manufacturer. A proper inspection of safety equipment should always be done in accordance with the manufacturer's information and any additional directives that exceed those recommendations. In addition to your own pre-use inspection, many companies also require that safety gear be inspected by supervision or management on a regular basis.

2.6.1 Belts and Harnesses

Belts and harnesses are made using a variety of materials. Different materials show signs of damage or wear in different ways. Inspect belts and harness as follows:

- Look for the manufacturers label to be affixed and legible.
- Check all webbing material for evidence of fraying or stress. Grab the webbing material between your hands, about a foot apart, and bend the belt up toward you, forming an upside-down U to detect broken fibers. Look for cuts, frayed edges, or chemical attack. Be sure to inspect both sides in the same manner.
- Inspect all sewing and lacing to ensure it is undamaged (*Figure 21*).
- Check any leather components for signs of deterioration.
- Look for any holes or parts that are not original; absolutely no fall protection or arrest components, including D-rings, should be added to a harness after its initial fabrication by the manufacturer. Check grommets for signs of stress and make sure they are secure.
- Ensure that all D-rings, buckles, or other metal parts are in good condition and show no signs of corrosion or distortion. D-rings should pivot freely and any attaching hardware should be sound and flush with the surface of the belt.

Figure 21 Stitching inspection.

- Ensure that the belt or harness design provides a proper fit for the individual user.

2.6.2 Ropes and Lanyards

Ropes and lanyards have a higher potential for damage than the harness itself, due to their use. All ropes and lanyards should be checked as follows:

- Ensure that lanyards have the manufacturer's label attached and that the label is legible.
- Check lanyard webbing to ensure it is relatively clean, and is not frayed or damaged. Grab the webbing material between your hands, about a foot apart, and bend the lanyard up towards you, forming an upside-down U (*Figure 22*). Inspecting the webbing while it is bent in this manner will encourage broken fibers or strands to stand out. Excess heat or flame usually causes the material to melt, leaving a hard, shiny, darkened spot. Again, remember to inspect both sides.
- Inspect ropes to ensure there is no fraying or tearing anywhere along the length.
- Ensure that all termination connections are sound and show no sign of damage or failure.
- Deceleration and/or shock-absorbing lanyards should show no sign of extension or decompression from previous use. A warning flag is sometimes exposed when the lanyard's shock-absorbing feature has been deployed.
- Check wire and cable lanyards for signs of corrosion or broken wires. Any D-rings attached should be checked for signs of corrosion or deformation (*Figure 23*).

Figure 22 Lanyard webbing inspection.

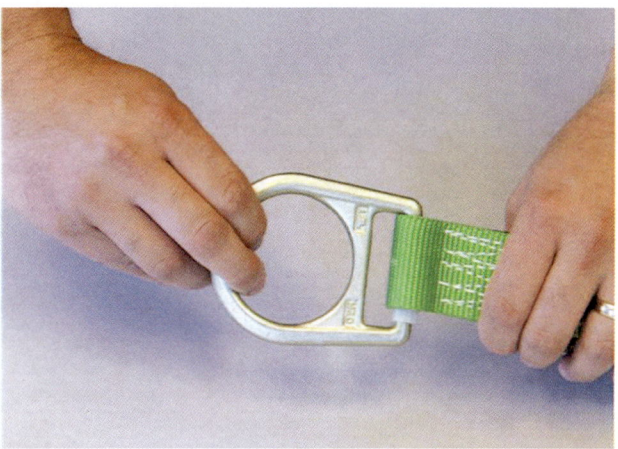

Figure 23 D-ring inspection.

2.6.3 Self-Retracting Lanyards

The internal mechanism of self-retracting lanyards cannot be inspected internally and should not be field-disassembled for this purpose. Inspect them as follows:

- Check the SRL housing or casing (*Figure 24*) for any sign of physical damage and to ensure that all assembly hardware is in place.
- Inspect the line itself regularly for any sign of concealed damage.
- Pull the line out slowly several feet, then hold it lightly as it retracts back into the housing.
- Test the braking mechanism by grabbing the line above the impact indicator and pull it sharply downward (*Figure 25*). This should engage the brakes. Once you begin to release tension, the brakes should disengage and the line should retract.

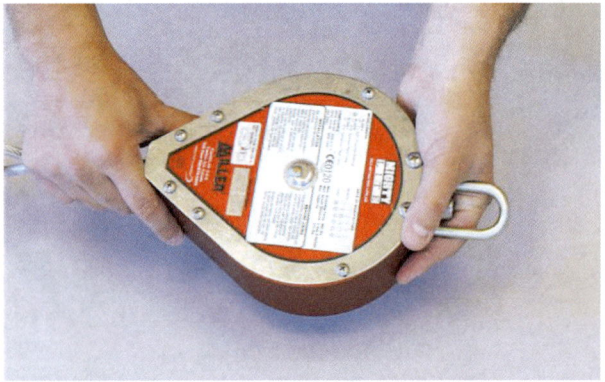

Figure 24 SRL casing inspection.

58103-11_F24.EPS

Figure 25 Braking mechanism test.

58103-11_F25.EPS

2.6.4 Hooks, Carabiners, and Cable/Rope Grabs

Cable and rope grabs have more moving parts than hooks and carabiners. Inspect your hardware as follows:

- Check hooks to be used as part of the PFAS for proper load capacity information stamped into the metal. The unit of measure is in pounds, and those used for fall arrest connections should be rated at 5,000 pounds.
- Inspect carabiners for the presence of appropriate force capacity information stamped into the metal. The unit of measure is in thousands of Newtons, or kN, and they should be rated at a minimum of 22kN.
- Check hooks and carabiners for any sign of corrosion, sharp edges, or shape distortion.
- Check the load indicator on hooks for an indication that the hook has been heavily loaded, if they are equipped with this feature (*Figure 26*). The connecting pin between the hook body and swivel connection will change position as shown.

58103-11_F26.EPS

Figure 26 Hook load indicator.

- Ensure that the safety gates or closures on both hooks and carabiners operate freely, lock, and are undamaged in any way.
- Inspect cable and rope grabs for any sign of stress, corrosion, or missing parts. It is also essential to ensure that the grabs be specifically designed for the type and diameter of the rope or cable in use.
- Each climber must take personal responsibility for the gear; it is your life that it is designed to protect.
- When not in use, safety equipment and harnesses should be properly stored in a dry location away from sunlight and chemicals that could cause parts to deteriorate.

3.0.0 HAZARDS OF THE ENVIRONMENT

It should be self-evident that the primary consideration in climbing safety is to avoid a fall. Like many other occupations though, a wide array of injuries can occur that have little to do with climbing at all, such as bee stings or cuts from sharp metal surfaces. Such relatively minor nuisances become highly magnified in the elevated environment and can quickly transform from a common nuisance to a serious and dangerous matter. In addition, there are other hazards associated with high-elevation work that you may never have considered from the safety of the ground.

3.1.0 Noise and Communication

The potential noise that any work on the turbine will create should be considered before leaving the ground. Proper hearing protection should be on hand. Noise can also interfere with communications. Portable radios are a common communication tool, but excessive noise, as well as RF interference, may prevent or inhibit their use. A second means of communication should be considered a necessity. Even a third means of communication, such as hand signals or the use of a code based on whistle blows, should be considered to ensure that some means of communicating with the ground crew is available in the worst of situations.

3.2.0 Living Dangers

A variety of animals and insects, and even humans, can be potentially hazardous to the climber, each in its own unique way. It is important to note that the hazards discussed here may not be extremely dangerous directly, but any external influence that causes the climber to be distracted can prove disastrous. Trainees will likely spend a great deal of time as members of a ground crew while experienced technicians are assigned climbing and turbine repair responsibilities. Many of these threats are just as dangerous to non-climbing crew members.

3.2.1 Birds

Birds can be a nuisance as well as a direct hazard. Although the average sparrow can inflict very little damage and has little incentive to do so, birds of prey are another matter. Lattice towers, especially, represent a unique habitat for some birds, such as the osprey (*Figure 27*) and other predatory birds known as raptors. In an attempt to protect the nest, birds can be very territorial. The beak and claws of these birds can inflict damage and injury. Although a tower free of nests represents a much smaller potential for bird attacks, attacks can and do happen anyway. The height of most utility-scale turbines on tubular towers is higher than most birds prefer for nesting, and the tubular tower itself offers little or no refuge for birds.

Bird droppings can be both smelly and slippery. Some birds of prey, like the vulture, possess highly corrosive digestive juices which enable them to digest putrid meat that would be lethal to man and other animals. When threatened, the contents of the vulture's stomach are used as a means of defense through projectile vomiting. This is not the best way to start a climber's day. Bird droppings can also carry a variety of diseases. Contact with droppings and other bird discharges should always be minimized.

3.2.2 Insects

Towers, transformers, and other wind site furnishings are common locations for the nests of stinging insects, which should be avoided. Although some experienced climbers may have developed their own solutions to the problem, avoidance is always best. Paper wasps (*Figure 28*) seem particularly drawn to communications and power transmission towers, although the reason is not clear.

Spiders are also a concern to the climber. It is very rare that a poisonous species is found on the higher structures, but some people tend to overreact to the mere sight of a spider or similar creature. For most insects, the potential danger created by

Figure 27 Osprey, or fish eagle.

Figure 28 Paper wasps.

the climber's reaction is far more serious than the damage the insect itself can do.

3.2.3 Snakes

Snakes are generally considered hazards to the ground crew only. Most snakes that would climb a short distance up a lattice tower are non-poisonous. Nevertheless, hawks and eagles prey on snakes, and it is not unheard of to find live snakes that have been dropped by their assailant onto a lattice tower. In general though, it is the ground crew that must keep an eye out for snakes, especially in remote areas.

3.2.4 Humans

One would not generally think of humans as being a potential hazard for the climber. However, humans can represent a significant danger. Agitated homeowners and others who may disapprove of the tall structures have been known to target climbers. Vandals and thieves can damage important structural supports or steal copper grounding components. Probably the most direct threat from humans comes from an unskilled or compromised member of a climbing team. Improper handling of tools and equipment by both climbers and ground crew members can be extremely dangerous.

3.3.0 Weather

Weather conditions and the potential dangers they represent must be considered well beyond the present situation. A clear day (*Figure 29*) when a climb begins can quickly become a stormy one. There are a number of ways to deal with these issues while safely on the ground, but the climber does not share that same advantage. It is also im-

58103-11_F29.EPS

Figure 29 Sunrise.

portant to remember that temperatures and wind conditions can be very different at a higher elevation than on the ground.

In all matters related to weather, climbers must be familiar with the guidelines and regulations of their immediate employer, and follow them without deviation. A brief discussion of common weather-related guidelines for climbers follows.

3.3.1 Sunshine

Tubular tower climbers would only need to concern themselves with sun-related issues if they are working in an open nacelle. Lattice tower climbing places the climber in a far more exposed situation.

The sun can cause discomfort in a variety of ways. Outdoor climbers should always be equipped with sunglasses to protect their eyes. In one part of the day, the sun's glare may not be a factor, but the glare may increase as the sun shifts. Climbers must also protect themselves from burning rays with proper clothing and sunscreen. On warmer days, water should either be carried by the climber or provided by the ground crew to prevent the possibility of dehydration. It is essential that climbers remain properly hydrated to prevent a loss of strength or agility during the descent, especially after many hours in or on a nacelle.

3.3.2 Rain, Snow, and Ice

Water makes almost any surface more slippery, and steel towers are no exception. It is very difficult to avoid wet surfaces completely, as a climber will often be called upon to climb while dew or the remnants of recent rain are still present. The greater concern comes from the other forms water often takes—ice and snow (*Figure 30*). Tubular tower climbers are typically well protected against all forms of precipitation, but accidents often happen before the climb ever begins.

Slipping and sliding are the most obvious hazards of ice. It may be difficult to negotiate an icy surface with both feet on the ground. Imagine the challenge at 100 feet in the air on a steel lattice structure. Remember that temperatures and conditions will be different at a higher elevation. Water at ground level may become ice as you ascend. Under the right conditions, a lattice tower can be wet from top to bottom during the climb, and then freeze while work is being conducted and attention is diverted from the weather.

Although there are several ways for wind turbine designers to avoid ice buildup on the blades, it can and does happen. While some turbines

58103-11_F30.EPS

Figure 30 Antarctic turbines.

simply furl the rotor and shut down when ice build-up is detected, others may use heaters in the blades to deter ice forming. Even certain blade coatings can help prevent ice formation. The location is an obvious factor in the developer's choice of turbine ice prevention.

Climbers should not approach a functioning tower with obvious ice buildup, as deadly projectiles are formed when pieces break off and fall to earth. Any time ice is present or is anticipated to form during a lattice tower climb, the work should not be considered safe. Work outside a nacelle should be avoided when towers are covered with ice and snow, but nacelle access is often permitted for tubular towers.

The hazards of snow can vary, due to differences in its consistency. In most cases, snow is considered as dangerous as ice, and lattice tower access is not be permitted without written exception by a site supervisor or equal. Again, nacelle access is generally permitted for tubular towers in the presence of snow.

3.3.3 Temperature

Temperature extremes can affect a climber's agility and stamina, especially in extreme cold. Cold temperatures are often accompanied by ice, which changes the situation dramatically. When temperatures below 10°F (–12°C) are encountered, site supervision generally evaluates the importance of the task and considers any risks involved before allowing a climber to proceed. Physical entry into the rotor hub is not typically permitted below –10°F (–23°C) for any purpose.

3.3.4 Lightning

A steel structure high in the air is an obvious target for a lightning strike (*Figure 31*). Protecting the structure from lightning damage is of the utmost importance in its design and construction, but personnel safety always ranks higher.

The view from an elevated environment is an advantage to the climber. From this perspective, storms may be visible at a great distance, allowing

58103-11_F31.EPS

Figure 31 Lightning.

sufficient time to react. Far-away lightning can often be seen without the thunder that accompanies it being heard. Generally, climbers are at a safe distance from the hazard under these circumstances. However, if thunder can be heard, lightning is present and close enough to be of concern. Under no circumstances should a climber ascend a tower, or remain on a tower at work, if thunder is audible.

Although guidelines for lightning safety differ among employers and wind sites, many U.S. sites consider a distance of 50 miles (80 km) to be the distance that leads to an informational alert. Personnel are notified that lightning is near and they should be prepared to leave elevated work areas and seek shelter if conditions worsen. At the 30-mile (48 km) mark, the turbine work area should be cleared immediately. Once the lightning is confirmed as being more than 48 km away for 30 consecutive minutes, work may resume.

3.3.5 Wind

Ironically, wind is undoubtedly an enemy of the climber. At a minimum, it creates noise, frustration, and a sense of urgency by its presence. At its worst, it can separate the climber from an exposed structure.

Wind can have a profound effect on the tower itself, often creating movement that is very disconcerting and distracting. Since towers are well above any type of wind break that protects you at ground level, acceptable wind conditions near the ground are often found to be extreme farther up. The towers themselves are designed to withstand the flexing and other stresses, but the climber may not enjoy the experience.

As is the case with lightning, different sites and maintenance groups have somewhat different guidelines regarding climbing and maintenance during high winds. Please note that all wind speed values in the list below are based on 10-minute averages rather than single readings. Some general guidelines:

- No restriction of wind turbine maintenance or climbing is considered necessary at wind speeds below 15 meters per second (m/s), or 33 miles per hour (mph).
- At wind speeds between 15 and 19 m/s (33 and 44 mph), climbing lattice towers is prohibited. For tubular towers with internal ladders, climbing and work may continue inside the nacelle only. Work cannot be conducted inside the rotor hub itself. Tasks requiring crane lifts to replace major components are prohibited as well.

- Once wind speeds exceed 20 m/s (45 mph), climbing all towers is prohibited.

Exceptions at most sites can be made for critical functions following a risk assessment by site or regional supervisors.

4.0.0 CLIMB PREPARATIONS

A few things need to be done in preparation for any climb. The work site and the structure itself need to be assessed; crew members should be aware of the location and availability of medical and rescue/escape kits; and everyone needs to know what tasks are to be done on the turbine system itself. These are all points of discussion in a proper pre-climb meeting.

Pre-climb meetings should take place on a daily basis, even on the same job site, and they must be taken seriously. All information exchanged and issues discussed should be documented as well.

4.1.0 Site and Tower Assessment

A site and tower assessment must be conducted prior to a pre-climb meeting so that the results can be presented to the group. A thorough assessment of the site includes observing the sky as well as the ground. Hazards associated with the environment have already been reviewed, including weather issues. During the site assessment, climbers put this learning to work. Look for the various types of wildlife and insects that may be a distraction, and deal with them if possible. For insects, be prepared to distribute repellents. Take a walk around the entire perimeter of the work area and structure, looking for hazards as you go. Monitor weather forecasts, and look for weather issues yourself. If it is a sunny day, workers need sunscreen and eye protection. Record all this information for discussion during the pre-climb meeting.

Lattice towers and tubular towers require slightly different approaches, since the tubular tower offers significant protection. Lattice towers must be examined carefully, with the known task and work location on the tower in mind. First, think about the big picture in your preliminary analysis of the tower. Ensure that the tower is standing up straight and appears normal. Look for any evidence of twists or turns in the structure that may indicate a problem. Then begin a more detailed visual inspection, looking for loose or missing pieces of the structure, rust and corrosion, missing hardware, or any signs that vandals have been at work. Pay particular attention to power and ground cables and other electrical devices, as they are a favorite target of thieves.

If the tower is guyed, you are likely to encounter several different styles of attachment and adjustment. Here are some general guidelines for examining them:

- Inspect the anchors where the guy wire attaches to the ground, especially at the point where the anchor shaft meets the ground or the concrete footing. Sometimes known as the point of daylight, this location is the first to show signs of deterioration and chemical reactions.
- If turnbuckles are in place for adjustment of guy wire tension, examine them for loose or missing hardware or damage. Both male threaded shafts should be completely through the adjusting nut, and they should be roughly equal in their length through. Turnbuckles for this application have no built-in shock absorbing properties.
- Inspect other guy wire connecting components as well. Ensure that all connections, thimbles, and terminations appear to be in sound condition and show no evidence of slippage or movement.
- All guy wires should be inspected to ensure that they are not loose. They should always be taut. A loose guy wire is an indication that something has changed. Determine the reason for a loose guy wire and correct it before climbing. Consider that when a loose guy wire is noted, a guy wire on the opposite side is likely to be too tight. This indicates a tower that is no longer tuned properly and should be evaluated further before climbing.

After examining the overall condition of a lattice structure, climbers should begin evaluating the climbing route. Inspect whatever step bolts or other climbing devices are permanently installed on the structure that can be seen. Those out of sight must be evaluated as the climb progresses to ensure their integrity. For tubular towers with internal ladders, examine all components of the ladder and lifeline system that you can see. Look up the ladder to see if anything looks unusual. Look over any safe climbing assist systems and make sure the proper size and style of rope or cable grab is available. Think about and evaluate where you need to work, plan the climb, and make adjustments as you progress. Plan how to deal with any obstacles in your path, or plan to avoid them altogether. Imagine yourself actively climbing and what you will likely use as the next anchor point on the route.

For offshore wind turbines like the ones shown in *Figure 32*, you obviously need to modify your site inspection procedure somewhat. You can

Figure 32 Offshore wind turbines.

safely eliminate snakes and wasps from your list of concerns here. Although there are no offshore sites active at this time in the U.S., development of the first site is in progress off the coast of Nantucket, MA.

4.2.0 The Pre-Climb Meeting

The pre-climb meeting should occur on site daily and be well documented. One important part of the pre-climb meeting is the job briefing. An employer-provided form is typically used to document the discussion and important data that should be readily available to all site workers. OSHA requires that a job briefing be conducted prior to the start of each job, and the following five topics should be covered:

- Hazards of the job
- Specific work procedures
- Special precautions required
- Required PPE
- Energy source controls

A thorough discussion of the task itself is needed. Required tools and materials and their location should be identified, and the order in which tasks will be undertaken should be reviewed, so that all climbers and ground crew members know and understand their respective responsibilities. The following are some general topics that should be covered in addition to the specifics of the work to be performed:

- *First aid* – OSHA requires employers to provide adequate access and availability to medical care. If a hospital or emergency room is nearby, that is certainly a bonus. Because the term *nearby* is subjective, it is best to follow the alternate ruling of OSHA, which requires that a person adequately trained in first aid be available on the site. Each crew member should know the location of first aid supplies, and everyone should know who is properly trained to administer it and who is not.

- *Emergency medical facilities* – The phone number and location of the nearest emergency medical facility should be clearly written on the pre-job briefing form and made available to all crew members. Workers should have knowledge of the site location in order to guide medical responders to the scene if necessary. GPS coordinates and the physical address of the site should be recorded on the pre-job briefing form. Being able to provide turn-by-turn directions may be better depending upon the location and terrain. Proactive wind energy sites may contact local medical facilities that would generally be dispatched to a site emergency and provide them with maps of the facility. On a wind farm, every turbine usually looks the same and it can be difficult to navigate. If any hazardous materials are involved with the job, the related MSDS, material handling procedures, and emergency response plans should be reviewed.

- *Rescue planning* – A rescue plan must be in place in the event that a climber becomes incapacitated. This is another OSHA requirement and a good common sense rule. All parties should know who is designated as the rescue climber if one or more climbers need help, and what form of rescue equipment is going to be employed. It is recommended that active climbers receive tower rescue training every twelve months.

- *Site and tower assessment* – The results of the site and tower assessment completed prior to the meeting should be reviewed. Make sure the team knows which hazards are present and any weather conditions that should be monitored. Discuss how to deal with hazards and adverse conditions.

- *Electrical hazards* – The hazards of electrical energy on the tower must be discussed during the meeting. Induced voltage from nearby energized circuits can be lethally hazardous. Follow approved work practices to avoid the hazards of induced voltage.

- *Climbing policies and guidelines* – During the pre-climb meeting, the meeting leader must confirm each worker is familiar with the policies for climbing safety. Although policies will differ among sites and employers, some widely accepted guidelines for wind turbine climbing should be acknowledged and followed unless alternatives are provided by site management. Common policies and guidelines are discussed in the next section.

- *Equipment inspection* – In some cases, this may be the chosen time for climbers to inspect their personal gear. This is best done as a group, so that everyone participates in full view.

If you are to take responsibility for the pre-climb meeting or any portion of it, take it seriously and always be on the lookout for changes in the environment or working conditions. Once the pre-climb meeting is complete, it is time to put the job in motion (*Figure 33*).

4.3.0 Common Climbing Policies and Guidelines

The personal protection equipment and climbing gear normally required has been discussed previously. Depending upon the employer, minor variations in those requirements may occur. Specific rescue gear may also be required by individual climbers if they are to play a specific role. The important subject of rescue is discussed in a later section. It is important to ensure that all climbers are aware of these employer- or site-specific requirements during the pre-climb meeting.

During the pre-climb meeting, it is essential that the meeting leader confirm each worker is familiar with the standard policies for climbing safety. Although policies differ among sites, some widely-accepted guidelines for wind turbine climbing should be acknowledged and followed unless alternatives are provided by site management:

- Before the climb begins, the wind turbine must be taken out of service. There are only a few cases where operation of the rotor and turbine should be permitted with personnel climbing to, or already in, the nacelle area. One possible example of such a situation is the commissioning, or startup, process of a new turbine. This generally requires someone in the nacelle during operation. However, in general, never climb while the rotor is turning.

58103-11_F33.EPS

Figure 33 Entering the turbine tower.

- The location of any rescue and escape gear assigned to the tower must be identified. Generally, having this equipment at the base of the tower is acceptable if the turbine is not functional. However, if the turbine is operating, the kit must be at least at the height of the work in progress.
- All services that involve climbing the tower require two people without exception. Each organization and wind site may have different policies regarding the qualifications and rescue skills of the two workers. In some cases, both climbers are required to have rescue skills. First aid and CPR certifications may also be required. Company policies must be followed without deviation. Only an individual with clear authority to make safety decisions in emergencies can authorize a departure from policy.
- If policies allow, both climbers may go up to the work area together if the task calls for it. One example would be when work is required in the hub. One worker remains in the nacelle while the hub is occupied.
- One climber can climb alone while the second climber remains on the ground in support. Good communication between the two parties is essential. Any communication problems should be resolved immediately. Some sites may allow a single downtower worker to provide support for as many as three towers, but that is often impractical. If a worker has a significant problem that requires physical assistance or rescue in one tower, the other climbers would be left without support.
- There are times when service personnel climbing the tower are not fully qualified as climbers. For example, a new technician may be a competent climber but has not yet completed any rescue training. In this case, at least one of the two qualified climbers present must be either above or at the same height as the non-qualified person.

During ascent or descent, the following guidelines should be recognized. These provisions should also be discussed during the pre-climb meeting when any crew member present is not fully familiar with them:

- Before beginning the ascent, shoes should be checked and cleaned as necessary. Both hands should be free and empty. Any items being carried up must be safely secured.

- Never stand under a climber as they ascend. For tubular towers, it is suggested that anyone inside the tower base move to the opposite side of the ladder from the climber or exit.
- Two people can ascend a tubular tower at the same time as long as there is always at least one closed hatch door separating them at all times. This provision assumes that the lifeline system—powered or non-powered—is rated at a sufficient weight capacity for the connected load.
- The first climber up should inspect the ladder system and tower for structural integrity or unusual conditions during ascent. Clean or wipe anything off the ladder that may create a hazard for the next climber.
- If a climber stops at any height that presents a fall hazard (above 6 feet), the fall arrest connection must be made in addition to any fall prevention or positioning lanyards. Stow unused lanyards on the harness per its guidelines, rather than leave an unused lanyard connected to a harness D-ring.
- Deck hatch doors should be closed once the climber has passed through, regardless of the number of climbers.
- Two climbers can descend at the same time, again providing that the lifeline system has a sufficient capacity rating and that at least one closed hatch door is between the two climbers.

Some additional provisions often apply to work in or near the nacelle area:

- The full body harness with needed lanyards should be worn at all times with few exceptions. If a technician is working near rotating shafts, then both the harness and lanyards can be removed to prevent them from being a hazard rather than a lifesaver. Otherwise, they should be worn. Lanyards may be removed when they present a hazard, but the harness should remain on.
- The nacelle's top safety rail is considered a personnel anchor point on many turbines, but it will likely be sufficiently rated for only one or two people. Ensure that all climbers are aware of the maximum capacity for the nacelle rail and any other equipped railings, as well as the capacity of the nacelle roof itself.
- Since working inside the rotor hub assembly when it is energized presents additional hazards, two qualified climbers must both be up in the nacelle area for one to enter the hub. If two workers need to enter the hub, a third qualified climber needs to be present in the nacelle.

5.0.0 BASIC CLIMBING SKILLS

Due to the extreme variety of situations and climbing tasks, it is virtually impossible to list every possible guideline or tip that could be associated with climbing. As a climber, you are subject to the regulations and policies in place by your employer, federal and state agencies, and possibly those of the equipment owner. There is no shortage of rules to follow as you engage in this line of work. Learning the rules is only a part of the process. In order to learn how to climb, you must practice. The physical process of climbing can be explained to some degree, but much of the learning must come from ascending a structure in the presence of a skilled and experienced climbing instructor.

Climbing safety is not a subject you learn once, take a test, and perform forever based on that single experience. It is a continual learning process that requires climbers to pay close attention to every successful climb, and even closer attention to climbs that did not go well or resulted in an injury. Not only must you be vigilant about every aspect of the climbing process, but it is also quite likely you will be working on or near live, high-voltage electrical systems. The distractions created by the electrical hazards and the work at hand cannot be allowed to interfere with attention to climbing safety. Full attention must be given to personal safety and fall avoidance, then to the work to be done. The task cannot be done if there is a failure to focus on personal safety first.

5.1.0 Preliminary Considerations

Safe work habits for the individual climber come into play before arrival at the job site. The following basic advice should be considered by each individual:

- Recognize that not everyone was born to climb and work at heights. Some will take to it naturally, others will work their way into proficiency, while still others will never be comfortable with climbing. If you are unable to conquer the fear of climbing to significant heights and working in a relaxed and focused manner, you should not be climbing wind towers.
- Understand that the most important safety factor is you, and assume that your own safety and the successful completion of the job rely solely on you. Although it is essential that team members be able to rely on each other, the best possible outcome is when all crew members and climbers are thinking the same thing— that the safe execution of the climb relies solely on them.

- Actively participate in all training programs that are offered to you. Consider safety a lifestyle rather than something you practice only at the job site.
- Follow the policies and procedures outlined by your employer and/or client. If you are not in agreement with them, feel free to discuss such issues with supervision before the climb. You will not generally be released from adhering to the guidelines and policies simply because you disagree, but changes and improvements in procedures are usually the result of discussion and mutual understanding.
- Climbing is a profession with significant physical and mental demands. As a professional, you must practice proper maintenance of yourself as well as your equipment. Endeavor to maintain good physical condition, eat properly, hydrate properly, and allow plenty of time for proper sleep.
- Maintain a drug-free lifestyle. Do not climb while under the influence of drugs or alcohol in any amount. Do not allow yourself to climb if you feel your health or condition is impaired in any way.

5.2.0 Ascent

Climbing is a very physical process and it should be done in a relaxed, unrushed, and confident manner. There is no award for speed, except possibly in a rescue attempt. Climbers often forget the physical exertion involved with climbing and find themselves suddenly tired. If you tire, stop and attach a short positioning lanyard for a brief rest while remaining connected to the lifeline. Decks in tubular towers may offer a good spot for a rest, but the hatch should be closed so that other climbers below may continue to ascend if they like.

- As you climb, you must maintain three points of contact at all times. Hopping or jumping from point-to-point is not the way to go. Move one foot or hand at a time, always remembering to keep three points of the body in contact with the structure. If you must provide your own fall arrest on the way up (no safe climb system installed), then disconnect and reconnect to anchor points while continuing to maintain three-point contact, and always ensure at least one leg of the Y fall arrest lanyard is connected. The Y-lanyard is designed specifically for this purpose.
- Climbing is to be done with the legs, shifting weight from one leg to another smoothly and gradually when necessary. Do not use your arms to pull yourself up—allow your legs to do the hard work. Move your legs up first, and then reposition each hand higher to prepare for the next step up.
- Connect your front chest D-ring to the cable or rope slide if such a system is installed. Otherwise, connect a shock-absorbing lanyard to your back D-ring. Never trust the clicking sound of hardware as a means of determining connection. Regardless of what you hear or expect to hear, and regardless of how well you think you know your personal equipment, you must visually and physically ensure that every connection made is sound and secure. Sound may help your confidence, but it is no guarantee a connection was made.
- If safe climb systems are not in place, then you should be planning to use other structures for anchor points as you ascend. Step bolts or any similar protrusion do not qualify as anchor points because they do not generally meet the load requirements, and lanyards or hooks can slip off too easily. They are convenient, but are not valid points of connection. You can alternate between fall restraint use (very short effective lanyard length) and fall arrest on your choice of anchor points, or alternate between fall restraint connections alone.
- Look up as you climb, not down. You should be assessing the route ahead of you, looking for any structural issues or obstructions, and planning your series of anchor points if necessary. If your back D-ring is connected to a fall arrest lanyard, climb with any slack in the lanyard collecting behind you.
- Step bolts or steps that are provided may not always be at perfectly consistent distances nor perfectly oriented to each other. Think about your foot position as you climb and evaluate each new step on its own.
- Climb at a speed and rhythm that is smooth and free of exaggerated movements to help prevent lanyards, harnesses and equipment from swinging wildly back and forth.

5.3.0 Maneuvering and Positioning

- Once you arrive at the work location, secure yourself with a positioning strap. Never disconnect from any safe climbing system until you have attached to another anchor point. Evaluate your situation and determine what else needs to be done to secure your safety in the needed work position.
- Remember that your fall restraint lanyards, when used for positioning, should leave no more than 2 feet of movement and must be attached to a 3,000-pound anchor point at least.

Your fall arrest lanyards should be attached to a 5,000-pound anchor point and should be higher than your shoulders whenever possible to minimize the fall distance and the swing zone. For each climber attached to a single point, the load capacity is increased another 5,000 pounds. In other words, for three workers to share a fall arrest anchor point, it must be able to handle a 15,000-pound load. Take a good look at it and consider if you would suspend three or four pickup trucks from the anchor and expect it to hold.

- Lattice towers generally have many diagonals in their structure. Always think twice about the result when connecting lanyards to diagonals. Remember that there will always be a tendency for a choked strap to slide down the diagonal to the next structural intersection. Visualize your connections here and consider what they will do to avoid a problem that either causes a fall or creates a very challenging disconnect situation.

- When attaching your PFAS system and using deceleration lanyards, remember that there is a 6-foot fall and roughly 3-foot deceleration distance, at a minimum. When self-retracting lanyards that are of significant length are used, do not forget that their convenience has a price. It may save a number of connects and disconnects, but always be mindful of, and respect, your swing zone and potential fall distance when using them. Consider the SRL anchoring position carefully, and it is quite likely that you will need to climb to your chosen anchor point using fall restraint only until you have it securely in place and connected to your back D-ring.

5.4.0 Descent

- If you are using a safe climbing system, return to it to begin your descent. Do not disconnect fall restraint lanyards until you are connected to the rope or cable grabs.
- Make a final check to ensure tools and equipment will not interfere with your descent and you are well balanced.
- During the descent, the hands are moved down first, and then the legs are moved. The hands and arms should still not be supporting your weight in any way, nor should your grip be too tight. The legs continue to do the heavy work, but they are not pushed down. Simply allow the weight of your body and the leg itself to lower it into position.
- Do not skip step bolts or try to descend too rapidly.

- Do not be tempted to rappel or slide down the tower. This is forbidden.
- Remember to be aware of your rhythm, just as you were during the ascent. Most climbers tend to descend a bit quicker than they ascend, and with more rhythm in the step which can cause all of your equipment to swing. Any displays of exuberance about finally being down from a difficult aerial position all day should be saved for ground contact. Keep your mind on the descent and your fall arrest and restraint system use.
- Once the climb is complete, look over all climbing equipment and stow it away in an organized manner. Look for any damage that might have occurred. The climber knows what happened up there and where a lanyard might have taken some abuse better than anyone does.

6.0.0 RESCUE AT HEIGHT

You have finally arrived as a climber. You have taken the training, performed perfectly in every simulated climbing situation, and now you have made it to the top of the first tower you will service with your colleagues. You open the upper nacelle doors, connect your fall arrest lanyard to the anchor point, and step out. What a view, until a gust of wind out of nowhere causes you to lose your footing and off the side you go. Fortunately, your fall arrest equipment worked, but now you are dangling over the side like a puppet, terrified and embarrassed. It is a long way down and you need some help.

Rescue at height is a skill you need to master and practice, but prefer never to use. In the case above, you would be dependent upon the training and skill of your co-workers. Some minor situations where a slip causes a climber to fall just a few feet may involve self-rescue. However, rescuing a fallen worker at height who may be unconscious or seriously injured is another matter altogether.

6.1.0 Suspension Trauma

Another good reason for the prompt rescue of a fall victim who is suspended in a harness is the potential for serious injury or even death from suspension trauma. Suspension trauma causes a variety of symptoms in the victim, including light-headedness, dizziness, nausea, headache, sweating, and heart palpitations. Being suspended vertically in a harness (*Figure 34*) for an excessive period can cause the death of an otherwise healthy and uninjured worker. It is essential that rescue actions begin immediately. The longer

Figure 34 Suspended technician.

58103-11_F34.EPS

the worker is suspended, the greater the danger from suspension trauma. A timely rescue is absolutely necessary.

As the human body is suspended in the vertical position with the groin straps tight around the upper legs, blood begins to pool and collect in the legs. Since blood-supplying arteries are located deeper inside the leg than the veins that return blood to the heart, the veins are most affected by the straps. Blood can go into the legs, especially with the assistance of gravity, but it is unable to return to the upper portion of the body.

The body initially responds to the apparent blood shortage by speeding up the heart to try and move more blood to the brain. Eventually, when this is ineffective, the body then slows the heart dramatically. Sudden fainting often occurs at this point. When on the ground, the body will simply fall to a horizontal position, which then helps the body return blood flow to normal. However, the worker suspended in a harness remains in the vertical position after losing consciousness.

For this reason, rescue needs to be as prompt as possible. A Joint Safety and Health Information Bulletin issued in 2004 by the U.S. Department of Labor, OSHA, and other agencies indicates that death from suspension trauma alone can occur in as little as 30 minutes. If other injuries prior to the suspension are experienced, or there are other weaknesses present in internal organs, the time can be even shorter. Workers should be trained to encourage a fall victim to use and move the leg muscles frequently, and to use a suspension trauma strap if it is provided. If the victim has been in the suspended position for an extended period, they should not be immediately placed into the horizontal position following rescue. This can cause death even though a successful rescue has been made, due to the abrupt change in blood flow through the heart and to the brain. Instead, try to keep the worker's upper body vertical until medical personnel are on the scene to monitor vital signs. Many medical professionals suggest that the victim take as long as 45 minutes to transition from a vertical position to fully horizontal. The victim should assume kneeling and sitting positions through this time period.

6.2.0 The Rescue Plan

It is the responsibility of the employer and operator of the wind turbine facility to create and adopt a rescue plan. This is an OSHA requirement. In addition, every rescue plan has to address speed in its provisions. OSHA requires that victims be rescued promptly. Common decency and concern for the well-being of your colleague is a better reason than regulations.

Rescue plans should be specifically designed for the facility and type(s) of turbine that are in service. No single plan or set of rescue equipment fits every turbine design perfectly. It is also best to stick with a limited number of devices in the rescue plan to avoid confusion when the equipment must be placed into service by an excited and possibly fatigued rescue climber.

Remember that not every rescue involves a worker who is suspended in climbing gear. The nacelle itself can be a dangerous area as well, and a stroke or heart attack can happen anywhere, any time.

Rescue plans should be detailed and provide guidance for every conceivable situation. They must also be realistic and practical. A good plan considers rescues from the rotor hub, nacelle, and suspended positions off the nacelle or tower. The rescue of both conscious and unconscious victims must be carefully thought out.

As a new member of a turbine service crew, you need to trust that your employer has done a good job of creating a rescue plan for workers at height. More importantly, it is essential that you learn and embrace the rescue plan in place, and take advantage of all opportunities for both climbing and rescue training.

ANSI has recently taken on the challenge of developing comprehensive standards for every aspect of the fall protection industry. Although the work is still in progress as of 2011, a number of standards have already been developed and shared with the wind industry.

The personnel and their related skill levels named in a rescue plan are very important to understand. Although ANSI and possibly other groups are working to create standards for identifying personnel with specific skills related to safety and rescue, the terminology varies at this time. As you review the rescue plans provided by your organization, be sure that you understand the identifying terms for personnel skilled in rescue, rescue planning, and climbing. The rescue plan should identify those individuals who can provide guidance and information related to the subject.

These individuals are identified in an organization's written rescue plan and should be able to provide you with information and training related to it.

6.3.0 Basic Rescue Equipment

Rescue equipment and systems can be permanently installed in the wind tower nacelle, carried to the nacelle by the first climber, or hoisted up as soon as the first climber reaches the nacelle area. The storage and arrangement for deployment of the equipment depends upon the company and wind site.

Most equipment in the category of rescue equipment for climbers falls into one of two classifications: automatic descent control equipment and manual descent control equipment. As the names imply, automatic descent equipment provides for the descent of a worker without the need for manual control over the speed of the descent. Certain functions can be manually controlled, but speed control is provided automatically.

One example of automatic descent control equipment is the Rollgliss® R500 System from DBI SALA (*Figure 35*). This unit comes as a kit with the option of manual rescue lifting integrated with descent control. An automatic descent control unit is selected based on the need for optional features and the anticipated maximum length of the descent. This particular model is available with up to 150 meters (500 feet) of descent capability.

The device lowers one or two people simultaneously from height, or can be used for multiple consecutive descents. The model shown here has a small hand crank to enable the rescuer to lift the victim enough to disconnect the fall arrest gear from the harness and allow descent. Once the descent begins, speed control is fully automatic. Although a rapid descent is often crucial, the speed must not be so fast as to create another serious injury at the base of the descent. The descent speed for this unit is typically between 60 and 90 centimeters/second (2 to 3 feet/second) for a single individual. The descent can easily be stopped by the rescuer or the victim (if conscious) by simply grabbing the ascending section of rope. When overhead anchor points are available and a clear line of sight exists between the victim and the anchor point, this type of equipment provides excellent service.

As you might expect, manual rescue equipment requires more attention and participation from the rescuer. On the other hand, they may also offer a bit more versatility. The rate of descent must be continually controlled by the operator. Systems like the Rollgliss® R350 from DBI SALA (*Figure 36*) are known as block and tackle systems. They allow the rescuer to quickly ascend or descend to the victim, transfer the victim's weight over to the device, disconnect their fall arrest lanyard, then raise or lower both parties to safety. This particular unit also has different pulley modules to allow for quick changing of the lift ratio from 2:1 to as high as 5:1.

The final choice of equipment remains with the responsible organization. Rescue training will be offered to you at the appropriate point of your career as a wind turbine technician, and it will certainly benefit everyone on your team, including yourself, to take advantage of the opportunity. Even the best rescue equipment is of no value unless a qualified rescuer is on hand with the skill and knowledge to place the rescue plan into action.

Figure 35 Automatic descent control unit.

Figure 36 Manual rescue system.

Summary

For those that possess the skill and aptitude for climbing, opportunities in the renewable energy industry will continue to expand. Safe climbing begins with a thorough understanding of the equipment that will become both your connection to security and your best friend. You must inspect and care for your equipment with the knowledge that, at any moment, it may save your life.

Climbers are faced with a number of challenges unique to the environment. Some factors can, and should, prevent a climbing event altogether, while others present a simple distraction that can lead to an accident. Preparation for a climb in the form of a site and tower inspection will help reveal hazards and challenges in the general area and on the tower itself. Pre-climb meetings and job briefings should precede each climb.

Consistent, ongoing training is essential to climbers for continued safety and proficiency. Although learning the basics of moving up and around a tower from a textbook is an appropriate start, climbing is a physical skill that must be learned and practiced.

1. Which of the following items are *not* considered by OSHA to be acceptable fall prevention systems for working at heights above 6 feet?

   a. A guardrail system
   b. A safety net system
   c. The two-climber buddy system
   d. A personal fall arrest system

2. Fall arrest systems _____.

   a. stop or take control of a fall in progress
   b. prevent falls from ever occurring
   c. ensure that every fall is reported to the proper authorities
   d. ensure that the climber does not swing from side-to-side during a fall

3. OSHA requires that a PFAS limit the maximum force imparted to the body during fall arrest to _____.

   a. 500 pounds
   b. 1,200 pounds
   c. 1,800 pounds
   d. 2,500 pounds

4. The free fall distance allowed by a PFAS should be no more than _____.

   a. 3½ feet
   b. 6 feet
   c. 10 feet
   d. 12 feet

5. The three primary components of a personal fall arrest system are _____.

   a. anchor points, a body harness, and a vertical lifeline
   b. anchor points, a body harness, and connecting devices
   c. anchor points, connecting devices, and a shock-absorbing lanyard
   d. a body harness, connecting devices, and a shock-absorbing lanyard

6. When an anchor point rated at 5,000 pounds is not available, the climber must then ensure that the chosen anchor position can at least withstand the maximum load that could be placed on it during a fall arrest, multiplied by a factor of _____.

   a. 1.5
   b. 2
   c. 3
   d. 5

7. An anchor point that is used as a fall restraint anchor, and not as a fall arrest anchor, must be load-rated to handle _____.

   a. 2,500 pounds
   b. 3,000 pounds
   c. 5,000 pounds
   d. 7,500 pounds

8. Unless a climber is on a ladder climbing system, the personal fall arrest harness must be connected to the anchor point using _____.

   a. at least two lanyards
   b. the chest D-ring only
   c. the back D-ring only
   d. both the chest and back D-rings together

9. Groin straps are an integral part of the full body harness and not an optional accessory.

   a. True
   b. False

10. A properly connected positioning lanyard should not allow a climber to fall more than _____.

    a. 18 inches
    b. 2 feet
    c. 3½ feet
    d. 6 feet

11. Self-retracting lanyards _____.

    a. are limited to a length of 20 feet
    b. are available in lengths up to 200 feet
    c. should stop a climber before the fall distance reaches 10 feet
    d. should never be used without adding a shock-absorbing lanyard

12. It is suggested that climbers wear _____.

    a. long-sleeve shirts with a collar
    b. short-sleeve shirts without a collar
    c. long-sleeve shirts without a collar
    d. short-sleeve shirts whenever weather allows

13. Personal safety equipment that has been involved in a fall arrest must be _____.

    a. taken out of service
    b. tagged and used only for fall restraint
    c. field-inspected by the manufacturer's rep before reusing
    d. field-tested at the appropriate load rating before reusing

14. Adding an additional D-ring to a body harness is often necessary for wind turbine technicians.

    a. True
    b. False

15. Broken fibers in a lanyard can be more easily detected by _____.

    a. black-light inspection
    b. soaking it in water first
    c. bending the lanyard in an upside-down U
    d. stretching it on a lanyard inspection bench

16. Carabiners used as part of the PFAS must have a minimum load capacity of _____.

    a. 22 kN
    b. 50 kN
    c. 3,000 pounds
    d. 5,000 pounds

17. New climbers on the job should expect to _____.

    a. be first up the tower
    b. be assigned the most challenging climbs for practice
    c. spend a significant amount of time as part of the ground crew
    d. be allowed to spend the majority of their time simply watching

18. Wind tower climbing should never be attempted when _____.

    a. the tower is wet
    b. lightning is in the area
    c. winds are above 5 knots
    d. thunderstorms are in the forecast

19. At wind speeds between 33 and 44 mph (15 and 19 m/s), _____.

    a. climbing of any wind tower is prohibited
    b. climbing lattice towers only may continue
    c. climbing and maintenance on tubular tower nacelles only may continue
    d. an additional authorized rescue climber must be added to the team for all towers

20. Turnbuckles provide shock absorption for tower guy wires in high winds.

    a. True
    b. False

21. It is recommended that climbers complete a course in tower rescue procedures _____.

    a. once each year
    b. every 6 months
    c. once a month when actively climbing
    d. only once if they are to be assigned rescue responsibilities

22. One possible situation where a technician may need to be in the nacelle while the turbine rotor is turning would be _____.

    a. during a gearbox oil change
    b. during the commissioning process
    c. when peak power is needed but maintenance work is scheduled
    d. when heavy rain is present and the technician cannot safely descend

23. When you are an experienced climber, if you do not agree with a particular climbing policy of the employer, you should _____.

    a. do what you feel is safer
    b. report your feelings to ANSI
    c. discuss it openly with the supervisor
    d. document your disagreement in writing to legally allow deviation from the policy until final review

24. While at height, a climber should always have _____.

    a. Three points of contact
    b. Visual contact with the horizon
    c. Visual contact with the ground below
    d. Two or more lanyard connections to anchor points

25. The maximum number of climbers that can be connected to a single anchor point is based on the _____.

    a. type of harnesses used by each climber
    b. load rating of the anchor point
    c. load rating of the carabiner
    d. load rating of the chosen shock-absorbing lanyards

Trade Terms Quiz

Fill in the blank with the correct term that you learned from your study of this module.

1. A _____ is a chain-link shaped connecting device that is usually rated in Newtons, since it is tested by impact to determine its strength.

2. A flexible rope, woven strap, or wire rope that attaches PFAS components to structures is called a(n) _____.

3. If you rapidly descend a vertical surface by sliding down a rope and performing a series of short backward leaps to control your descent, then you know how to _____.

4. Components that ride along the length of a cable smoothly when moved casually, but lock onto the cable sharply when downward movement becomes too rapid or violent are known as _____.

5. The collection of anchor points, full body harness, and various connecting devices comprises a complete _____.

6. A(n) _____ is a system of straps and rings worn on the body as a single unit with the intent of distributing weight and force applied evenly across the shoulders, chest, waist, thighs, and pelvic area.

7. One pound of force = 4.45 _____.

8. A place on a structure designed or authorized for the attachment of fall protection lanyards and equipment is called a(n) _____.

9. A means of stopping or controlling a fall in progress with minimal or no injury to the climber describes _____ systems or equipment.

10. The government organization that works to ensure employers provide safe workplaces for their employees is known as _____.

11. You can use one of several types of _____ to connect the PFAS and positioning belts to anchor and positioning points.

12. A(n) _____ provides a means of preventing a fall from occurring, or aids a worker in holding a position.

13. The area in space where the body moves back and forth due to the momentum of a fall is called the _____.

14. The _____ is where guy wire anchors and tower structural components meet the soil and are no longer exposed to daylight.

15. _____ defines a number of physical symptoms that result from gravitational effects, lack of muscle movement to help pump blood back to the heart, and the restrictive nature of PFAS groin straps.

Trade Terms

Anchor point	Connecting devices	Occupational Safety and Health	Personal fall arrest system (PFAS)	Suspension trauma
Body harness	Fall arrest	Administration	Point of daylight	Swing zone
Cable grabs	Fall restraint	(OSHA)	Rappel	
Carabiner	Lanyards			
	Newtons			

Robert E. DeGraw, Jr.

Director of Renewable Energy
Miller-Motte Technical College
Madison, Tennessee

How did you get started in the construction industry?

Growing up I enjoyed climbing trees and building forts and tree houses in the woods. This was followed by creating landscapes and buildings for model railroads. I also enjoyed making model airplanes, cars, boats, and more. My dad had his own construction business when I was growing up, and he was a good carpenter, plumber, and electrician. Much of what I know I learned from him.

As a teenager I began racing motorcycles, and I enjoyed taking them apart and doing my own repairs and customizations. I was also an avid writer, and I enjoyed writing science fiction and adventure stories. However, in high school, my guidance counselor convinced my parents that I should pursue a trade, and in my junior year I enrolled in the electrical program at our local vocational-technical school. I studied residential and commercial electricity through my senior year, after which I joined the U.S. Army.

After serving four years in the military, I returned home and enrolled at the State University of New York, where I studied architectural design. Upon graduating, I worked designing solar home additions and constructing site models for IBM. Later, I went to work for a company that built uninterruptible power systems for IBM's 415Hz mainframe computers. As a designer with this company, I was initially assigned to assist with the development of the company's 50kW wind turbines. Later, I began designing the company's 1,200A through 3,000A DC disconnect switches, and after streamlining the design and manufacturing processes for these, I was asked to redesign most of the company's products, including 400A through 4,000A switchboard and switchgear, 200A thru 5,000A AC disconnects, 12kVA through 150kVA inverters, battery systems, and more.

I was an early adopter of CAD systems, and I was successful in convincing the company to migrate from drafting machines to computer-assisted 2D and 3D design tools. My proposals for CNC machinery were realized, and I wrote my own code so that our CAD systems could communicate with early CAM software and with our CNC machines. Fortunately, this automation came just in time. A downturn in our industry resulted in the company's first layoffs. However, the product streamlining and the automation adopted earlier allowed us to remain viable and competitive, and within a few years we had regained our position in the marketplace, and were employing more people than ever before.

Who inspired you to enter the industry? Why?

If we're talking about the construction industry, then it was my dad. What came after that was the result of external influences, which steered me away from my aspirations as a writer, and towards a trade. Interestingly enough, I spent a good part of my career doing technical writing.

What do you enjoy most about your job?

What I enjoyed the most about my career in the power industry was designing and improving processes. I was responsible for the implementation of document and data control systems and engineering change control processes, and I wrote nearly every policy, procedure, and work instruction while leading the company's ISO 9001 initiative.

In my current job as an educator, I thoroughly enjoy sharing what I know with others, and helping them to realize their own dreams and aspirations.

Do you think training and education are important in construction? If so, why?

Yes, absolutely! Most employers look for semi-skilled to skilled labor, and for people who have made an investment in themselves. For students, training is a way to develop skills that others may not possess, thereby making themselves more attractive and valuable to employers.

How important are NCCER credentials to your career?

As an educator invested in the NCCER curriculum, I believe NCCER credentials are invaluable.

How has training/construction impacted your life and your career?

Since the age of 12, I have always had a job. I have always made my own way. I can't point to a time when I didn't work for what I wanted. That said, I used education as a way to advance myself, and to explore new and more meaningful careers.

How do you define craftsman?

I am tempted to use my dictionary; however, my gut feels that a craftsman is akin to an artisan; someone who can shape an idea into a near art form by using hand and eye coordination. What was a pile of lumber becomes beautiful cabinetry; what was mere stone becomes a magnificent fireplace; what was steel becomes a perfectly machined part for a wind turbine.

The difference between a craftsman and someone merely doing their job is caring. And when a job is done right, it shows.

Trade Terms Introduced in This Module

Anchor point: The location of attachments on a structure for all types of climbing and/or rigging systems.

Body harness: A system of straps and rings worn on the body with the intent of distributing weight and force applied evenly across the shoulders, chest, waist, thighs, and pelvic area. The body harness must not be confused with a body belt, which is simply worn around the waist and is not approved as a fall arrest device.

Cable grabs: Components that ride along the length of a cable smoothly when moved casually, but lock onto the cable sharply when downward movement becomes too rapid or violent.

Carabiner (Kare-uh-BEAN-er): A chain-link shaped device which can be opened on one side for the insertion of a line, then closed securely. It is usually rated in Newtons, since it is tested by impact to determine its strength.

Connecting devices: Devices used to connect the PFAS and positioning belts to anchor points and positioning points.

Fall arrest: A means of stopping or controlling a fall in progress with minimal or no injury to the climber.

Fall restraint: A means of preventing a fall from occurring or aiding in worker positioning.

Lanyards: Flexible ropes, woven straps, and wire ropes that attach PFAS components to structures or other people.

Newtons: A measure of force applied, equal to the amount of force required to accelerate a mass of one kilogram at a rate of one meter per second, per second. One pound of force = 4.45 Newtons.

Occupational Safety and Health Administration (OSHA): A government organization that works to ensure that employers provide a safe workplace for their employees.

Personal fall arrest system (PFAS): System consisting of anchor points, full body harness, and connecting devices.

Point of daylight: The point where guy wire anchor assemblies or tower components meet the soil. Below this point, they are no longer exposed to daylight.

Rappel: Descent of a vertical surface by sliding down a rope, typically while facing the surface and performing a series of short backward leaps to control the descent.

Suspension trauma: The development of physical symptoms due to blood accumulation in the lower extremities from being suspended in a harness. This trauma results from gravitational effects, lack of muscle movement to help pump blood back to the heart, and the restrictive nature of the groin straps; also known as orthostatic intolerance.

Swing zone: The area in space where the momentum and inertia of a fall would cause the body or protected object to swing until a center of gravity is stabilized (hanging straight down).

Additional Resources

This module presents thorough resources for task training. The following resource material is suggested for further study.

OSHA Standard 1926, Safety and Health Regulations for Construction, available at www.osha.gov.

Figure Credits

Paul Anderson, Module opener

Brian Shultz, Figures 1, 33, and 34

Fall protection materials provided courtesy of Miller Fall Protection, Franklin, PA., Figures 2–7, 9, 11–15, 17, 18, and 22–26

Photo courtesy of Capital Safety (DBI-SALA and PROTECTA), Figures 8, 16, 19, 35, and 36

Courtesy of Snap-on Industrial – Tools at Height Program, SA01

Topaz Publications, Inc., Figures 10 and 21

Pict. Hailo, Germany, SA03

Petzl America, Figure 20

NASA, Figure 27

Matt Young, Figure 28

National Weather Service Central Region Headquarters, Figures 29 and 31

Courtesy: National Science Foundation, Figure 30

Courtesy of DOE/NREL, Figure 32

Centers for Disease Control and Prevention, National Institute for Occupational Safety and Health, SA04

NCCER CURRICULA — USER UPDATE

NCCER makes every effort to keep its textbooks up-to-date and free of technical errors. We appreciate your help in this process. If you find an error, a typographical mistake, or an inaccuracy in NCCER's Curricula, please fill out this form (or a photocopy), or complete the online form at **www.nccer.org/olf**. Be sure to include the exact module number, page number, a detailed description, and your recommended correction. Your input will be brought to the attention of the Authoring Team. Thank you for your assistance.

Instructors – If you have an idea for improving this textbook, or have found that additional materials were necessary to teach this module effectively, please let us know so that we may present your suggestions to the Authoring Team.

NCCER Product Development and Revision

13614 Progress Blvd., Alachua, FL 32615

Email: curriculum@nccer.org
Online: www.nccer.org/olf

❏ Trainee Guide   ❏ AIG   ❏ Exam   ❏ PowerPoints   Other _____

Craft / Level: _____   Copyright Date: _____

Module Number / Title: _____

Section Number(s): _____

Description: _____

_____

_____

_____

_____

Recommended Correction: _____

_____

_____

_____

_____

Your Name: _____

Address: _____

_____

Email: _____   Phone: _____

Introduction to Electrical Circuits

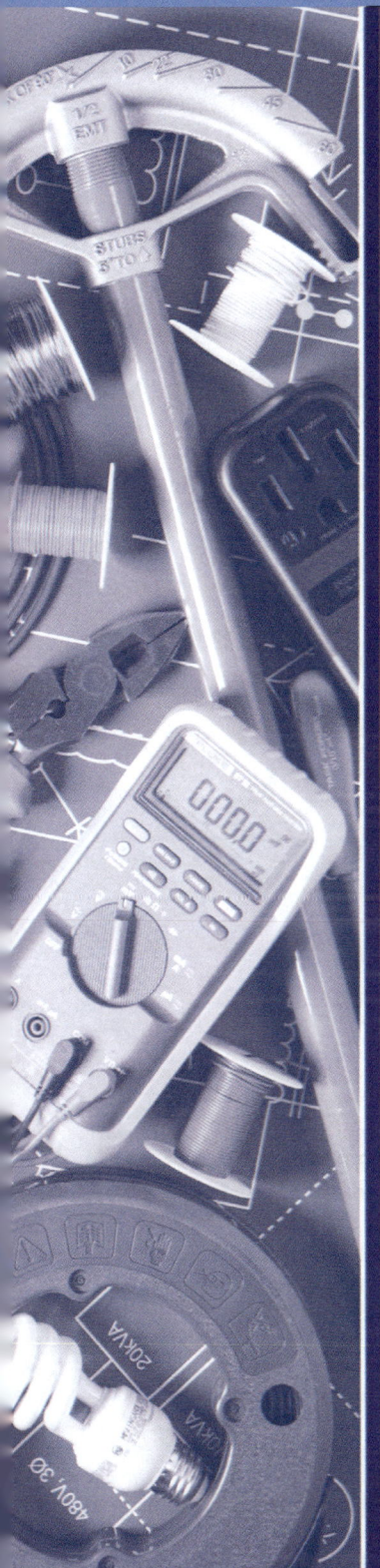

Reno Transportation Rail Access Corridor

The Reno Transportation Rail Access Corridor (ReTRAC) is the largest public works project ever undertaken in Nevada and includes the design and construction of a 2.1-mile long, 54-foot wide, 33-foot deep train trench running through downtown Reno. ReTRAC was built using the design-build method, resulting in shorter construction times, lessened traffic impacts, and lower costs.

26103-11

Trainees with successful module completions may be eligible for credentialing through NCCER's National Registry. To learn more, go to **www.nccer.org** or contact us at **1.888.622.3720.** Our website has information on the latest product releases and training, as well as online versions of our *Cornerstone* newsletter and Pearson's Contren® product catalog.

Your feedback is welcome. You may email your comments to **curriculum@nccer.org,** send general comments and inquiries to **info@nccer.org,** or use the User Update form at the back of this module.

 V.1 6/11

Objectives

When you have completed this module, you will be able to do the following:

1. Define voltage and identify the ways in which it can be produced.
2. Explain the difference between conductors and insulators.
3. Define the units of measurement that are used to measure the properties of electricity.
4. Identify the meters used to measure voltage, current, and resistance.
5. Explain the basic characteristics of series and parallel circuits.

Performance Tasks

This is a knowledge-based module. There are no performance tasks.

Trade Terms

Ammeter	Matter	Schematic
Ampere (A)	Mega	Series circuit
Atom	Neutrons	Solenoid
Battery	Nucleus	Transformer
Circuit	Ohm (Ω)	Valence shell
Conductor	Ohmmeter	Volt (V)
Coulomb	Ohm's law	Voltage
Current	Power	Voltage drop
Electron	Protons	Voltmeter
Insulator	Relay	Watt (W)
Joule (J)	Resistance	
Kilo	Resistor	

Required Trainee Materials

1. Paper and pencil
2. Appropriate personal protective equipment
3. Calculator

Note: *NFPA 70*®, *National Electrical Code*®, and *NEC*® are registered trademarks of the National Fire Protection Association, Inc., Quincy, MA 02269. All *National Electrical Code*® and *NEC*® references in this module refer to the 2011 edition of the *National Electrical Code*®.

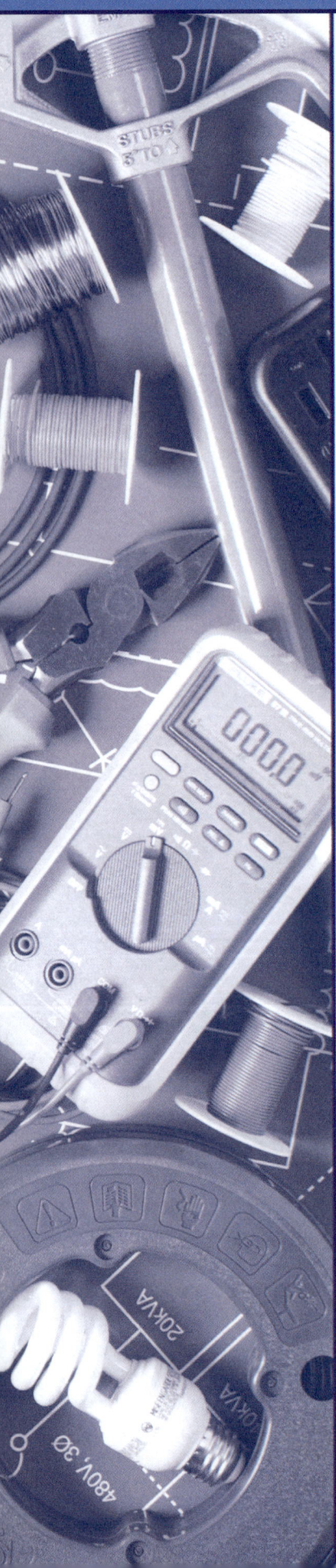

Contents ———————————————————

Topics to be presented in this module include:

1.0.0 Introduction 1
2.0.0 Atomic Theory 1
   2.1.0 The Atom 1
      2.1.1 The Nucleus 2
      2.1.2 Electrical Charges 2
   2.2.0 Conductors and Insulators 2
   2.3.0 Magnetism 3
3.0.0 Electrical Power Generation and Distribution 4
4.0.0 Electric Charge and Current 5
   4.1.0 Current Flow 6
   4.2.0 Voltage 6
   4.3.0 Resistance 7
      4.3.1 Characteristics of Resistance 7
5.0.0 Ohm's Law 8
6.0.0 Schematic Representation of Circuit Elements 9
7.0.0 Resistors 10
   7.1.0 Resistor Color Codes 12
8.0.0 Electrical Circuits 13
   8.1.0 Series Circuits 13
   8.2.0 Parallel Circuits 13
   8.3.0 Series-Parallel Circuits 15
9.0.0 Electrical Measuring Instruments 15
   9.1.0 Measuring Current 16
   9.2.0 Measuring Voltage 16
   9.3.0 Measuring Resistance 16
   9.4.0 Voltage Testers 17
10.0.0 Electrical Power 19
   10.1.0 Power Equation 20
   10.2.0 Power Rating of Resistors 21

Figures and Tables ———————————

Figure 1   Basic electrical circuit 1
Figure 2   Hydrogen atom 1
Figure 3   Law of electrical charges 3
Figure 4   Valence shell and electrons 3
Figure 5   Magnetism 3
Figure 6   Electromagnet 4
Figure 7   Electrical power distribution 4
Figure 8   Internal power distribution 5
Figure 9   Potential difference causing electric current 6
Figure 10 Ohm's law circle 9
Figure 11 Electrical circuit 9
Figure 12 Standard schematic symbols 10
Figure 13 Common resistors 11
Figure 14 Symbols used for variable resistors 12

Figure 15 Resistor color codes 12
Figure 16 Sample color codes on a fixed resistor 13
Figure 17 Types of circuits 13
Figure 18 Digital and analog meters 15
Figure 19 Clamp-on ammeter 15
Figure 20 Clamp-on ammeter in use 16
Figure 21 In-line ammeter test setup 16
Figure 22 Voltmeter connection 16
Figure 23 Ohmmeter connection for continuity testing 17
Figure 24 Continuity tester 17
Figure 25 Voltage tester 17
Figure 26 One watt 19
Figure 27 Expanded Ohm's law circle 21

Table 1 Conductor Properties 8
Table 2 Conversion Table 20

1.0.0 INTRODUCTION

Electricity is a form of energy that can be used by electrical devices such as motors, lights, TVs, heaters, and numerous other devices to perform work. Electricity is also used to control non-electrical devices that perform work. For example, your car is driven by a gasoline engine, but you wouldn't be able to start it or turn it off without the electrical system. In order to work with electricity, you need to know how it is produced and how it acts in electrical circuits.

You will hear the term *circuit* throughout your training. An electrical circuit contains, at minimum, a voltage source, a load, and conductors (wires) to carry the electrical current (*Figure 1*). The circuit should also have a means to stop and start the current, such as a switch.

Electricity is all about cause and effect. The presence of voltage (volts) in a closed circuit will cause current (amps) to flow. The more voltage you apply, the more current will flow. However, the amount of current flow is also determined by how much resistance (ohms) the load offers to the flow of current. In order to convert electrical energy into work, the load consumes energy. The amount of energy a device consumes is called power, and is expressed in watts (W). Volts (V), amps, ohms, and watts are related in such a way that if any one of them changes, the others are proportionally affected. This relationship can be seen using basic math principles that you will learn in this module. You will also learn how electricity is produced and how test instruments are used to measure electricity.

2.0.0 ATOMIC THEORY

In order to understand electrical theory, you must first understand the basic concepts of atomic theory. Atomic theory explains the construction and behavior of atoms, including the transfer of electrons that results in current flow.

2.1.0 The Atom

The atom is the smallest part of an element that enters into a chemical change, but it does so in the form of a charged particle. These charged particles are called ions, and are of two types—positive and negative. A positive ion may be defined as an atom that has become positively charged. A negative ion may be defined as an atom that has become negatively charged. One of the properties of charged ions is that ions of the same charge tend to repel one another, whereas ions of unlike charge will attract one another. The term *charge* can be taken to mean a quantity of electricity that is either positive or negative.

The structure of an atom is best explained by a detailed analysis of the simplest of all atoms, that of the element hydrogen. The hydrogen atom in *Figure 2* is composed of a nucleus containing one proton and a single orbiting electron. As the electron revolves around the nucleus, it is held in this orbit by two counteracting forces. One of these forces is called centrifugal force, which is the force that tends to cause the electron to fly outward as it travels around its circular orbit. The second force acting on the electron is electrostatic force. This force tends to pull the electron in toward the

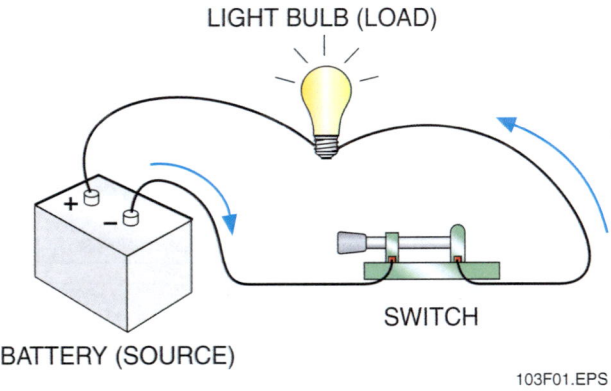

LIGHT BULB (LOAD)

SWITCH

BATTERY (SOURCE)

103F01.EPS

Figure 1   Basic electrical circuit.

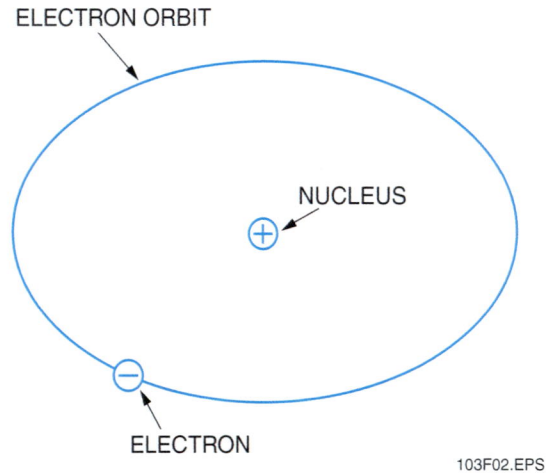

ELECTRON ORBIT

NUCLEUS

ELECTRON

103F02.EPS

Figure 2   Hydrogen atom.

Why Bother Learning Theory?

Many trainees wonder why they need to bother learning the theory behind how things operate. They figure, why should I learn how it works as long as I know how to install it? The answer is, if you only know how to install something (e.g., run wire, connect switches, etc.), that's all you are ever going to be able to do. For example, if you don't know how your car operates, how can you troubleshoot it? The answer is, you can't. You can only keep changing out the parts until you finally hit on what is causing the problem. (How many times have you seen people do this?) Remember, unless you understand not only how things work but why they work, you'll only be a parts changer. With theory behind you, there is no limit to what you can do.

nucleus and is caused by the mutual attraction between the positive nucleus and the negative electron. At some given radius, the two forces will balance each other, providing a stable path for the electron.

- A proton (+) repels another proton (+).
- An electron (−) repels another electron (−).
- A proton (+) attracts an electron (−).

Basically, an atom contains three types of subatomic particles that are of interest in electricity: electrons, protons, and neutrons.

The protons and neutrons are located in the center, or nucleus, of the atom, and the electrons travel around the nucleus in orbits.

Because protons are relatively heavy, the repulsive force they exert on one another in the nucleus of an atom has little effect.

The attracting and repelling forces on charged materials occur because of the electrostatic lines of force that exist around the charged materials. In a negatively charged object, the lines of force of the excess electrons combine to produce an electrostatic field that has lines of force coming into the object from all directions. In a positively charged object, the lines of force of the excess protons combine to produce an electrostatic field that has lines of force going out of the object in all directions. The electrostatic fields either aid or oppose each other to attract or repel.

2.1.1 The Nucleus

The nucleus is the central part of the atom. It is made up of heavy particles called protons and neutrons. The proton is a charged particle containing the smallest known unit of positive electricity. The neutron has no electrical charge. The number of protons in the nucleus determines how the atom of one element differs from the atom of another element.

Although a neutron is actually a particle by itself, it is generally thought of as an electron and proton combined, and is electrically neutral. Since

Electrical Charges

Think about the things you come in contact with every day. Where do you see or find examples of electrostatic attraction?

neutrons are electrically neutral, they are not considered relevant to the electrical nature of atoms.

2.1.2 Electrical Charges

The negative charge of an electron is equal but opposite to the positive charge of a proton. The charges of an electron and a proton are called electrostatic charges. The lines of force associated with each particle produce electrostatic fields. Because of the way these fields act together, charged particles can attract or repel one another. The Law of Electrical Charges states that particles with like charges repel each other and those with unlike charges attract each other. This is shown in *Figure 3*.

2.2.0 Conductors and Insulators

The difference between atoms, with respect to chemical activity and stability, depends on the number and position of the electrons included within the atom. In general, the electrons reside in groups of orbits called shells. The shells are arranged in steps that correspond to fixed energy levels.

The outer shell of an atom is called the valence shell, and the electrons contained in this shell are called valence electrons (*Figure 4*). The number of valence electrons determines an atom's ability to gain or lose an electron, which in turn determines the chemical and electrical properties of the atom. An atom that is lacking only one or two electrons from its outer shell will easily gain electrons to complete its shell, but a large amount of energy is required to

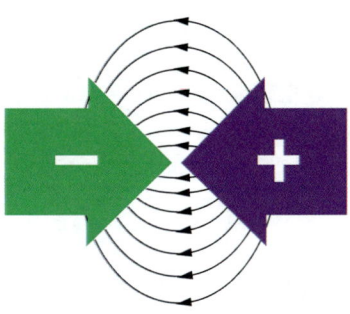

UNLIKE CHARGES ATTRACT

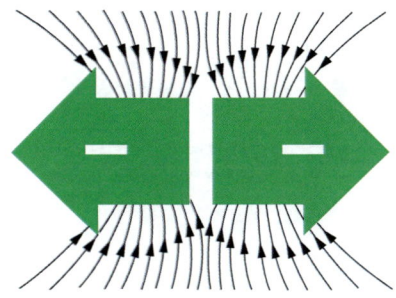

LIKE CHARGES REPEL

103F03.EPS

Figure 3 Law of electrical charges.

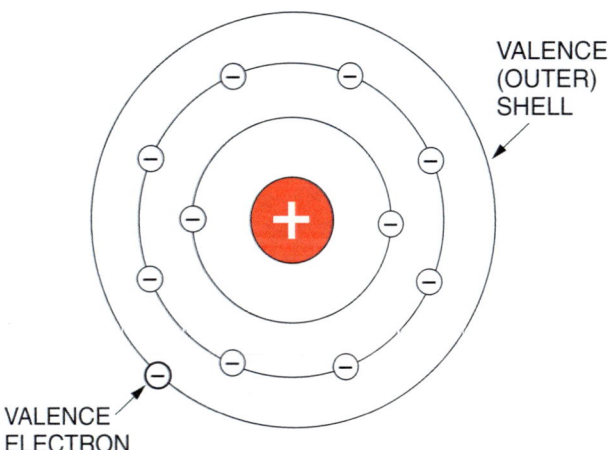

VALENCE
(OUTER)
SHELL

VALENCE
ELECTRON

103F04.EPS

Figure 4 Valence shell and electrons.

free any of its electrons. An atom having a relatively small number of electrons in its outer shell in comparison to the number of electrons required to fill the shell will easily lose these valence electrons.

It is the valence electrons that we are most concerned with in electricity. These are the electrons that are easiest to break loose from their parent atom. Normally, a conductor has three or less valence electrons, an insulator has five or more valence electrons, and semiconductors usually have four valence electrons.

All the elements of which matter is made may be placed into one of three categories: conductors, insulators, and semiconductors.

Conductors, for example, are elements such as copper and silver that will conduct a flow of electricity very readily. Because of their good conducting abilities, they are formed into wire and used whenever it is desired to transfer electrical energy from one point to another.

Insulators, on the other hand, do not conduct electricity to any great degree and are used when it is desirable to prevent the flow of electricity. Compounds such as porcelain and plastic are good insulators.

Materials such as germanium and silicon are not good conductors but cannot be used as insulators either, since their electrical characteristics fall between those of conductors and those of insulators. These in-between materials are classified as semiconductors. As you will learn later in your training, semiconductors play a crucial role in electronic circuits.

2.3.0 Magnetism

The operation of many electrical components relies on the power of magnetism. Motors, relays, transformers, and solenoids are examples. Magnetized iron generates a magnetic field consisting of magnetic lines of force, also known as magnetic flux lines (*Figure 5*). Magnetic objects within the field will be attracted or repelled by the magnetic field. The more powerful the magnet, the more powerful the magnetic field around it. Each magnet has a north pole and a south pole. Opposing poles attract each other; like poles repel each other.

Electricity also produces magnetism. Current flowing through a conductor produces a small magnetic field around the conductor. If the conductor is coiled around an iron bar, the result is an electromagnet (*Figure 6*) that attracts and repels other magnetic objects just like an iron magnet. This is the basis on which electric motors and other components operate.

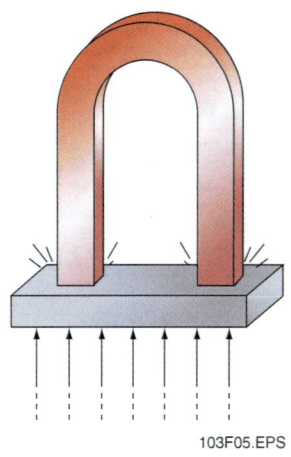

103F05.EPS

Figure 5 Magnetism.

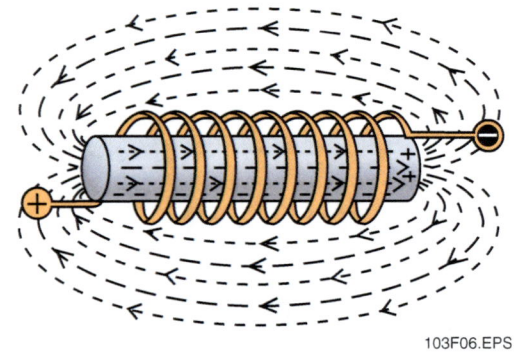

Figure 6 Electromagnet.

3.0.0 ELECTRICAL POWER GENERATION AND DISTRIBUTION

Electricity comes from electrical generating plants (*Figure 7*) operated by utilities like your local power company. Steam from coal-burning or nuclear power plants is used to power huge generators called turbines, which generate electricity. There are also hydroelectric power plants, solar power generating plants, and wind-driven turbines.

The electrical power that travels through long-distance transmission lines may be as high as 750,000 volts (V). Devices known as transformers are used to step the voltage down to lower levels as it reaches electrical substations and eventually homes, offices, and factories. The voltage you receive at home is usually about 240V. At the wall outlet where you plug in small appliances such as televisions and toasters, the voltage is about 120V (*Figure 8*). Electric stoves, clothes dryers, water heaters, and central air conditioning systems usually require the full 240V. Commercial buildings and factories may receive anywhere from 208V to 575V. This depends on the amount of power their machines consume.

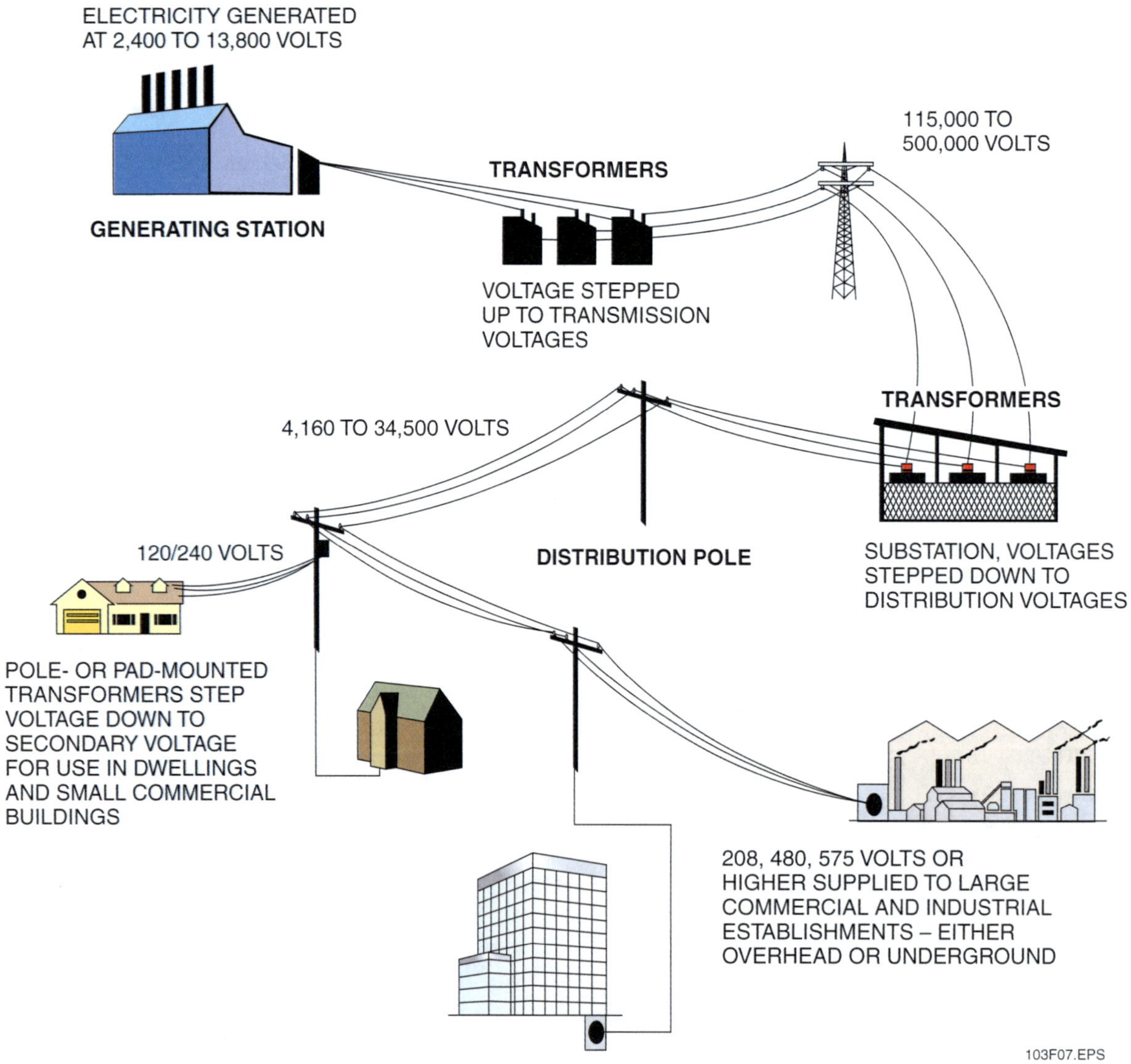

ELECTRICITY GENERATED
AT 2,400 TO 13,800 VOLTS

GENERATING STATION

TRANSFORMERS

VOLTAGE STEPPED
UP TO TRANSMISSION
VOLTAGES

115,000 TO
500,000 VOLTS

TRANSFORMERS

SUBSTATION, VOLTAGES
STEPPED DOWN TO
DISTRIBUTION VOLTAGES

4,160 TO 34,500 VOLTS

DISTRIBUTION POLE

120/240 VOLTS

POLE- OR PAD-MOUNTED
TRANSFORMERS STEP
VOLTAGE DOWN TO
SECONDARY VOLTAGE
FOR USE IN DWELLINGS
AND SMALL COMMERCIAL
BUILDINGS

208, 480, 575 VOLTS OR
HIGHER SUPPLIED TO LARGE
COMMERCIAL AND INDUSTRIAL
ESTABLISHMENTS – EITHER
OVERHEAD OR UNDERGROUND

103F07.EPS

Figure 7 Electrical power distribution.

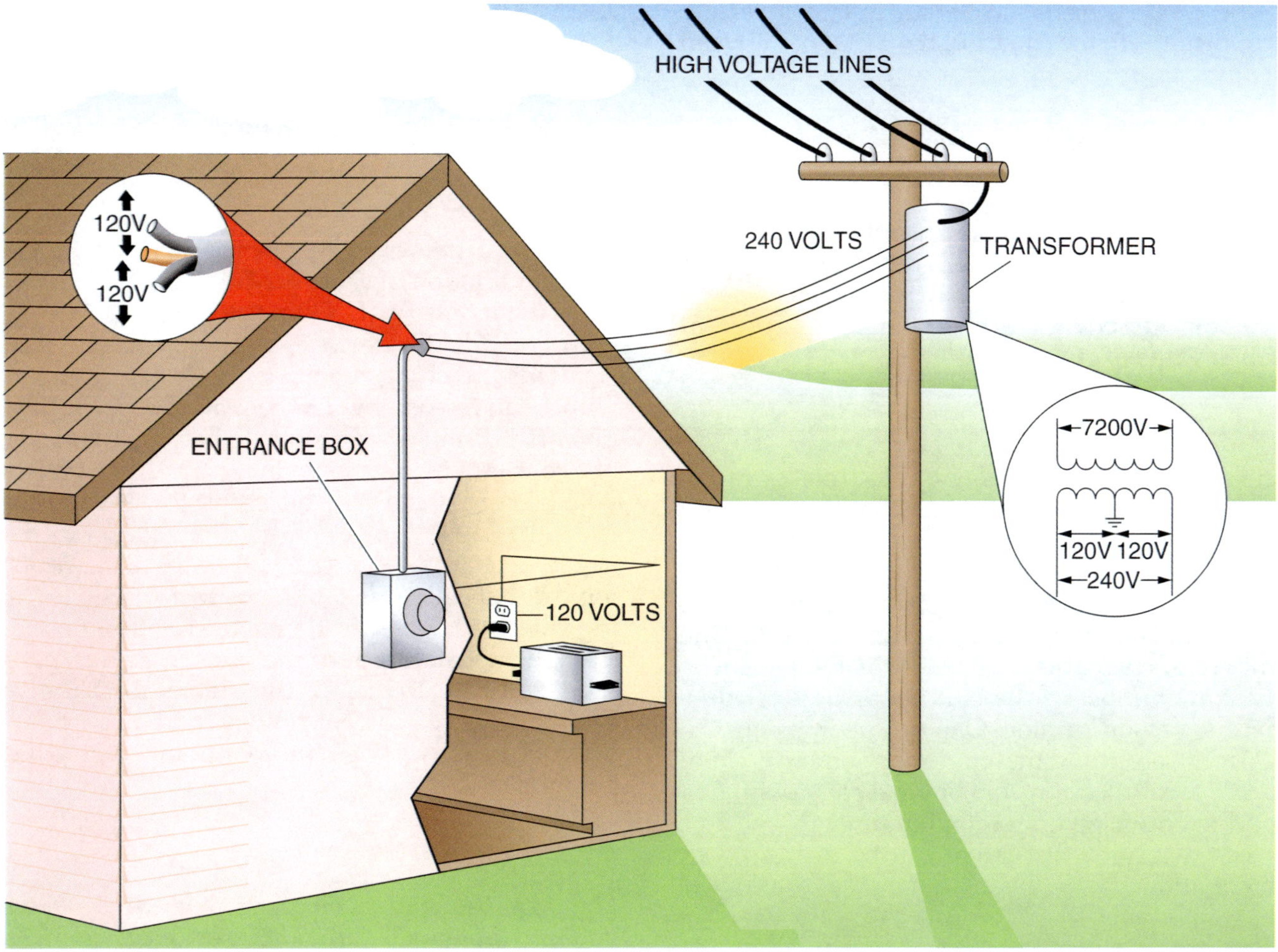

Figure 8 Internal power distribution.

Hydroelectric Plants

GOING GREEN

Hydroelectric plants use the power generated by water to drive turbines that produce electricity.

4.0.0 ELECTRIC CHARGE AND CURRENT

An electric charge has the ability to do the work of moving another charge by attraction or repulsion. The ability of a charge to do work is called its potential. When one charge is different from another, there is a difference in potential between them. The sum of the difference of potential of all the charges in the electrostatic field is referred to as electromotive force (emf) or voltage. Voltage is frequently represented by the letter E.

Electric charge is measured in **coulombs**. An electron has 1.6×10^{-19} coulombs of charge. Therefore, it takes 6.25×10^{18} electrons to make up one coulomb of charge, as shown below.

$$\frac{1}{1.6 \times 10^{-19}} = 6.25 \times 10^{18} \text{ electrons}$$

If two particles, one having charge Q_1 and the other charge Q_2, are a distance (d) apart, then the force between them is given by Coulomb's law, which states that the force is directly proportional to the product of the two charges and inversely proportional to the square of the distance between them:

$$Force = \frac{k \times Q_1 \times Q_2}{d^2}$$

If Q_1 and Q_2 are both positive or both negative, then the force is positive; it is repulsive. If Q_1 and Q_2 are of opposite charges, then the force is negative; it is attractive. The letter k equals a constant with a value of 10^9.

4.1.0 Current Flow

The movement of the flow of electrons is called current. To produce current, the electrons are moved by a potential difference. Current is represented by the letter *I*. The basic unit in which current is measured is the **ampere (A)**, also called the amp. The symbol for the ampere is *A*. One ampere of current is defined as the movement of one coulomb past any point of a conductor during one second of time. One coulomb is equal to 6.25×10^{18} electrons; therefore, one ampere is equal to 6.25×10^{18} electrons moving past any point of a conductor during one second of time.

The definition of current can be expressed as an equation:

Transformers

Large distribution transformers at power substations step down the power to the level required for local distribution. Pole transformers like the one shown here step it down further to the voltages needed for homes and businesses.

103SA02.EPS

$$I = \frac{Q}{T}$$

Where:

I = current (amperes)
Q = charge (coulombs)
T = time (seconds)

Charge differs from current in that charge (Q) is an accumulation of charge, while current (I) measures the intensity of moving charges.

In a conductor, such as copper wire, the free electrons are charges that can be forced to move with relative ease by a potential difference. If a potential difference is connected across two ends of a copper wire, as shown in *Figure 9*, the applied voltage forces the free electrons to move. This current is a flow of electrons from the point of negative charge (–) at one end of the wire, moving through the wire to the positive charge (+) at the other end. The direction of the electron flow is from the negative side of the **battery**, through the wire, and back to the positive side of the battery. The direction of current flow is therefore from a point of negative potential to a point of positive potential.

4.2.0 Voltage

The force that causes electrons to move is called voltage, potential difference, or electromotive force (emf). One volt is the potential difference between two points for which one coulomb of electricity will do one **joule (J)** of work. A battery is one of several means of creating voltage. It chemically creates a large reserve of free electrons at the

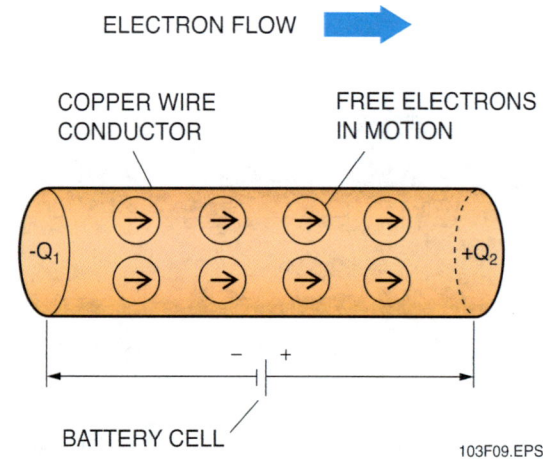

103F09.EPS

Figure 9 Potential difference causing electric current.

Units of Electricity and Volta

A disagreement with a fellow scientist over the twitching of a frog's leg eventually led 18th-century physicist Alessandro Volta to theorize that when certain objects and chemicals come into contact with each other, they produce an electric current. Believing that electricity came from contact between metals only, Volta coined the term *metallic electricity*. To demonstrate his theory, Volta placed two discs, one of silver and the other of zinc, into a weak acidic solution. When he linked the discs together with wire, electricity flowed through the wire. Thus, Volta introduced the world to the battery, also known as the Voltaic pile. Now Volta needed a term to measure the strength of the electric push or the flowing charge; the volt is that measure.

negative (–) terminal. The positive (+) terminal has electrons chemically removed and will therefore accept them if an external path is provided from the negative (–) terminal. When a battery is no longer able to chemically deposit electrons at the negative (–) terminal, it is said to be dead, or in need of recharging. Batteries are normally rated in volts. Large batteries are also rated in ampere-hours, where one ampere-hour is a current of one amp supplied for one hour.

4.3.0 Resistance

Resistance is directly related to the ability of a material to conduct electricity. All conductors have very low resistance; insulators have very high resistance.

Law of Electrical Force and de Coulomb

In the 18th century, a French physicist named Charles de Coulomb was concerned with how electric charges behaved. He watched the repelling forces electric charges exerted by measuring the twist in a wire. An object's weight acted as a turning force to twist the wire, and the amount of twist was proportional to the object's weight. After many experiments with opposing forces, de Coulomb proposed the Inverse Square Law, later known as the Law of Electrical Force.

The Magic of Electricity

The effect of the flow of electrons occurs at close to the speed of light, about 186,000 miles per second. How long does it take the light from the end of a flashlight to reach the floor? If you ran a light circuit from Maine to California and flipped the switch, how long would it take for the light to come on?

4.3.1 Characteristics of Resistance

Resistance can be defined as the opposition to current flow. To add resistance to a circuit, electrical components called resistors are used. A resistor is a device whose resistance to current flow is a known, specified value. Resistance is measured in ohms and is represented by the symbol R in equations. One ohm is defined as the amount of resistance that will limit the current in a conductor to one ampere when the voltage applied to the conductor is one volt. The symbol for an ohm is Ω.

The resistance of a wire is proportional to the length of the wire, inversely proportional to the cross-sectional area of the wire, and dependent upon the kind of material of which the wire is made. The relationship for finding the resistance of a wire is:

$$R = \rho \frac{L}{A}$$

Where:

R = resistance (ohms)

L = length of wire (feet)

A = area of wire (circular mils, CM, or cm^2)

ρ = specific resistance (ohm-CM/ft or microhm-CM)

A mil equals 0.001 inch; a circular mil is the cross-sectional area of a wire one mil in diameter.

The specific resistance is a constant that depends on the material of which the wire is made. *Table 1* shows the properties of various wire conductors.

Current Flow

Why do you need two wires to use electrical devices? Why can't current simply move to a lamp and be released as light energy?

Joule's Law

While other scientists of the 19th century were experimenting with batteries, cells, and circuits, James Joule was theorizing about the relationship between heat and energy. He discovered, contrary to popular belief, that work did not just move heat from one place to another; work, in fact, generated heat. Furthermore, he demonstrated that over time, a relationship existed between the temperature of water and electric current. These ideas formed the basis for the concept of energy. In his honor, the modern unit of energy was named the joule.

Table 1 shows that at 75°F, a one-mil diameter, pure annealed copper wire that is one foot long has a resistance of 10.351 ohms; while a one-mil diameter, one-foot-long aluminum wire has a resistance of 16.758 ohms. Temperature is important in determining the resistance of a wire. The hotter a wire, the greater its resistance.

Table 1 Conductor Properties

Metal	Specific Resistance (Resistance of 1 CM/ft in ohms)	
	32°F or 0°C	75°F or 23.8°C
Silver, pure annealed	8.831	9.674
Copper, pure annealed	9.390	10.351
Copper, annealed	9.590	10.505
Copper, hard-drawn	9.810	10.745
Gold	13.216	14.404
Aluminum	15.219	16.758
Zinc	34.595	37.957
Iron	54.529	62.643

103T01.EPS

5.0.0 OHM'S LAW

Ohm's law defines the relationship between current, voltage, and resistance. There are three ways to express Ohm's law mathematically.

- The current in a circuit is equal to the voltage applied to the circuit divided by the resistance of the circuit:

$$I = \frac{E}{R}$$

- The resistance of a circuit is equal to the voltage applied to the circuit divided by the current in the circuit:

$$R = \frac{E}{I}$$

The Visual Language of Electricity

Learning to read circuit diagrams is like learning to read a book—first you learn to read the letters, then you learn to read the words, and before you know it, you are reading without paying attention to the individual letters anymore. Circuits are the same way—you will struggle at first with the individual pieces, but before you know it, you will be reading a circuit without even thinking about it. Studying the table below will help you to understand the fundamental language of electricity.

What's Measured	Unit of Measurement	Symbol	Ohm's Law Symbol
Amount of current	Amp	A	I
Electrical power	Watt	W	P
Force of current	Volt	V	E
Resistance to current	Ohm	Ω	R

- The applied voltage to a circuit is equal to the product of the current and the resistance of the circuit:

$$E = I \times R = IR$$

Where:

$$I = \text{current (amperes)}$$
$$R = \text{resistance (ohms)}$$
$$E = \text{voltage or emf (volts)}$$

If any two of the quantities E, I, or R are known, the third can be calculated.

The Ohm's law equations can be memorized and practiced effectively by using an Ohm's law circle, as shown in *Figure 10*. To find the equation for E, I, or R when two quantities are known, cover the unknown third quantity. The other two quantities in the circle will indicate how the covered quantity may be found.

Example 1:

Find I when E = 120V and R = 30Ω.

$$I = \frac{E}{R}$$
$$I = \frac{120V}{30\Omega}$$
$$I = 4A$$

This formula shows that in a DC circuit, current (I) is directly proportional to voltage (E) and inversely proportional to resistance (R).

Example 2:

Find R when E = 240V and I = 20A.

$$R = \frac{E}{I}$$
$$R = \frac{240V}{20A}$$
$$R = 12\Omega$$

Example 3:

Find E when I = 15A and R = 8Ω.

$$E = I \times R$$
$$E = 15A \times 8\Omega$$
$$E = 120V$$

6.0.0 SCHEMATIC REPRESENTATION OF CIRCUIT ELEMENTS

The simple electric circuit shown earlier is shown in both pictorial and schematic forms in *Figure 11*. The schematic diagram is a shorthand way to draw an electric circuit, and circuits are usually represented in this way. In addition to the connecting wire, three components are shown symbolically: the battery, the switch, and the lamp. Note the positive (+) and negative (−) markings in both

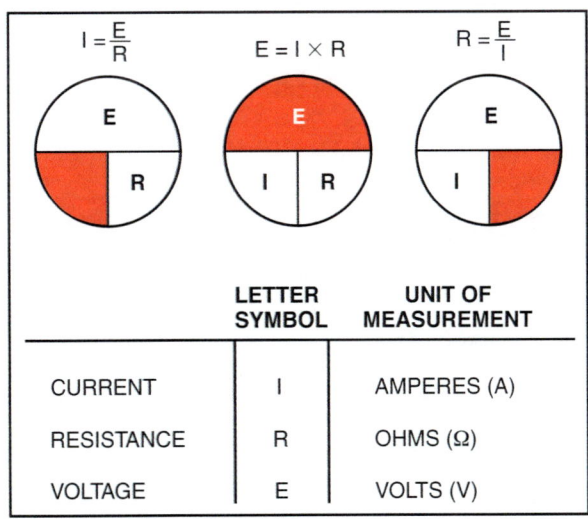

103F10.EPS

Figure 10 Ohm's law circle.

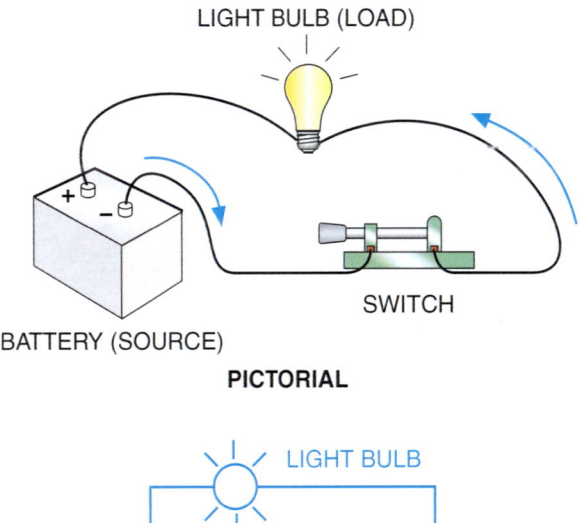

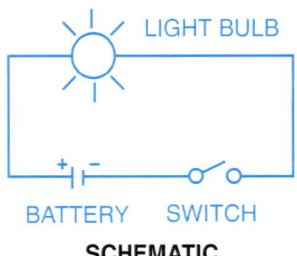

103F11.EPS

Figure 11 Electrical circuit.

Voltage Matters

Standard household voltage is different the world over, from 100V in Japan to 600V in Bombay, India. Many countries have no standard voltage; for example, France varies from 110V to 360V. If you were to plug a 120V hair dryer into England's 240V, you would burn out the dryer. Use basic electric theory to explain exactly what would happen to destroy the hair dryer.

Drawing a Schematic

Draw a schematic diagram showing a voltage source, switch, motor, and fuse.

7.0.0 RESISTORS

The function of a resistor is to offer a particular resistance to current flow. For a given current and known resistance, the change in voltage across the component, or voltage drop, can be predicted using Ohm's law. Voltage drop refers to a specific amount of voltage used, or developed, by that component. An example is a very basic circuit of a 10V battery and a single resistor in a series circuit. The voltage drop across that resistor is 10V because it is the only component in the circuit and all voltage must be dropped across that resistor. Similarly, for a given applied voltage, the current that flows may be predetermined by selection of the resistor value.

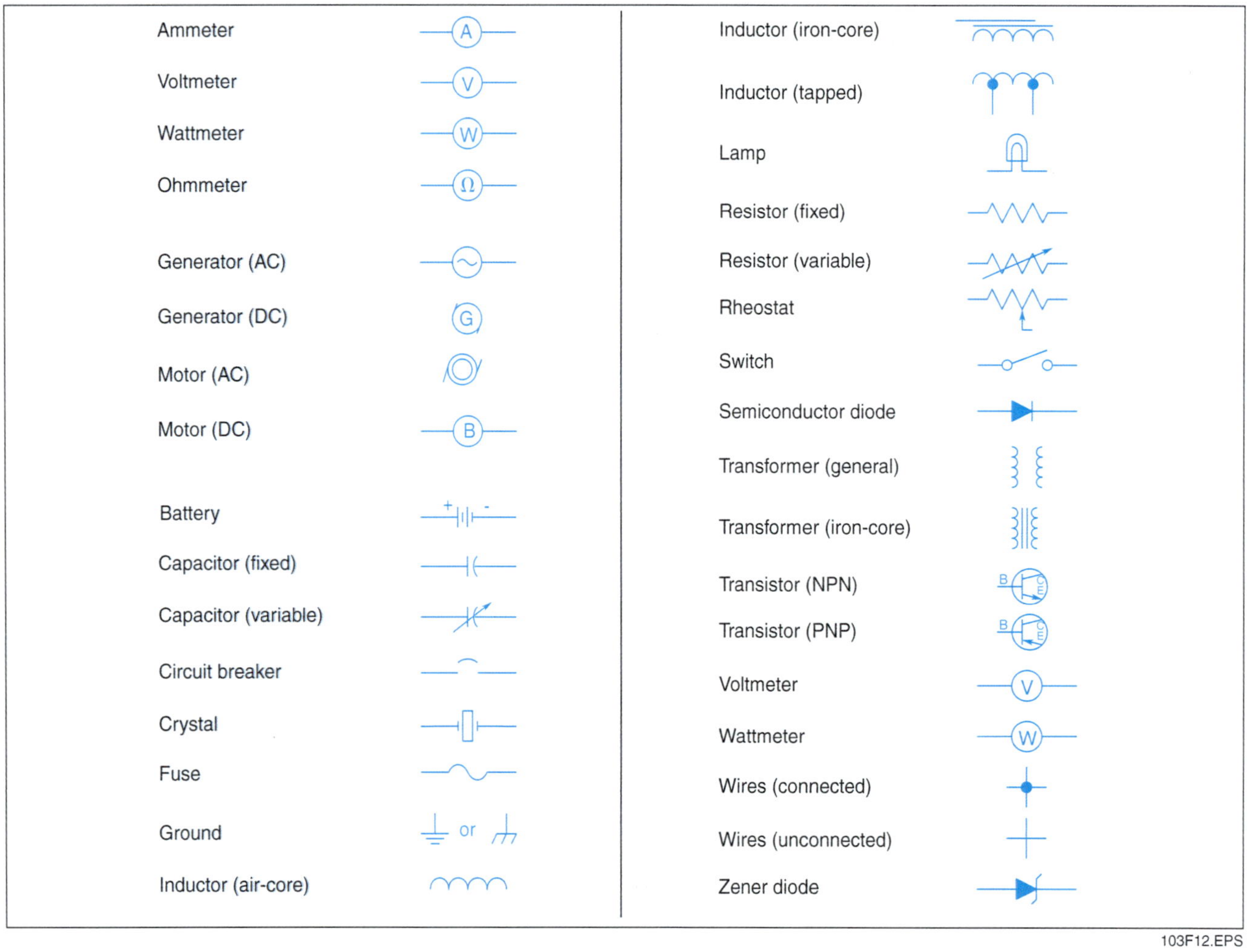

the pictorial and schematic representations of the battery. The schematic components represent the pictorial components in a simplified manner. A schematic diagram is one that shows, by means of graphic symbols, the electrical connections and functions of the different parts of a circuit.

The standard graphic symbols for commonly used electrical and electronic components are shown in *Figure 12*.

Ammeter	(A)	Inductor (iron-core)	
Voltmeter	(V)	Inductor (tapped)	
Wattmeter	(W)	Lamp	
Ohmmeter	(Ω)	Resistor (fixed)	
Generator (AC)	(~)	Resistor (variable)	
Generator (DC)	(G)	Rheostat	
Motor (AC)	(⊚)	Switch	
Motor (DC)	(B)	Semiconductor diode	
Battery	+|||-	Transformer (general)	
Capacitor (fixed)		Transformer (iron-core)	
Capacitor (variable)		Transistor (NPN)	
Circuit breaker		Transistor (PNP)	
Crystal		Voltmeter	(V)
Fuse		Wattmeter	(W)
Ground	⏚ or ⎓	Wires (connected)	
Inductor (air-core)		Wires (unconnected)	
		Zener diode	

103F12.EPS

Figure 12   Standard schematic symbols.

Using Your Intuition

Learning the meanings of various electrical symbols may seem overwhelming, but if you take a moment to study *Figure 12*, you will see that most of them are intuitive—that is, they are shaped (in a symbolic way) to represent the actual object. For example, the battery shows + and –, just like an actual battery. The motor has two arms that suggest a spinning rotor. The transformer shows two coils. The resistor has a jagged edge to suggest pulling or resistance. Connected wires have a black dot that reminds you of solder. Unconnected wires simply cross. The fuse stretches out in both directions as though to provide extra slack in the line. The circuit breaker shows a line with a break in it. The capacitor shows a gap. The variable resistor has an arrow like a swinging compass needle. As you learn to read schematics, take the time to make mental connections between the symbol and the object it represents.

The required power dissipation largely dictates the construction and physical size of a resistor.

The two most common types of electronic resistors are wire-wound and carbon composition construction. A typical wire-wound resistor consists of a length of nickel wire wound on a ceramic tube and covered with porcelain. Low-resistance connecting wires are provided, and the resistance value is usually printed on the side of the component. *Figure 13* illustrates the

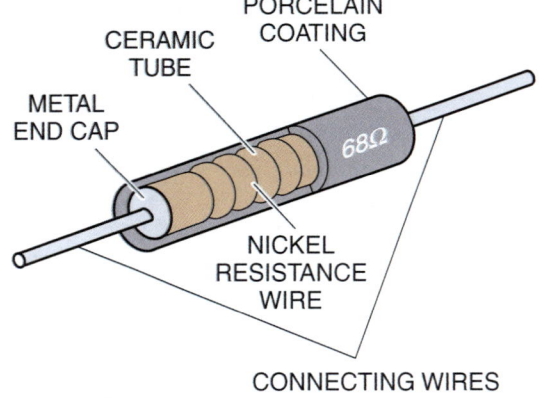

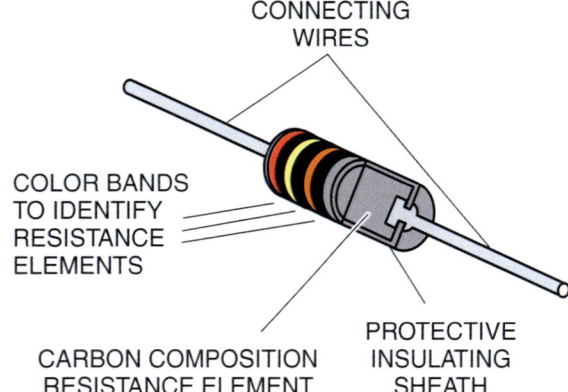

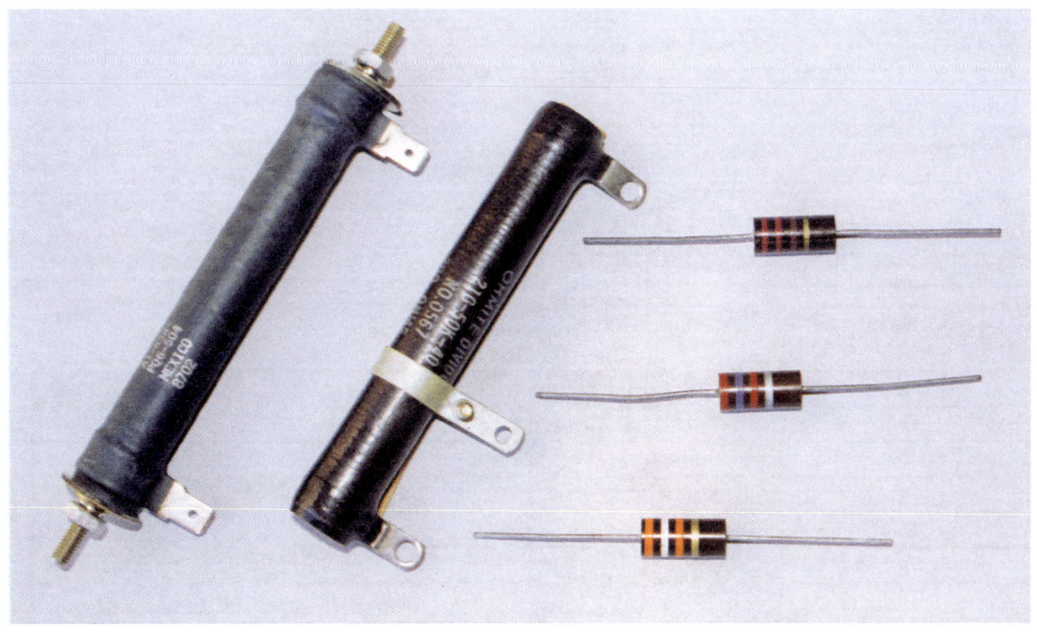

103F13.EPS

Figure 13   Common resistors.

construction of typical resistors. Carbon composition resistors are constructed by molding mixtures of powdered carbon and insulating materials into a cylindrical shape. An outer sheath of insulating material affords mechanical and electrical protection, and copper connecting wires are provided at each end. Carbon composition resistors are smaller and less expensive than the wire-wound type. However, the wire-wound type is the more rugged of the two and is able to survive much larger power dissipations than the carbon composition type.

Most resistors have standard fixed values, so they can be termed fixed resistors. Variable resistors, also known as adjustable resistors, are used a great deal in electronics. Two common symbols for a variable resistor are shown in *Figure 14*.

A variable resistor consists of a coil of closely wound insulated resistance wire formed into a partial circle. The coil has a low-resistance terminal at each end, and a third terminal is connected to a movable contact with a shaft adjustment facility. The movable contact can be set to any point on a connecting track that extends over one (uninsulated) edge of the coil.

Using the adjustable contact, the resistance from either end terminal to the center terminal may be adjusted from zero to the maximum coil resistance.

Another type of variable resistor is known as a decade resistance box. This is a laboratory component that contains precise values of switched series-connected resistors.

7.1.0 Resistor Color Codes

Because carbon composition resistors are physically small (some are less than 1 cm in length), it is not convenient to print the resistance value on the side. Instead, a color code in the form of colored bands is employed to identify the resistance value and tolerance. The color code is illustrated in *Figure 15*. Starting from one end of the resistor, the first two bands identify the first and second digits of the resistance value, and the third band indicates the number of zeros. An exception to this is when the third band is either silver or gold, which indicates a 0.01 or 0.1 multiplier, respectively. The fourth band is always either silver or gold, and in this position, silver indicates a ± 10% tolerance and gold indicates a ± 5% tolerance. Where no fourth band is present, the resistor tolerance is ± 20%.

You can put this information to practical use by determining the range of values for the carbon resistor in *Figure 16*.

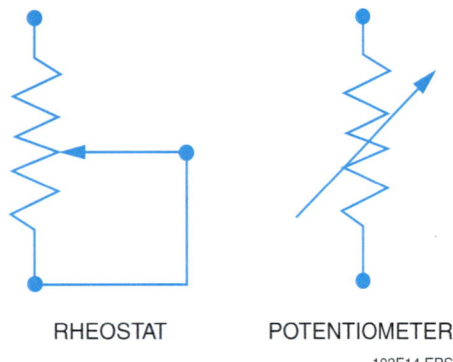

RHEOSTAT   POTENTIOMETER

103F14.EPS

Figure 14 Symbols used for variable resistors.

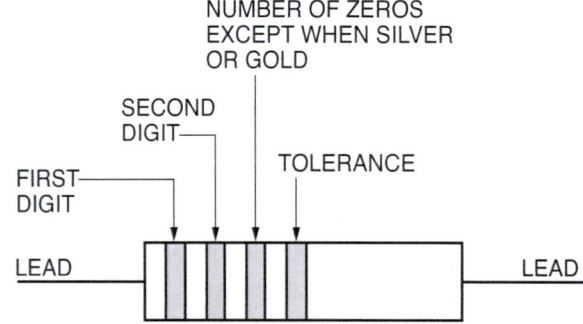

0	BLACK	7	VIOLET
1	BROWN	8	GRAY
2	RED	9	WHITE
3	ORANGE	0.1	GOLD
4	YELLOW	0.01	SILVER
5	GREEN	5%	GOLD – TOLERANCE
6	BLUE	10%	SILVER – TOLERANCE

103F15.EPS

Figure 15 Resistor color codes.

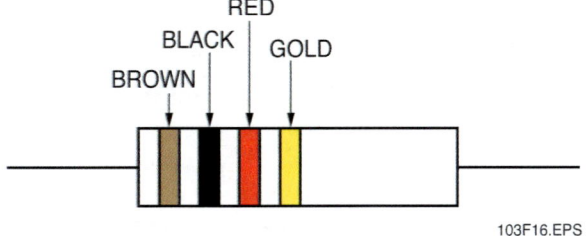

Figure 16　Sample color codes on a fixed resistor.

The color code for this resistor is as follows:

- Brown = 1, black = 0, red = 2, gold = a tolerance of ±5%
- First digit = 1, second digit = 0, number of zeros (2) = 1,000Ω

Since this resistor has a value of 1,000Ω ± 5%, the resistor can range in value from 950Ω to 1,050Ω.

8.0.0 ELECTRICAL CIRCUITS

You will often hear the terms *series circuit* and *parallel circuit* during your training. When you hear these terms, keep in mind that they refer to the way loads are connected in the circuit.

8.1.0 Series Circuits

A series circuit provides only one path for current flow and is a voltage divider. The total resistance of the circuit is equal to the sum of the individual resistances. The 12V series circuit in *Figure 17* has two 30Ω loads. The total resistance is therefore 60Ω. The amount of current flowing in the circuit is 0.2A.

$$I = \frac{E}{R} = \frac{12V}{60\Omega} = 0.2A$$

If there were five 30Ω loads, the total resistance would be 150Ω. The current flow is the same through all the loads. The voltage measured across any load (voltage drop) depends on the resistance of that load. The sum of the voltage drops equals the total voltage applied to the circuit. Circuits containing loads in series are uncommon. An important trait of a series circuit is that if the circuit is open at any point, no current will flow. For example, if you have five light bulbs connected in series and one of them blows, all five lights will go off.

8.2.0 Parallel Circuits

In a parallel circuit, each load is connected directly to the voltage source; therefore, the voltage drop is the same through all loads

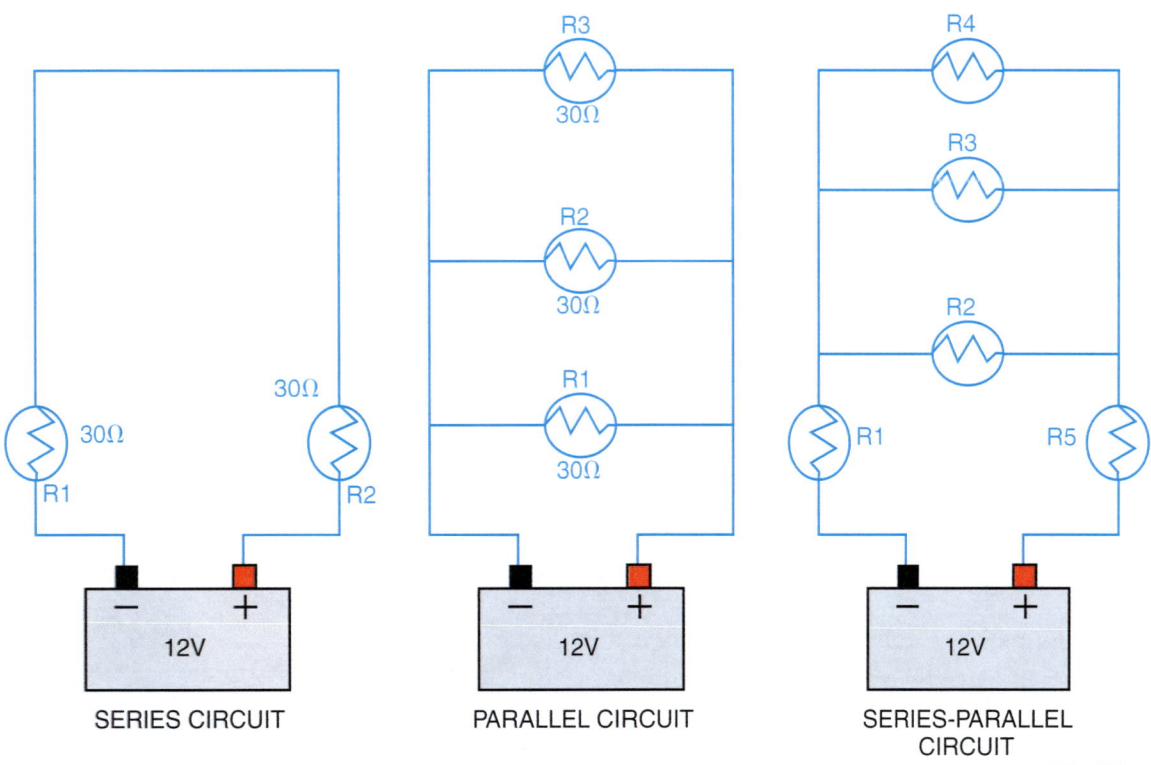

Figure 17　Types of circuits.

Is It a Series Circuit?

When the term *series circuit* is used, it refers to the way the loads are connected. The same is true for parallel and series-parallel circuits. You will rarely, if ever, find loads connected in series, or in a series-parallel arrangement. The simple circuit shown here illustrates this point. At first glance, you might think it is a series-parallel circuit. On closer examination, you can see that there are only two loads—the relay and the contactor—and they are connected in parallel. Therefore, it is a parallel circuit. The control devices are wired in series with the loads, but only the loads are considered in determining the type of circuit.

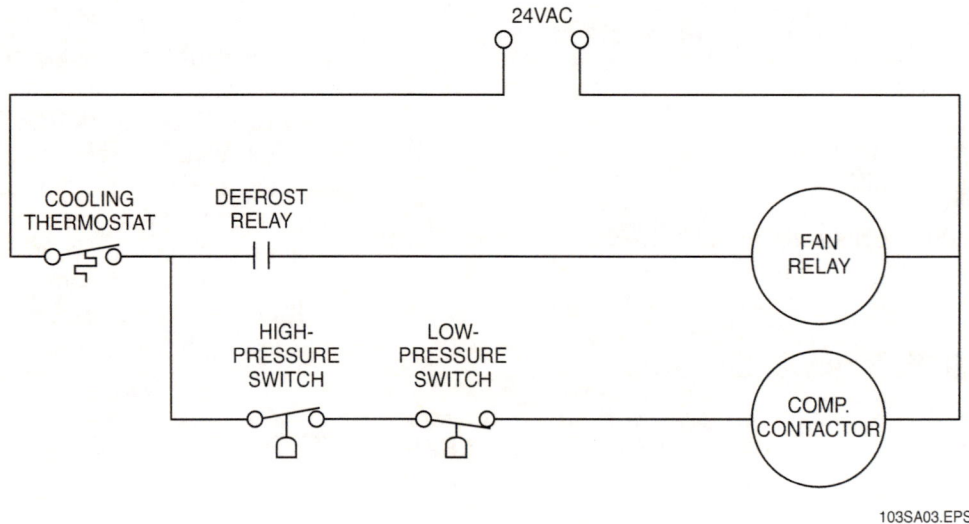

103SA03.EPS

and current is divided between the loads. The source sees the circuit as two or more individual circuits containing one load each. In the parallel circuit in *Figure 17*, the source sees three circuits, each containing a 30Ω load. The current flow through any load is determined by the resistance of that load. Thus the total current drawn by the circuit is the sum of the individual currents. The total resistance of a parallel circuit is calculated differently from that of a series circuit. In a parallel circuit, the total resistance is less than the smallest of the individual resistances.

For example, each of the 30Ω loads draws 0.4A at 12V; therefore, the total current is 1.2A:

$$I = \frac{E}{R} = \frac{12V}{30} = 0.4A \text{ per circuit}$$

0.4A per circuit × three circuits = 1.2A

Now, Ohm's law can be used again to calculate the total resistance:

$$R = \frac{E}{I} = \frac{12V}{1.2A} = 10Ω$$

This one was simple because all the resistances were the same value. The process is the same when the resistances are different, but the current calculation has to be done for each load. The individual currents are added to get the total current.

Unlike series circuits, parallel circuits continue working even if one circuit opens. Household circuits are wired in parallel. In fact, almost all the load circuits you encounter will be parallel circuits.

Either of the following formulas can be used to convert parallel resistances to a single resistance value. The first one is used when there are two resistances in parallel. The second is used when there are three or more.

$$\text{Total resistance} = \frac{R1 \times R2}{R1 + R2}$$

$$\text{Total resistance} = \frac{1}{\frac{1}{R1} + \frac{1}{R2} + \frac{1}{R3}}$$

Example:

1. The total resistance of the parallel circuit below is 6Ω.

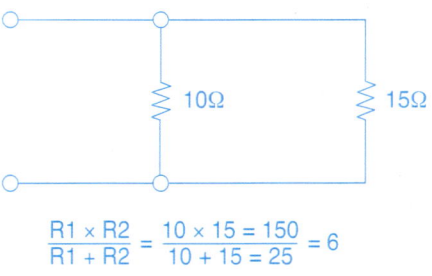

$$\frac{R1 \times R2}{R1 + R2} = \frac{10 \times 15 = 150}{10 + 15 = 25} = 6$$

UA0301.EPS

2. The total resistance of the parallel circuit below is 4.76Ω.

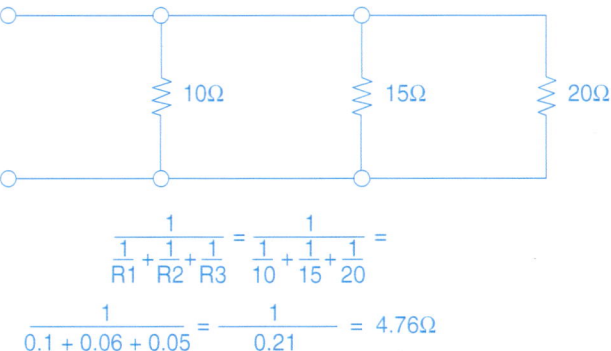

$$\frac{1}{\frac{1}{R1} + \frac{1}{R2} + \frac{1}{R3}} = \frac{1}{\frac{1}{10} + \frac{1}{15} + \frac{1}{20}} =$$

$$\frac{1}{0.1 + 0.06 + 0.05} = \frac{1}{0.21} = 4.76\Omega$$

UA0302.EPS

8.3.0 Series-Parallel Circuits

Electronic circuits often contain a hybrid arrangement known as a series-parallel circuit (*Figure 17*). It is unlikely, however, that you will ever have to determine the electrical characteristics of one of these circuits. If it becomes necessary, the parallel loads must be converted to their equivalent series resistance. Then the load resistances are added to determine total circuit resistance.

9.0.0 ELECTRICAL MEASURING INSTRUMENTS

Electricians frequently use test meters to measure voltage, current, and resistance. The most common test meter is the volt-ohm-milliammeter (VOM), also called a multimeter. *Figure 18* shows both digital and analog multimeters. The analog meter is so-called because the pointer moves in proportion to the value being measured. The person using the meter must then interpret the scale to determine the measured value. Digital meters display the result numerically on the screen.

Multimeters are commonly used to measure AC and DC voltage, DC current, and resistance. They can also be used to measure AC current in the milliamp

range. For larger current values, it is usually necessary to use a clamp-on **ammeter** (*Figure 19*).

> **WARNING!**
>
> Only qualified individuals may use these meters. Consult your company's safety policy for applicable rules.

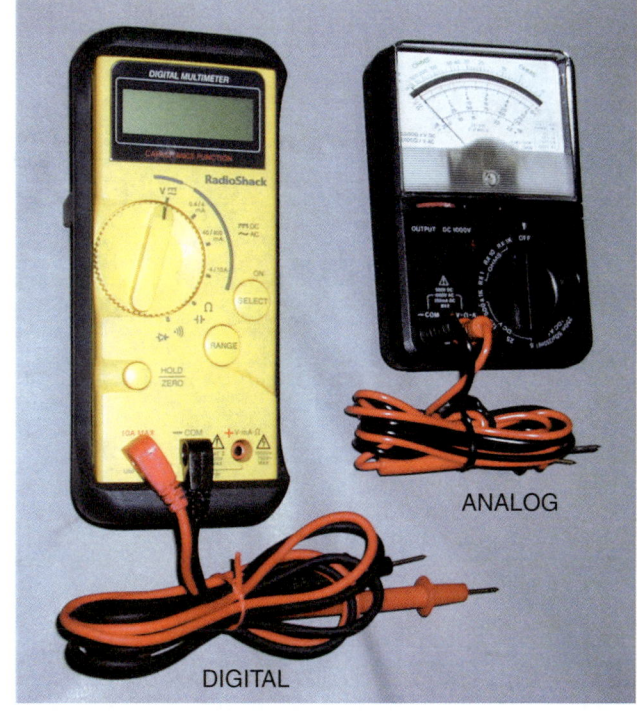

ANALOG

DIGITAL

103F18.EPS

Figure 18   Digital and analog meters.

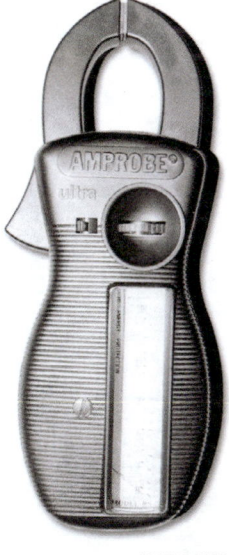

103F19.EPS

Figure 19   Clamp-on ammeter.

9.1.0 Measuring Current

A clamp-on ammeter is used to measure current. The jaws of the ammeter are placed around a single conductor (*Figure 20*). Current flowing through the wire creates a magnetic field, which induces a proportional current in the ammeter jaws. This current is read by the meter movement and appears as a direct readout or, on an analog meter, as a deflection of the meter needle.

In-line ammeters (*Figure 21*) are less common. This type of meter must be connected in series with the circuit, which means that the circuit must be opened.

Aside from following good safety practices, there are a few things to remember when measuring current:

- If the ammeter jaws are dirty or misaligned, a meter will not read correctly.
- When using an analog meter, always start at a high range and work down to avoid damaging the meter.
- Do not clamp the meter jaws around two different conductors at the same time, or an inaccurate reading will result.

9.2.0 Measuring Voltage

A **voltmeter** must be connected in parallel with (across) the component or circuit to be tested (*Figure 22*). If a circuit function is not operating, the voltmeter can be used to determine if the correct voltage is available to the circuit. Voltage must be checked with power applied.

9.3.0 Measuring Resistance

An **ohmmeter** contains an internal battery that acts as a voltage source. Therefore, resistance measurements are always made with the system power shut off. Sometimes, an ohmmeter is used to measure

103F20.EPS

Figure 20 Clamp-on ammeter in use.

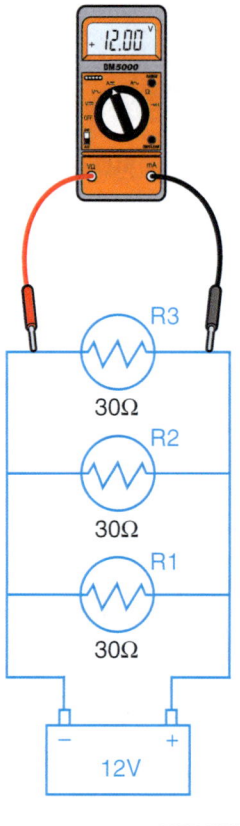

SERIES CIRCUIT

103F21.EPS

Figure 21 In-line ammeter test setup.

103F22.EPS

Figure 22 Voltmeter connection.

resistance in a load; motor windings are a good example. More often, an ohmmeter is used to check continuity in a circuit. A wire or closed switch offers negligible resistance. With the ohmmeter connected as in *Figure 23*, and the three switches closed, the current produced by the ohmmeter battery will flow unopposed and the meter will show zero resistance. The circuit has continuity; that is, it is continuous. If a switch is open, however, there is no path for current and the meter will see infinite resistance; that is, a lack of continuity.

A continuity tester (*Figure 24*) is a simple device consisting mainly of a battery and either an audible or visual indicator. It can be used in place of an ohmmeter to test the continuity of a wire and to identify individual wires contained in a conduit or other raceway. To test the continuity of a wire, strip the insulation off the end of the wire to be tested at one end of the conduit run, then connect (short) the wire to the metal conduit. At the other end of the conduit run, clip the alligator clip lead of the tester to the conduit and touch the probe to the end of the wire under test. If the tester audible alarm sounds or the indicator light comes on, there is continuity. Note that this only indicates that there is continuity between the two points being tested; it does not indicate the actual value of the resistance. If there is no indication, the wire is open.

To identify individual wires in a conduit run, touch the tester probe to the wires in the conduit one at a time until the tester audible alarm sounds or the indicator lights. Then, put matching identification tags on both ends of the wire.

103F24.EPS

Figure 24   Continuity tester.

Continue this procedure until all the wires have been identified.

9.4.0 Voltage Testers

Figure 25 shows one of the wide variety of devices available for checking for the presence of voltage. It can be used as a troubleshooting tool and as a safety device to make sure the voltage is turned off before touching any terminals or conductors. When the probes are touched to the circuit, the light on the instrument will turn

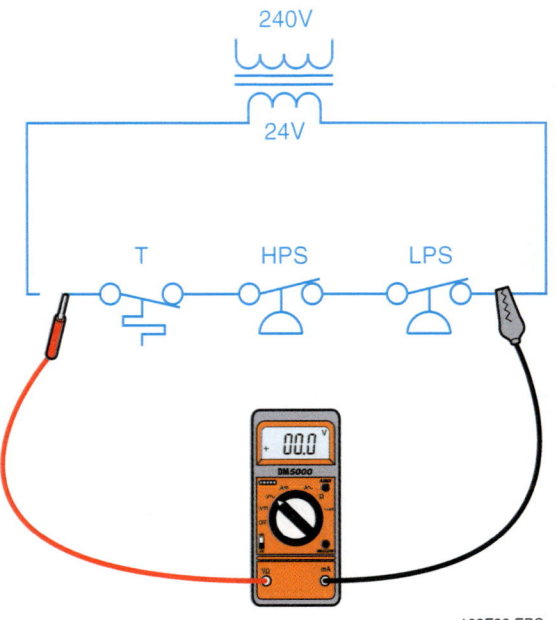

103F23.EPS

Figure 23   Ohmmeter connection for continuity testing.

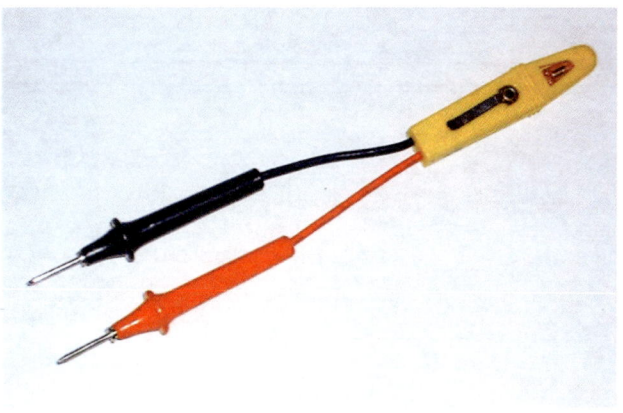

103F25.EPS

Figure 25   Voltage tester.

Test Instruments— Old and New

Early electricians used individual meters to test circuit parameters. Today, those instruments seem primitive given the availability of all-purpose instruments like the combination multimeter and clamp-on ammeter with its direct digital readout shown here.

103SA04.EPS

103SA05.EPS

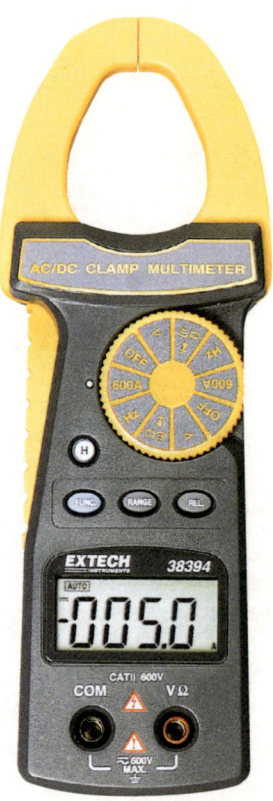

103SA06.EPS

on if a voltage is present. Instruments like these are available in several voltage ranges, so it is important to know something about the circuit you are checking.

10.0.0 ELECTRICAL POWER

Power is defined as the rate of doing work. This is equivalent to the rate at which energy is used or dissipated. Electrons passing through a resistance dissipate energy in the form of heat. In electrical circuits, power is measured in units called watts (W). The power in watts equals the rate of energy conversion. One watt of power equals the work done in one second by one volt of potential difference in moving one coulomb of charge. One coulomb per second is an ampere; therefore, power in watts equals the product of amperes times volts.

The work done in an electrical circuit can be useful work or it can be wasted work. In both cases, the rate at which the work is done is still measured in power. The turning of an electric motor is useful work. On the other hand, the heating of wires or resistors in a circuit is wasted work, since no useful function is performed by the heat.

The unit of electrical work is the joule. This is the amount of work done by one coulomb flowing through a potential difference of one volt. Thus, if five coulombs flow through a potential difference of one volt, five joules of work are done. The time it takes these coulombs to flow through the potential difference has no bearing on the amount of work done.

It is more convenient when working with circuits to think of amperes of current rather than coulombs. As previously discussed, one ampere equals one coulomb passing a point in one second. Using amperes, one joule of work is done in one second when one ampere moves through a potential difference of one volt. This rate of one joule of work in one second is the basic unit of power, and is called a watt. Therefore, a watt is the power used when one ampere of current flows through a potential difference of one volt, as shown in *Figure 26*.

Mechanical power is usually measured in units of horsepower (hp). To convert from horsepower to watts, multiply the number of horsepower by 746. To convert from watts to horsepower, divide the number of watts by 746. Conversions for common units of power are given in *Table 2*.

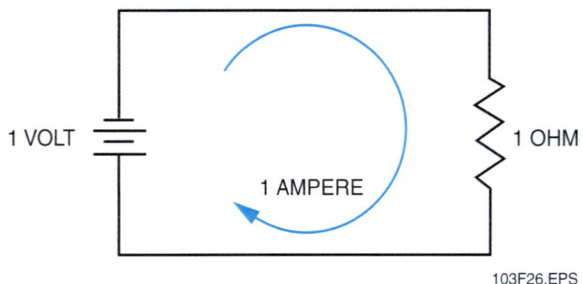

103F26.EPS

Figure 26   One watt.

Table 2 Conversion Table

1,000 watts (W)	= 1 kilowatt (kW)
1,000,000 watts (W)	= 1 megawatt (MW)
1,000 kilowatts (kW)	= 1 megawatt (MW)
1 watt (W)	= 0.00134 horsepower (hp)
1 horsepower (hp)	= 746 watts (W)

103T02.EPS

The kilowatt-hour (kWh) is commonly used for large amounts of electrical work or energy. (The prefix **kilo** means one thousand.) The amount is calculated simply as the product of the power in kilowatts multiplied by the time in hours during which the power is used. If a light bulb uses 300W or 0.3kW for 4 hours, the amount of energy is 0.3×4, which equals 1.2kWh.

Very large amounts of electrical work or energy are measured in megawatts (MW). (The prefix **mega** means one million.)

Electricity usage is figured in kilowatt-hours of energy. The power line voltage is fairly constant at 120V. Suppose the total load current in the main line equals 20A. Then the power in watts from the 120V line is:

$$P = 120V \times 20A$$
$$P = 2,400W \text{ or } 2.4kW$$

If this power is used for five hours, then the energy of work supplied equals:

$$2.4 \times 5 = 12kWh$$

10.1.0 Power Equation

When one ampere flows through a difference of two volts, two watts must be used. In other words, the number of watts used is equal to the number of amperes of current times the potential difference. This is expressed in equation form as:

$$P = I \times E \text{ or } P = IE$$

Where:

P = power used in watts

I = current in amperes

E = potential difference in volts

The equation is sometimes called Ohm's law for power, because it is similar to Ohm's law. This equation is used to find the power consumed in a circuit or load when the values of current and voltage are known. The second form of the equation is used to find the voltage when the power and current are known:

$$E = \frac{P}{I}$$

The third form of the equation is used to find the current when the power and voltage are known:

$$I = \frac{P}{E}$$

Using these three equations, the power, voltage, or current in a circuit can be calculated whenever any two of the values are already known.

Example 1:

Calculate the power in a circuit where the source of 100V produces 2A in a 50Ω resistance.

$$P = IE$$
$$P = 2 \times 100$$
$$P = 200W$$

This means the source generates 200W of power while the resistance dissipates 200W in the form of heat.

Example 2:

Calculate the source voltage in a circuit that consumes 1,200W at a current of 5A.

$$E = \frac{P}{I}$$
$$E = \frac{1,200}{5}$$
$$E = 240V$$

Example 3:

Calculate the current in a circuit that consumes 600W with a source voltage of 120V.

$$I = \frac{P}{E}$$
$$I = \frac{600}{120}$$
$$I = 5A$$

Components that use the power dissipated in their resistance are generally rated in terms of power. The power is rated at normal operating voltage, which is usually 120V. For instance, an appliance that draws 5A at 120V would dissipate 600W. The rating for the appliance would then be 600W/120V.

To calculate I or R for components rated in terms of power at a specified voltage, it may be convenient to use the power formula in different forms. There are actually three basic power formulas, but each can be rearranged into two other forms for a total of nine combinations:

$$P = IE \qquad P = I^2R \qquad P = \frac{E^2}{R}$$

$$I = \frac{P}{E} \qquad R = \frac{P}{I^2} \qquad R = \frac{E^2}{P}$$

$$E = \frac{P}{I} \qquad I = \sqrt{\frac{P}{R}} \qquad E = \sqrt{PR}$$

Note that all of these formulas are based on Ohm's law (E = IR) and the power formula (P = I × E). *Figure 27* shows all of the applicable power, voltage, resistance, and current equations.

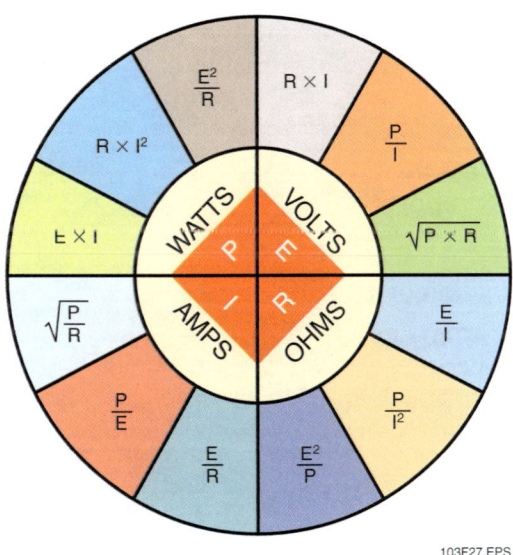

103F27.EPS

Figure 27 Expanded Ohm's law circle.

10.2.0 Power Rating of Resistors

If too much current flows through a resistor, the heat caused by the current will damage or destroy the resistor. This heat is caused by I²R heating, which is power loss expressed in watts. Therefore, every resistor is given a wattage, or power rating, to show how much I²R heating it can take before it burns out. This means that a resistor with a power rating of one watt will burn out if it is used in a circuit where the current causes it to dissipate heat at a rate greater than one watt.

If the power rating of a resistor is known, the maximum current it can carry is found by using an equation derived from P = I²R:

$$P = I^2R$$
$$I^2 = P\,R$$
$$I = \sqrt{P/R}$$

Using this equation, find the maximum current that can be carried by a 1Ω resistor with a power rating of 4W:

$$I = \sqrt{P/R} = \sqrt{4/1} = 2 \text{ amperes}$$

If such a resistor conducts more than 2 amperes, it will dissipate more than its rated power and burn out.

Power ratings assigned by resistor manufacturers are usually based on the resistors being mounted in an open location where there is free air circulation, and where the temperature is not higher than 104°F (40°C). Therefore, if a resistor is mounted in a small, crowded, enclosed space, or where the temperature is higher than 104°F, there is a good chance it will burn out even before its power rating is exceeded. Also, some resistors are designed to be attached to a chassis or frame that will carry away the heat.

Think About It

Putting It All Together

Notice the common electrical devices in the building you're in. What is their wattage rating? How much current do they draw? How would you test their voltage or amperage?

Summary

Electricians often test and troubleshoot electrical circuits. This work can be done safely and more effectively if you know the theory of electricity and the interrelationships of voltage, current, resistance, and power. The basic tool for understanding these relationships is Ohm's law.

Testing and troubleshooting of electrical circuits involves the use of test instruments such as the multimeter, or VOM. The multimeter combines the voltmeter, ammeter, and ohmmeter into a single instrument. In analog meters, a pointer moves across a scale in proportion to the current flowing though the meter. In a digital meter, the measured value is displayed directly on the screen of the meter.

Review Questions

1. An electrical circuit contains, at minimum, a(n) _____.

   a. voltage source, load, and switch
   b. ammeter, load, and voltage source
   c. voltage source, load, and conductors
   d. conductor, switch, and load

2. A type of subatomic particle with a positive charge is a(n) _____.

   a. proton
   b. neutron
   c. electron
   d. nucleus

3. Which of the following substances is considered an insulator?

   a. Gold
   b. Copper
   c. Silver
   d. Porcelain

4. The voltage commonly supplied to a residence by the local utility is _____.

   a. 120V
   b. 240V
   c. 480V
   d. 208V

5. Another term used for voltage is _____.

   a. emf
   b. coulomb
   c. current
   d. joule

6. In order to calculate the current flowing in a circuit, you would multiply voltage by resistance.

   a. True
   b. False

7. The color band that represents tolerance on a resistor is the _____.

   a. 4th band
   b. 3rd band
   c. 2nd band
   d. 1st band

8. In a parallel circuit, the total resistance is _____ the smallest resistance.

   a. greater than
   b. equal to
   c. less than
   d. proportional to

9. Circuit continuity is checked using the _____ function of a multimeter.

   a. ammeter
   b. voltmeter
   c. ohmmeter
   d. wattmeter

10. The power in a circuit with 120 volts and 5 amps is _____.

    a. 24 watts
    b. 600 watts
    c. 6,000 watts
    d. $\frac{1}{24}$ watt

Trade Terms Quiz

Fill in the blank with the correct term that you learned from your study of this module.

1. A(n) _____ is an instrument for measuring electrical current.

2. Measured in amperes, _____ is the flow of electrons in a circuit.

3. A(n) _____ is the force required to produce a current of one ampere through a resistance of one ohm.

4. Voltage is measured with a(n) _____.

5. One volt applied across one ohm of resistance causes a current flow of one _____.

6. One volt is the potential difference between two points for which one coulomb of electricity will do one _____ of work.

7. A(n) _____ is the common unit used for specifying the size of a given charge.

8. _____ is the driving force that makes current flow in a circuit.

9. The basic unit of measurement for electrical power is the _____.

10. The _____ is the smallest particle of an element that will still retain the properties of that element.

11. The _____ is the center of an atom.

12. Found in the nuclei of atoms, _____ are electrically positive particles and _____ are electrically neutral particles.

13. The outermost ring of electrons orbiting the nucleus of an atom is known as the _____.

14. A(n) _____ is a negatively charged particle that orbits the nucleus of an atom.

15. _____ is any substance that has mass and occupies space.

16. The prefix used to indicate one thousand is _____.

17. The prefix used to indicate one million is _____.

18. Consisting of two or more cells, _____ convert chemical energy into electrical energy.

19. A(n) _____ is a complete path for current flow.

20. A material through which it is relatively easy to maintain an electric current is a(n) _____.

21. A(n) _____ is a material through which it is difficult to conduct an electric current.

22. The basic unit of measurement for resistance is the _____.

23. The instrument that is used to measure resistance is called a(n) _____.

24. _____ is a statement of the relationship between current, voltage, and resistance in an electrical circuit.

25. _____ is the rate of doing work or the rate at which energy is used or dissipated.

26. Measured in ohms, _____ is the electrical property that opposes the flow of current through a circuit.

27. A(n) _____ is a component that normally opposes current flow in a DC circuit.

28. A(n) _____ is a drawing in which symbols are used to represent the components in a system.

29. A(n) _____ circuit has only one route for current flow.

30. The change in voltage across a component is called _____.

31. A(n) _____ is an electromechanical component used as a switching device.

32. A device containing one or more coils of wire wrapped around a common core is called a(n) _____.

33. An electromagnetic device used to control a mechanical device such as a valve is called a(n) _____.

Trade Terms

Ammeter
Ampere (A)
Atom
Battery
Circuit
Conductor
Coulomb
Current
Electron

Insulator
Joule (J)
Kilo
Matter
Mega
Neutrons
Nucleus
Ohm (Ω)
Ohmmeter

Ohm's law
Power
Protons
Relay
Resistance
Resistor
Schematic
Series circuit
Solenoid

Transformer
Valence shell
Volt (V)
Voltage
Voltage drop
Voltmeter
Watt (W)

1. An atom that is missing two or fewer electrons from its outer shell will _____.

2. Current is measured in units called _____.
    a. joules
    b. coulombs
    c. amperes
    d. volt-amperes

3. Joules are _____.
    a. units of work
    b. the potential difference between two points
    c. the difference between EMF and current
    d. the rate of current flow

4. All conductors have _____.
    a. current flow
    b. some resistance
    c. EMF
    d. voltage potential

5. To find voltage when both current and resistance are known, use the formula _____.
    a. $E = I \times R$
    b. $E = I \div R$
    c. $E = \dfrac{R}{I}$
    d. $E = I + R$

6. A resistor with a color code of yellow, orange, red, and silver has a tolerance of _____.

7. Ammeters measure _____.
    a. voltage
    b. resistance
    c. current
    d. power

8. True or False? When using an in-line ammeter, it is important to connect the meter in series.

9. If a toaster draws 6.2A of current at 120V, how many kilowatt-hours of energy will be used in 3.5 hours? _____

10. If a battery sends a current of 10A through a circuit for one hour, how many coulombs will flow through the circuit? _____

11. The charged particles of an atom are called _____.

12. Conductors have _____ or fewer valence electrons.

13. The sum of the difference in potential of all the charges in an electrostatic field is called _____

_____.

14. Electric charge is measured in _____.

15. Resistance is measured in _____.

16. Find the resistance when the voltage is 120V and the current is 6A. _____

17. Give the resistance value and tolerance of a resistor where the color bands are red, yellow, brown, and silver. _____

18. What is the power in a 120V circuit with a current of 12.5A?_____

19. What is the voltage in a 30Ω circuit with a power rating of 480W? _____

20. An ohmmeter is used to measure resistance and check for _____.

E. L. Jarrell

Associated Builders and Contractors

Eurlin Layne (E. L.) Jarrell is another prime example of a master electrician giving back to the electrical community by teaching and mentoring.

After serving in the United States Army, E. L. went to work for Cities Services, now known as CITGO. He stayed at CITGO for 38 years, finally retiring in 1995. It was during his employment at CITGO that he first received apprenticeship training in the electrical field.

While at CITGO, E. L. worked as a process unit operator before moving to the electrical department. While there, he worked as a trainee electrician for three years until he became a first-class electrician. A few years later, he was promoted to temporary supervisor, planning and scheduling shut-down maintenance. In 1983, he took and passed the Block Master Electrician test for the City of Lake Charles, Louisiana. In 1997, E. L. became involved with Associated Builders and Contractors (ABC).

E. L. is currently the Electrical Department Head for the ABC Training Center, where he works in the lab, overseeing students doing hands-on electrical work.

During his first semester teaching at the ABC Training Center, it became clear to E. L. that many students simply had no time to study because they worked 10-hour days, drove over 100 miles to work, and had family obligations. In response, E. L. began an in-class study guide. He encouraged students to form study groups, and he gave students time to study in class.

E. L. was an instrumental member of NCCER's Technical Review Committee, which completely rewrote all four levels of NCCER's Electrical curriculum. In addition, E. L. is currently a member of both NCCER's National Skills Assessment Written Test Committee and the Performance Verification Packet for Industrial Electricians Committee.

E. L. has decided to give back to the electrical community with his expertise and mentoring. Many of E. L.'s students have become his personal friends. He says, "At this point in my life, I just want to continue being the best electrical instructor that I can be and share some of my knowledge and experience with my students and hope that I can make a difference in their lives and careers."

Trade Terms Introduced in This Module

Ammeter: An instrument for measuring electrical current.

Ampere (A): A unit of electrical current. For example, one volt across one ohm of resistance causes a current flow of one ampere.

Atom: The smallest particle to which an element may be divided and still retain the properties of the element.

Battery: A DC voltage source consisting of two or more cells that convert chemical energy into electrical energy.

Circuit: A complete path for current flow.

Conductor: A material through which it is relatively easy to maintain an electric current.

Coulomb: An electrical charge equal to 6.25×10^{18} electrons or 6,250,000,000,000,000,000 electrons. A coulomb is the common unit of quantity used for specifying the size of a given charge.

Current: The movement, or flow, of electrons in a circuit. Current (I) is measured in amperes.

Electron: A negatively charged particle that orbits the nucleus of an atom.

Insulator: A material through which it is difficult to conduct an electric current.

Joule (J): A unit of measurement that represents one newton-meter (Nm), which is a unit of measure for doing work.

Kilo: A prefix used to indicate one thousand; for example, one kilowatt is equal to one thousand watts.

Matter: Any substance that has mass and occupies space.

Mega: A prefix used to indicate one million; for example, one megawatt is equal to one million watts.

Neutrons: Electrically neutral particles (neither positive nor negative) that have the same mass as a proton and are found in the nucleus of an atom.

Nucleus: The center of an atom. It contains the protons and neutrons of the atom.

Ohm (Ω): The basic unit of measurement for resistance.

Ohmmeter: An instrument used for measuring resistance.

Ohm's law: A statement of the relationships among current, voltage, and resistance in an electrical circuit: current (I) equals voltage (E) divided by resistance (R). Generally expressed as a mathematical formula: $I = E/R$.

Power: The rate of doing work or the rate at which energy is used or dissipated. Electrical power is the rate of doing electrical work. Electrical power is measured in watts.

Protons: The smallest positively charged particles of an atom. Protons are contained in the nucleus of an atom.

Relay: An electromechanical device consisting of a coil and one or more sets of contacts. Used as a switching device.

Resistance: An electrical property that opposes the flow of current through a circuit. Resistance (R) is measured in ohms.

Resistor: Any device in a circuit that resists the flow of electrons.

Schematic: A type of drawing in which symbols are used to represent the components in a system.

Series circuit: A circuit with only one path for current flow.

Solenoid: An electromagnetic coil used to control a mechanical device such as a valve.

Transformer: A device consisting of one or more coils of wire wrapped around a common core. It is commonly used to step voltage up or down.

Valence shell: The outermost ring of electrons that orbit about the nucleus of an atom.

Volt (V): The unit of measurement for voltage (electromotive force or difference of potential). One volt is equivalent to the force required to produce a current of one ampere through a resistance of one ohm.

Voltage: The driving force that makes current flow in a circuit. Voltage (E) is also referred to as electromotive force or difference of potential.

Voltage drop: The change in voltage across a component that is caused by the current flowing through it and the amount of resistance opposing it.

Voltmeter: An instrument for measuring voltage. The resistance of the voltmeter is fixed. When the voltmeter is connected to a circuit, the current passing through the meter will be directly proportional to the voltage at the connection points.

Watt (W): The basic unit of measurement for electrical power.

Additional Resources

This module presents thorough resources for task training. The following resource material is suggested for further study.

Electronics Fundamentals: Circuits, Devices, and Applications, Thomas L. Floyd. New York: Prentice Hall.

Principles of Electric Circuits, Thomas L. Floyd. New York: Prentice Hall.

Figure Credits

AGC of America, Module opener

Topaz Publications, Inc., Figure 13B (Bottom), Figure 18, Figure 19, Figure 25, 103SA02, 103SA04, 103SA05

Tim Dean, Module overview, Figure 20

Amprobe Instruments, Figure 24

U.S. Army Corps of Engineers, 103SA01

Extech Instruments, Inc., 103SA06

CONTREN® LEARNING SERIES — USER UPDATE

NCCER makes every effort to keep its textbooks up-to-date and free of technical errors. We appreciate your help in this process. If you find an error, a typographical mistake, or an inaccuracy in NCCER's Contren® materials, please fill out this form (or a photocopy), or complete the online form at www.nccer.org/olf. Be sure to include the exact module number, page number, a detailed description, and your recommended correction. Your input will be brought to the attention of the Authoring Team. Thank you for your assistance.

Instructors – If you have an idea for improving this textbook, or have found that additional materials were necessary to teach this module effectively, please let us know so that we may present your suggestions to the Authoring Team.

NCCER Product Development and Revision
3600 NW 43rd Street, Building G, Gainesville, FL 32606

Fax: 352-334-0932
Email: curriculum@nccer.org
Online: www.nccer.org/olf

☐ Trainee Guide ☐ AIG ☐ Exam ☐ PowerPoints Other _____

Craft / Level: _____ Copyright Date: _____

Module Number / Title: _____

Section Number(s): _____

Description: _____

_____

_____

_____

Recommended Correction: _____

_____

_____

_____

Your Name: _____

Address: _____

_____

Email: _____ Phone: _____

Electrical Theory

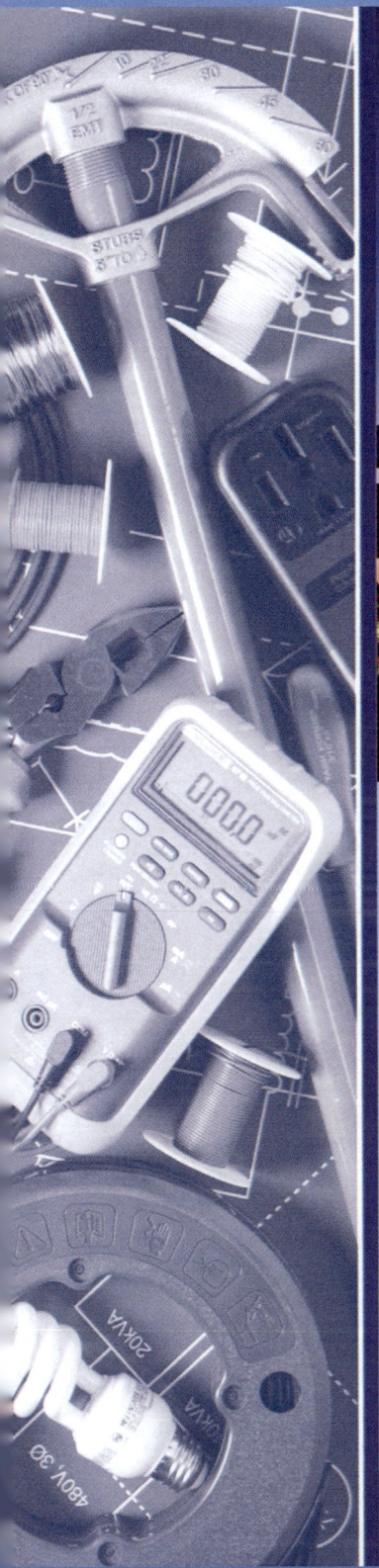

GM Lansing Delta Township Assembly Complex

This project consisted of converting 375 acres of 1,100 acres of farm land into a campus-style automotive assembly facility totaling 2.4 million square feet. In 20 months, a greenfield site was transformed into a major complex where raw steel enters at one end and finished vehicles are shipped to dealers at the opposite end. This project was the first design-build/guaranteed maximum price new manufacturing facility that General Motors has created.

26104-11

Trainees with successful module completions may be eligible for credentialing through NCCER's National Registry. To learn more, go to **www.nccer.org** or contact us at **1.888.622.3720**. Our website has information on the latest product releases and training, as well as online versions of our *Cornerstone* newsletter and Pearson's Contren® product catalog.

Your feedback is welcome. You may email your comments to **curriculum@nccer.org,** send general comments and inquiries to **info@nccer.org,** or use the User Update form at the back of this module.

 V.1 6/11

Objectives

When you have completed this module, you will be able to do the following:

1. Explain the basic characteristics of combination circuits.
2. Calculate, using Kirchhoff's voltage law, the voltage drop in series, parallel, and series-parallel circuits.
3. Calculate, using Kirchhoff's current law, the total current in parallel and series-parallel circuits.
4. Using Ohm's law, find the unknown parameters in series, parallel, and series-parallel circuits.

Performance Tasks

This is a knowledge-based module. There are no Performance Tasks.

Trade Terms

Kirchhoff's current law
Kirchhoff's voltage law

Parallel circuits
Series circuits

Series-parallel circuits

Required Trainee Materials

1. Paper and pencil
2. Appropriate personal protective equipment

Note: *NFPA 70*®, *National Electrical Code*®, and *NEC*® are registered trademarks of the National Fire Protection Association, Inc., Quincy, MA 02269. All *National Electrical Code*® and *NEC*® references in this module refer to the 2011 edition of the *National Electrical Code*®.

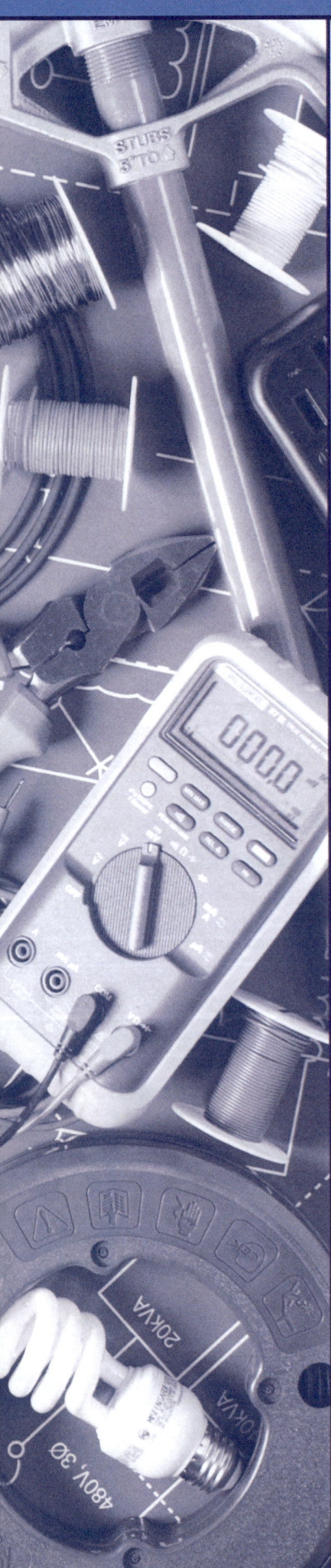

Contents

Topics to be presented in this module include:

1.0.0 Introduction 1
2.0.0 Resistive Circuits 1
  2.1.0 Resistances in Series. 1
  2.2.0 Resistances in Parallel. 1
    2.2.1 Simplified Formulas 2
  2.3.0 Series-Parallel Circuits 3
    2.3.1 Reducing Series-Parallel Circuits 4
  2.4.0 Applying Ohm's Law. 5
    2.4.1 Voltage and Current in Series Circuits 5
    2.4.2 Voltage and Current in Parallel Circuits 6
    2.4.3 Voltage and Current in Series-Parallel Circuits 7
3.0.0 Kirchhoff's Laws. 9
  3.1.0 Kirchhoff's Current Law. 9
  3.2.0 Kirchhoff's Voltage Law. 9
  3.3.0 Loop Equations 10

Figures and Tables

Figure 1   Series circuit 1
Figure 2   Total resistance 1
Figure 3   Parallel branch 2
Figure 4   Equal resistances in a parallel circuit 3
Figure 5   Series, parallel, and series-parallel circuits 4
Figure 6   Redrawing a series-parallel circuit 4
Figure 7   Reducing a series-parallel circuit 6
Figure 8   Calculating voltage drops 6
Figure 9   Parallel circuit 7
Figure 10 Solving for an unknown current 7
Figure 11 Series-parallel circuit 8
Figure 12 Simplified series-parallel circuit 8
Figure 13 Kirchhoff's current law 9
Figure 14 Application of Kirchhoff's current law 10
Figure 15 Kirchhoff's voltage law 10
Figure 16 Loop equation 11
Figure 17 Applying Kirchhoff's voltage law 11

1.0.0 INTRODUCTION

Ohm's law was explained in the module *Introduction to Electrical Circuits*. This fundamental concept is now going to be used to analyze more complex series circuits, parallel circuits, and combination series-parallel circuits. This module will explain how to calculate resistance, current, and voltage in these complex circuits. Ohm's law will be used to develop a new law for voltage and current determination. This law, called Kirchhoff's law, will become the new foundation for analyzing circuits.

2.0.0 RESISTIVE CIRCUITS

Resistance is calculated in different ways, depending on whether it is a series or parallel circuit. Resistance is calculated in ohms.

2.1.0 Resistances in Series

A series circuit is a circuit in which there is only one path for current flow. Resistance is measured in ohms (Ω). In the series circuit shown in *Figure 1* the current (I) is the same in all parts of the circuit. This means that the current flowing through R_1 is the same as the current flowing through R_2 and R_3, and it is also the same as the current supplied by the battery.

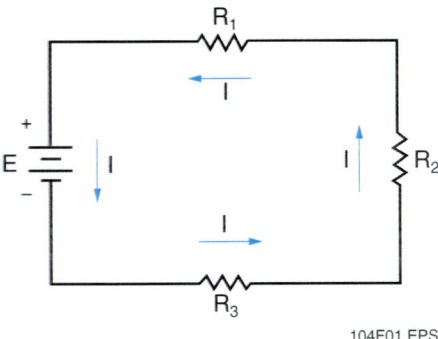

Figure 1   Series circuit.

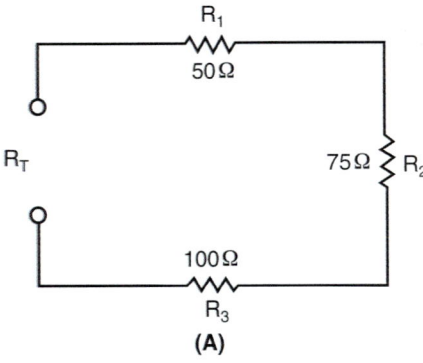

When resistances are connected in series as in this example, the total resistance in the circuit is equal to the sum of the resistances of all the parts of the circuit:

$$R_T = R_1 + R_2 + R_3$$

Where:

R_T = total resistance
$R_1 + R_2 + R_3$ = resistances in series

Example 1:

The circuit shown in *Figure 2(A)* has 50Ω, 75Ω, and 100Ω resistors in series. Find the total resistance of the circuit.

Add the values of the three resistors in series:

$$R_T = R_1 + R_2 + R_3 = 50 + 75 + 100 = 225\Omega$$

Example 2:

The circuit shown in *Figure 2(B)* has three lamps connected in series with the resistances shown. Find the total resistance of the circuit.

Add the values of the three lamp resistances in series:

$$R_T = R_1 + R_2 + R_3 = 20 + 40 + 60 = 120\Omega$$

2.2.0 Resistances in Parallel

The total resistance in a parallel resistive circuit is given by the formula:

$$R_T = \dfrac{1}{\dfrac{1}{R_1} + \dfrac{1}{R_2} + \dfrac{1}{R_3} + \dfrac{1}{R_n}}$$

Where:

R_T = total resistance in parallel
R_1, R_2, R_3, and R_n = branch resistances

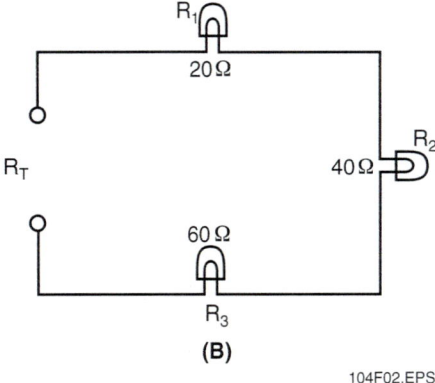

(A)   **(B)**

Figure 2   Total resistance.

Series Circuits

Simple series circuits are seldom encountered in practical wiring. The only simple series circuit you may recognize is older strands of Christmas lights, in which the entire string went dead when one lamp burned out. Think about what the actual wiring of a series circuit would look like in household receptacles. How would the circuit physically be wired? What kind of illumination would you get if you wired your household receptacles in series and plugged half a dozen lamps into those receptacles?

Example 1:

Find the total resistance of the 2Ω, 4Ω, and 8Ω resistors in parallel shown in *Figure 3*.

Write the formula for the three resistances in parallel:

$$R_T = \frac{1}{\dfrac{1}{R_1} + \dfrac{1}{R_2} + \dfrac{1}{R_3}}$$

Substitute the resistance values:

$$R_T = \frac{1}{\dfrac{1}{2} + \dfrac{1}{4} + \dfrac{1}{8}}$$

$$R_T = \frac{1}{0.5 + 0.25 + 0.125}$$

$$R_T = \frac{1}{0.875}$$

$$R_T = 1.14\Omega$$

Note that when resistances are connected in parallel, the total resistance is always less than the resistance of any single branch.

In this case:

$$R_T = 1.14\Omega < R_1 = 2\Omega, R_2 = 4\Omega, \text{ and } R_3 = 8\Omega$$

Example 2:

Add a fourth parallel resistor of 2Ω to the circuit in *Figure 3*. What is the new total resistance, and what is the net effect of adding another resistance in parallel?

Write the formula for four resistances in parallel:

$$R_T = \frac{1}{\dfrac{1}{R_1} + \dfrac{1}{R_2} + \dfrac{1}{R_3} + \dfrac{1}{R_4}}$$

Substitute values:

$$R_T = \frac{1}{\dfrac{1}{2} + \dfrac{1}{4} + \dfrac{1}{8} + \dfrac{1}{2}}$$

$$R_T = \frac{1}{0.5 + 0.25 + 0.125 + 0.5}$$

$$R_T = \frac{1}{1.375}$$

$$R_T = 0.73\Omega$$

The net effect of adding another resistance in parallel is a reduction of the total resistance from 1.14Ω to 0.73Ω.

2.2.1 Simplified Formulas

The total resistance of equal resistors in parallel is equal to the resistance of one resistor divided by the number of resistors:

$$R_T = \frac{R}{N}$$

Where:

R_T = total resistance of equal resistors in parallel
R = resistance of one of the equal resistors
N = number of equal resistors

If two resistors with the same resistance are connected in parallel, the equivalent resistance is half of that value, as shown in *Figure 4*.

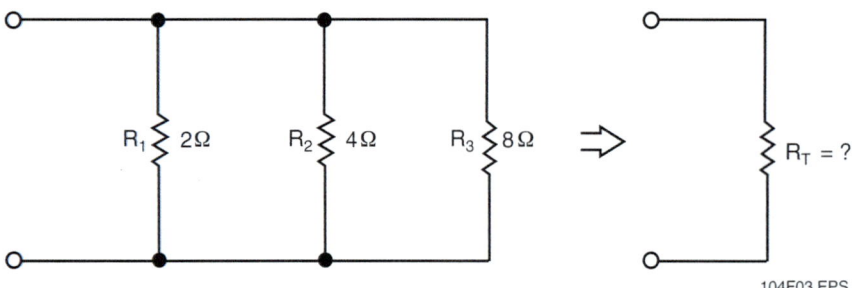

104F03.EPS

Figure 3   Parallel branch.

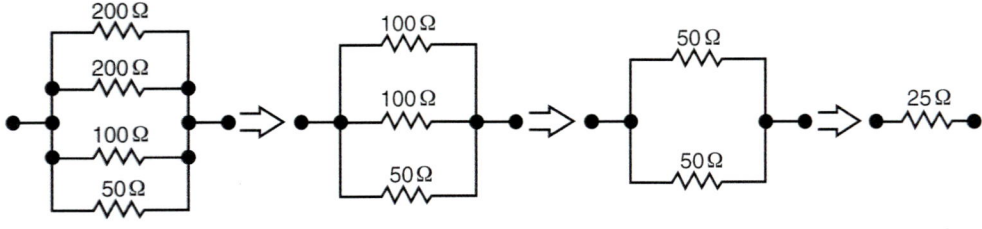

Figure 4 Equal resistances in a parallel circuit.

104F04.EPS

The two 200Ω resistors in parallel are the equivalent of one 100Ω resistor; the two 100Ω resistors are the equivalent of one 50Ω resistor; and the two 50Ω resistors are the equivalent of one 25Ω resistor.

When any two unequal resistors are in parallel, it is often easier to calculate the total resistance by multiplying the two resistances and then dividing the product by the sum of the resistances:

$$R_T = \frac{R_1 \times R_2}{R_1 + R_2}$$

Where:

R_T = total resistance of unequal resistors in parallel

R_1, R_2 = two unequal resistors in parallel

Example 1:

Find the total resistance of a 6Ω (R_1) resistor and an 18Ω (R_2) resistor in parallel:

$$R_T = \frac{R_1 \times R_2}{R_1 + R_2} = \frac{6 \times 18}{6 + 18} = \frac{108}{24} = 4.5Ω$$

Example 2:

Find the total resistance of a 100Ω (R_1) resistor and a 150Ω (R_2) resistor in parallel:

$$R_T = \frac{R_1 \times R_2}{R_1 + R_2} = \frac{100 \times 150}{100 + 150} = \frac{15,000}{250} = 60Ω$$

2.3.0 Series-Parallel Circuits

To find current, voltage, and resistance in series circuits and parallel circuits is fairly easy. When working with either type, use only the rules that apply to that type. In a series-parallel circuit, some parts of the circuit are series connected and other parts are parallel connected. Thus, in some parts the rules for series circuits apply, and in other parts, the rules for parallel circuits apply. To analyze or solve a problem involving a series-parallel

Parallel Circuits

An interesting fact about circuits is the drop in resistance in a parallel circuit as more resistors are added. But this fact does not mean that you can add an endless number of devices, such as lamps, in a parallel circuit. Why not?

circuit, it is necessary to recognize which parts of the circuit are series connected and which parts are parallel connected. This is obvious if the circuit is simple. Many times, however, the circuit must be redrawn, putting it into a form that is easier to recognize.

In a series circuit, the current is the same at all points. In a parallel circuit, there are one or more points where the current divides and flows in separate branches. In a series-parallel circuit, there are both separate branches and series loads. The easiest way to find out whether a circuit is a series, parallel, or series-parallel circuit is to start at the negative terminal of the power source and trace the path of current through the circuit back to the positive terminal of the power source. If the current does not divide anywhere, it is a series circuit. If the current divides into separate branches, but there are no series loads, it is a parallel circuit. If the current divides into separate branches and there are also series loads, it is a series-parallel circuit. *Figure 5* shows electric lamps connected in series, parallel, and series-parallel circuits.

After determining that a circuit is series-parallel, redraw the circuit so that the branches and the series loads are more easily recognized. This is especially helpful when computing the total resistance of the circuit. *Figure 6* shows resistors connected in a series-parallel circuit and the equivalent circuit redrawn to simplify it.

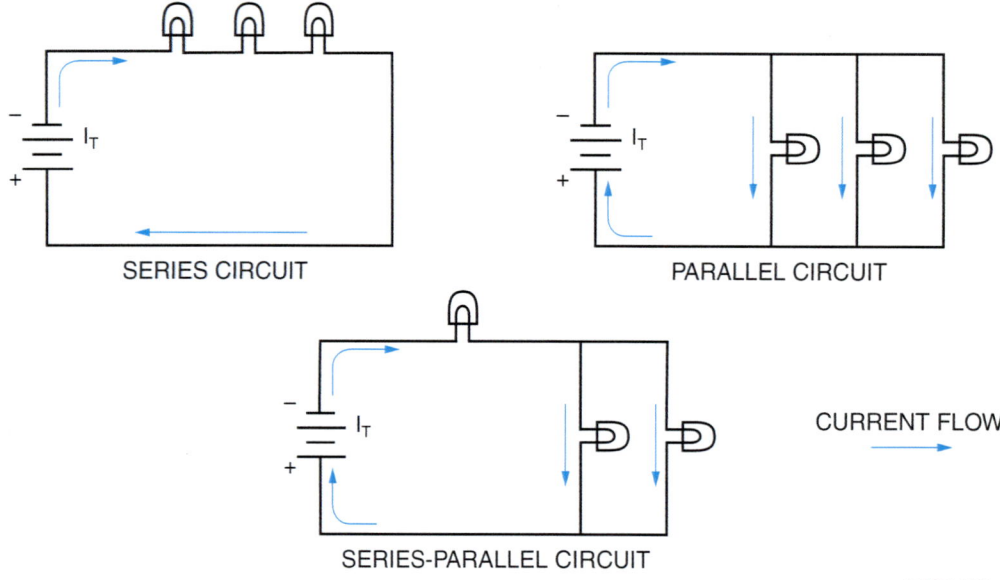

Figure 5 Series, parallel, and series-parallel circuits.

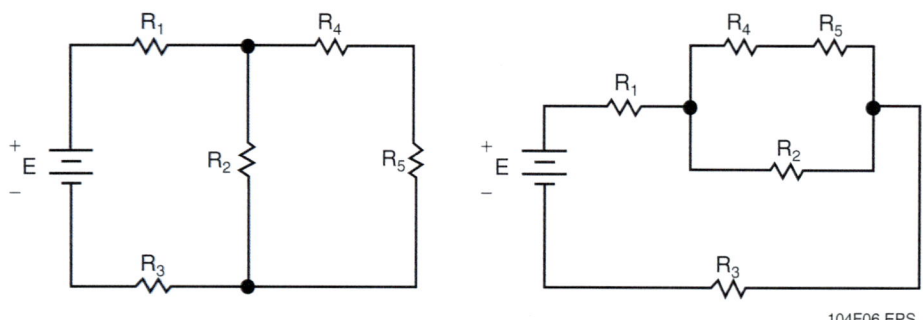

Figure 6 Redrawing a series-parallel circuit.

2.3.1 *Reducing Series-Parallel Circuits*

Very often, all that is known about a series-parallel circuit is the applied voltage and the values of the individual resistances. To find the voltage drop across any of the loads or the current in any of the branches, the total circuit current must usually be known. But to find the total current, the total resistance of the circuit must be known. To find the total resistance, reduce the circuit to its simplest form, which is usually one resistance that forms a series circuit with the voltage source. This simple series circuit has the equivalent resistance of the series-parallel circuit it was derived from, and also has the same total current. There are four basic steps in reducing a series-parallel circuit:

- If necessary, redraw the circuit so that all parallel combinations of resistances and series resistances are easily recognized.

- For each parallel combination of resistances, calculate its effective resistance.
- Replace each of the parallel combinations with one resistance whose value is equal to the effective resistance of that combination. This provides a circuit with all series loads.
- Find the total resistance of this circuit by adding the resistances of all the series loads.

Examine the series-parallel circuit shown in *Figure 7* and reduce it to an equivalent series circuit.

Think About It

Series-Parallel Circuits

Explain *Figure 6*. Which resistors are in series and which are in parallel?

Parallel Circuits

Most practical circuits are wired in parallel, like the pole lights shown here.

104SA01.EPS

In this circuit, resistors R_2 and R_3 are connected in parallel, but resistor R_1 is in series with both the battery and the parallel combination of R_2 and R_3. The current I_T leaving the negative terminal of the voltage source travels through resistor R_1 before it is divided at the junction of resistors R_1, R_2, and R_3 (Point A) to go through the two branches formed by resistors R_2 and R_3.

Given the information in *Figure 7*, calculate the resistance of R_2 and R_3 in parallel and the total resistance of the circuit, R_T.

The total resistance of the circuit is the sum of R_1 and the equivalent resistance of R_2 and R_3 in parallel. To find R_T, first find the resistance of R_2 and R_3 in parallel. Because the two resistances have the same value of 20Ω, the resulting equivalent resistance is 10Ω. Therefore, the total resistance (R_T) is 15Ω (5Ω + 10Ω).

2.4.0 Applying Ohm's Law

In resistive circuits, unknown circuit parameters can be found by using Ohm's law and the techniques for determining equivalent resistance.

2.4.1 Voltage and Current in Series Circuits

Ohm's law may be applied to an entire series circuit or to the individual parts of the circuit. When it is used on a particular part of a circuit, the voltage across that part is equal to the current in that part multiplied by the resistance of that part.

For example, given the information in *Figure 8*, calculate the total resistance (R_T) and the total current (I_T).

To find R_T:

$$R_T = R_1 + R_2 + R_3$$
$$R_T = 20 + 50 + 120$$
$$R_T = 190\Omega$$

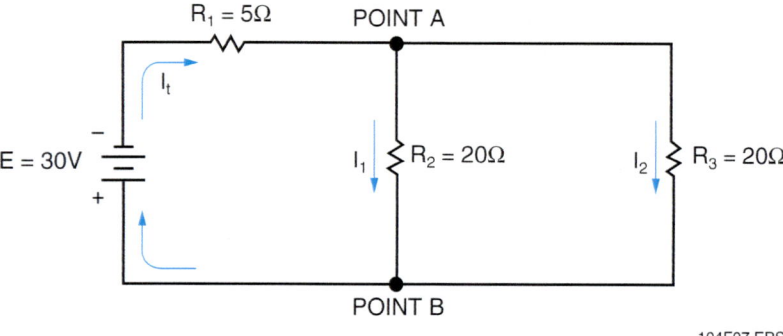

Figure 7 Reducing a series-parallel circuit.

To find I_T using Ohm's law:

$$I_T = \frac{E_T}{R_T}$$

$$I_T = \frac{95}{190}$$

$$I_T = 0.5A$$

Find the voltage across each resistor. In a series circuit, the current is the same; that is, I = 0.5A through each resistor:

$$E_1 = IR_1 = 0.5(20) = 10V$$
$$E_2 = IR_2 = 0.5(50) = 25V$$
$$E_3 = IR_3 = 0.5(120) = 60V$$

The voltages E_1, E_2, and E_3 found for *Figure 8* are known as voltage drops or IR drops. Their effect is to reduce the voltage that is available to be applied across the rest of the components in the circuit. The sum of the voltage drops in any

series circuit is always equal to the voltage that is applied to the circuit. The total voltage (E_T) is the same as the applied voltage and can be verified in this example ($E_T = 10 + 25 + 60$ or 95V).

2.4.2 Voltage and Current in Parallel Circuits

A parallel circuit is a circuit in which two or more components are connected across the same voltage source, as illustrated in *Figure 9*. The resistors R_1, R_2, and R_3 are in parallel with each other and with the battery. Each parallel path is then a branch with its own individual current. When the total current I_T leaves the voltage source E, part I_1 of the current I_T will flow through R_1, part I_2 will flow through R_2, and the remainder I_3 will flow through R_3. The branch currents I_1, I_2, and I_3 can be different. However, if a voltmeter is connected across R_1, R_2, and R_3, the respective voltages E_1, E_2, and E_3 will be equal to the source voltage E.

The total current I_T is equal to the sum of all branch currents.

This formula applies for any number of parallel branches whether the resistances are equal or unequal.

Using Ohm's law, each branch current equals the applied voltage divided by the resistance between the two points where the voltage is

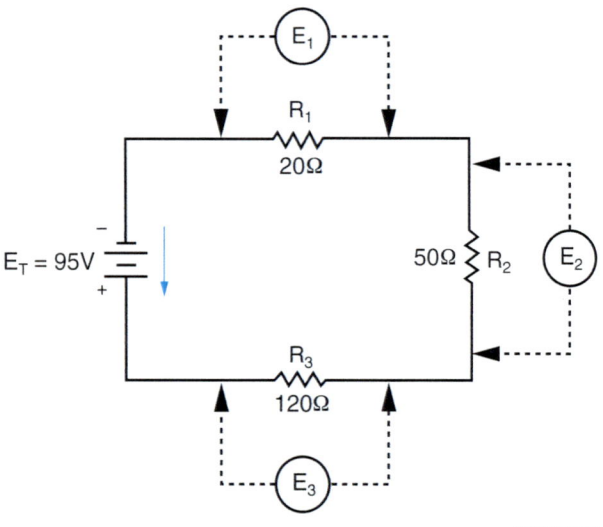

Figure 8 Calculating voltage drops.

Think About It

Voltage Drops

Calculating voltage drops is not just a schoolroom exercise. It is important to know the voltage drop when sizing circuit components. What would happen if you sized a component without accounting for a substantial voltage drop in the circuit?

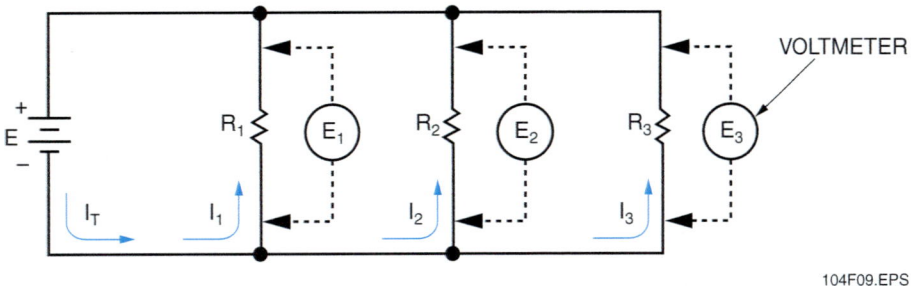

Figure 9 Parallel circuit.

applied. Hence, for each branch in *Figure 9* we have the following equations:

$$\text{Branch 1: } I_1 = \frac{E_1}{R_1} = \frac{E}{R_1}$$

$$\text{Branch 2: } I_2 = \frac{E_2}{R_2} = \frac{E}{R_2}$$

$$\text{Branch 3: } I_3 = \frac{E_3}{R_3} = \frac{E}{R_3}$$

With the same applied voltage, any branch that has less resistance allows more current through it than a branch with higher resistance.

Example 1:

The two branches R_1 and R_2, shown in *Figure 10(A)* across a 110V power line draw a total line current of 20A. Branch R_1 takes 12A. What is the current I_2 in branch R_2?

Transpose to find I_2 and then substitute given values:

$$I_T = I_1 + I_2$$
$$I_2 = I_T - I_1$$
$$I_2 = 20 - 12 = 8A$$

Example 2:

As shown in *Figure 10(B)* the two branches R_1 and R_2 across a 240V power line draw a total line current of 35A. Branch R_2 takes 20A. What is the current I_1 in branch R_1?

Transpose to find I_1 and then substitute given values:

$$I_T = I_1 + I_2$$
$$I_1 = I_T - I_2$$
$$I_1 = 35 - 20 = 15A$$

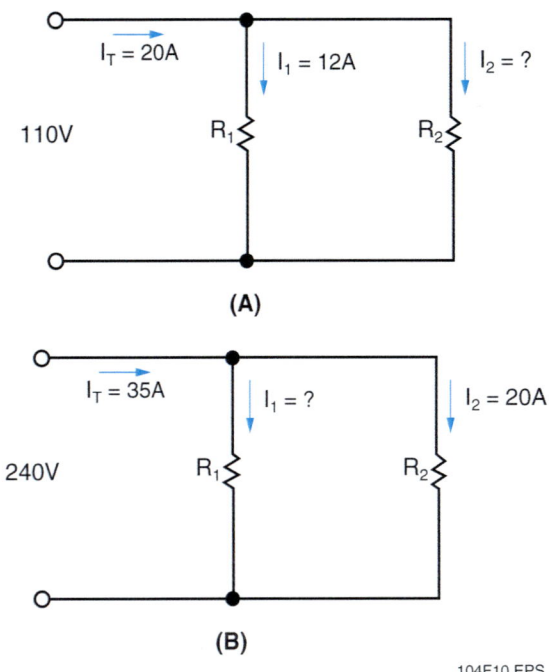

Figure 10 Solving for an unknown current.

2.4.3 Voltage and Current in Series-Parallel Circuits

Series-parallel circuits combine the elements and characteristics of both the series and parallel configurations. By properly applying the equations and methods previously discussed, the values of individual components of the circuit can be determined. *Figure 11* shows a simple series-parallel circuit with a 1.5V battery.

The current and voltage associated with each component can be determined by first simplifying the circuit to find the total current, and then working across the individual components.

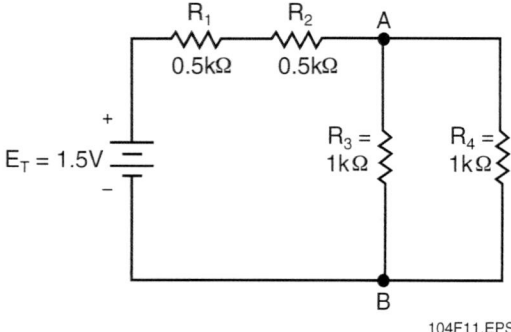

Figure 11 Series-parallel circuit.

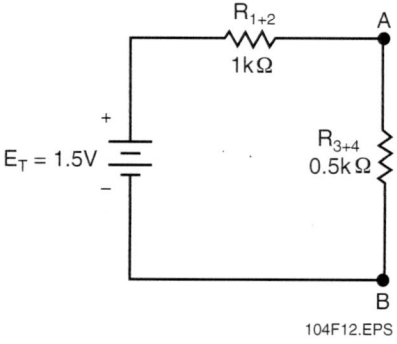

Figure 12 Simplified series-parallel circuit.

This circuit can be broken into two components: the series resistances R_1 and R_2, and the parallel resistances R_3 and R_4.

R_1 and R_2 can be added together to form the equivalent series resistance R_{1+2}:

$$R_{1+2} = R_1 + R_2$$
$$R_{1+2} = 0.5k\Omega + 0.5k\Omega$$
$$R_{1+2} = 1k\Omega$$

R_3 and R_4 can be totaled using either the general reciprocal formula or, since there are two resistances in parallel, the product over sum method. Both methods are shown below.

$$R_{3+4} = \cfrac{1}{\cfrac{1}{R_3} + \cfrac{1}{R_4}} = \cfrac{1}{\cfrac{1}{1k\Omega} + \cfrac{1}{1k\Omega}}$$

$$= \cfrac{1}{\cfrac{2}{1,000\Omega}} = \frac{1}{0.002} = 0.5k\Omega$$

$$R_{3+4} = \frac{R_3 \times R_4}{R_3 + R_4} = \frac{1k\Omega \times 1k\Omega}{1k\Omega + 1k\Omega}$$

$$= \frac{1,000,000\Omega}{2,000\Omega} = 0.5k\Omega$$

The equivalent circuit containing the R_{1+2} resistance of $1k\Omega$ and the R_{3+4} resistance of $0.5k\Omega$ is shown in *Figure 12*.

Using the Ohm's law relationship that total current equals voltage divided by circuit resistance, the circuit current can be determined. First, however, total circuit resistance must be found. Since the simplified circuit consists of two resistances in series, they are simply added together to obtain total resistance.

$$R_T = R_{1+2} + R_{3+4}$$
$$R_T = 1k\Omega + 0.5k\Omega$$
$$R_T = 1.5k\Omega$$

Applying this to the current/voltage equation:

$$I_T = \frac{E_T}{R_T}$$

$$I_T = \frac{1.5V}{1.5k\Omega}$$

$$I_T = 1mA \text{ or } 0.001A$$

Now that the total current is known, voltage drops across individual components can be determined:

$$E_{R1} = I_T R_1 = 1mA \times 0.5K\Omega = 0.5V$$
$$E_{R2} = I_T R_2 = 1mA \times 0.5K\Omega = 0.5V$$

Since the total voltage equals the sum of all voltage drops, the voltage drop from A to B can be determined by subtraction:

$$E_T = E_{R1} + E_{R2} + E_{A+B}$$
$$E_T - E_{R1} - E_{R2} = E_{A+B}$$
$$1.5V - 0.5V - 0.5V = E_{A+B} = 0.5V$$

Since R_3 and R_4 are in parallel, some of the total current must pass through each resistor. R_3 and R_4 are equal, so the same current should flow through each branch. Using the relationship:

$$I = \frac{E}{R}$$

$$I_{R3} = \frac{E_{R3}}{R_3} \qquad\qquad I_{R4} = \frac{E_{R4}}{R_4}$$

$$I_{R3} = \frac{0.5V}{1k\Omega} \qquad\qquad I_{R4} = \frac{0.5V}{1k\Omega}$$

$$I_{R4} = 0.5mA \qquad\qquad I_{R3} = 0.5mA$$

$$0.5mA + 0.5mA = 1mA$$

Therefore, the total current for the circuit passes through R_1 and R_2 and is evenly divided between R_3 and R_4.

3.0.0 KIRCHHOFF'S LAWS

Kirchhoff's laws provide a simple, practical method of solving for unknown parameters in a circuit.

3.1.0 Kirchhoff's Current Law

In its most general form, Kirchhoff's current law states that at any point in a circuit, the total current entering that point must equal the total current leaving that point. For parallel circuits, this implies that the current in a parallel circuit is equal to the sum of the currents in each branch.

When using Kirchhoff's laws to solve circuits, it is necessary to adopt conventions that determine the algebraic signs for current and voltage terms. A convenient system for current is to consider all current flowing into a branch point as positive, and all current directed away from that point as negative.

As an example, in *Figure 13* the currents can be written as:

$$I_A + I_B - I_C = 0$$

or

$$5A + 3A - 8A = 0$$

Currents I_A and I_B are positive terms because these currents flow into P, but I_C, directed out of P, is negative.

For a circuit application, refer to Point C at the top of the diagram in *Figure 14*. The 6A I_T into Point C divides into the 2A I_3 and 4A $I_{4/5}$, both directed out. Note that $I_{4/5}$ is the current through R_4 and R_5. The algebraic equation is:

$$I_T - I_3 - I_{4/5} = 0$$

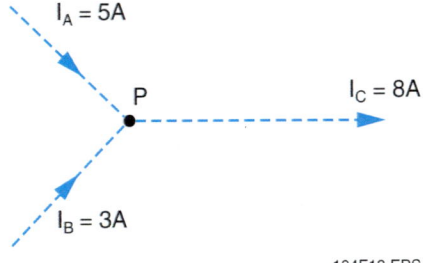

$I_A = 5A$

P

$I_C = 8A$

$I_B = 3A$

104F13.EPS

Figure 13   Kirchhoff's current law.

Substituting the values for each current:

$$6A - 2A - 4A = 0$$

For the opposite direction, refer to Point D at the bottom of *Figure 14*. Here, the branch currents into Point D combine to equal the mainline current I_T returning to the voltage source. Now, I_T is directed out from Point D, with I_3 and $I_{4/5}$ directed in. The algebraic equation is:

$$I_3 + I_{4/5} - I_T = 0$$
$$2A + 4A - 6A = 0$$

Note that at either Point C or Point D, the sum of the 2A and 4A branch currents must equal the 6A total line current. Therefore, Kirchhoff's current law can also be stated as:

$$I_{IN} = I_{OUT}$$

For *Figure 14*, the equations for current can be written as shown below.

At Point C:

$$6A = 2A + 4A$$

At Point D:

$$2A + 4A = 6A$$

Kirchhoff's current law is really the basis for the practical rule in parallel circuits that the total line current must equal the sum of the branch currents.

3.2.0 Kirchhoff's Voltage Law

Kirchhoff's voltage law states that the algebraic sum of the voltages around any closed path is zero.

Referring to *Figure 15*, the sum of the voltage drops around the circuit must equal the voltage applied to the circuit:

$$E_A = E_1 + E_2 + E_3$$

Where:

E_A = voltage applied to the circuit

E_1, E_2, and E_3 = voltage drops in the circuit

Another way of stating this law is that the algebraic sum of the voltage rises and voltage drops must be equal to zero. A voltage source is considered a voltage rise; a voltage across a resistor is a voltage drop. (For convenience in labeling, letter subscripts are shown for voltage sources and numerical subscripts are used for voltage drops.)

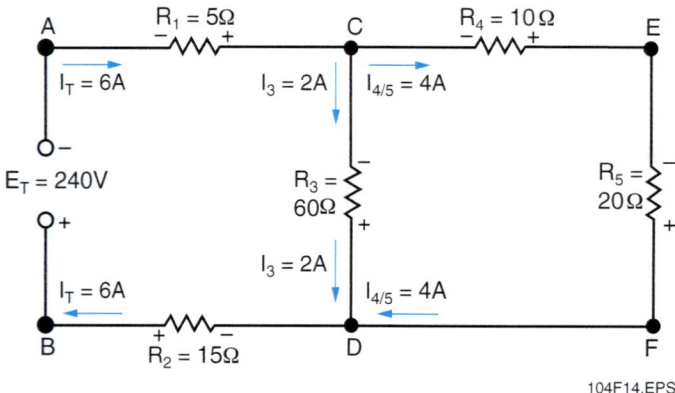

Figure 14   Application of Kirchhoff's current law.

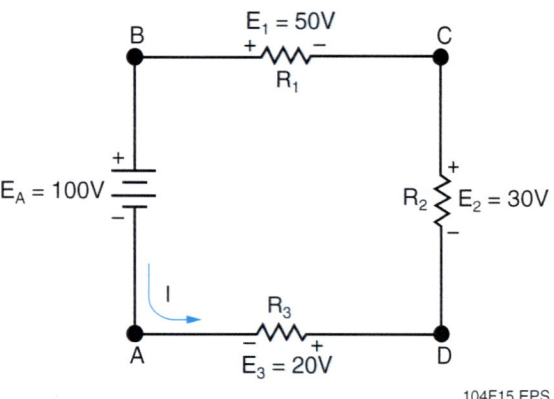

Figure 15   Kirchhoff's voltage law.

This form of the law can be written by transposing the right members to the left side:

Voltage applied – sum of voltage drops = 0

Substitute letters:

$$E_A - E_1 - E_2 - E_3 = 0$$
$$E_A - (E_1 + E_2 + E_3) = 0$$

3.3.0 Loop Equations

Any closed path for current flow is called a loop. A loop equation specifies the voltages around the loop. Refer to *Figure 16*.

Consider the inside loop through A, C, D, and B. This includes the voltage drops E_1, E_3, and E_2, and the source E_T. In a clockwise direction, starting at Point A, the algebraic sum of the voltages is:

$$- E_1 - E_3 - E_2 + E_T = 0$$

or

$$- 30V - 120V - 90V + 240V = 0$$

Voltages E_1, E_3, and E_2 have a negative value, because there is a decrease in voltage seen across each of the resistors in a clockwise direction. However, the source E_T is a positive term because an increase in voltage is seen in that same direction.

For the opposite direction, going counterclockwise in the same loop from Point B, E_T is negative while E_1, E_2, and E_3 have positive values. Therefore:

$$- E_T + E_2 + E_3 + E_1 = 0$$

or

$$- 240V + 90V + 120V + 30V = 0$$

When the negative term is transposed, the equation becomes:

$$240V = 90V + 120V + 30V$$

In this form, the loop equation shows that Kirchhoff's voltage law is really the basis for the practical rule in series circuits that the sum of the voltage drops must equal the applied voltage.

For example, determine the voltage E_B for the circuit shown in *Figure 17*. The direction of the

NCCER — *Contren® Learning Series*   26104-11

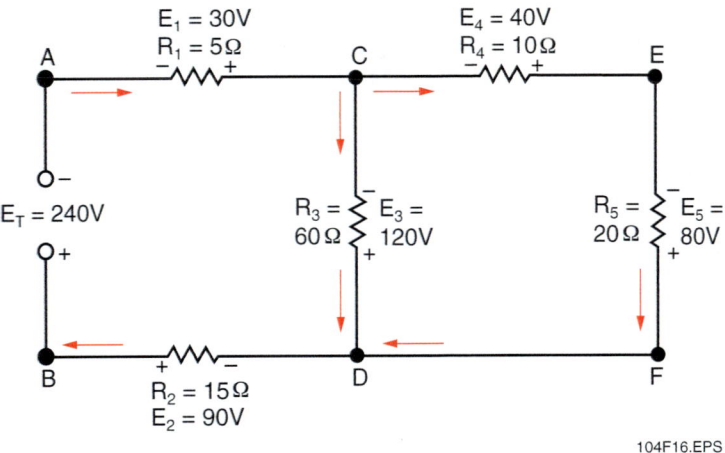

Figure 16 Loop equation.

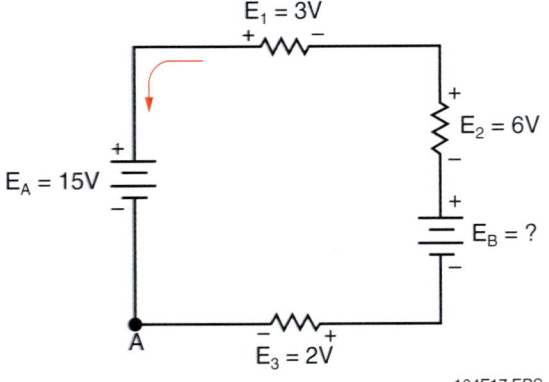

Figure 17 Applying Kirchhoff's voltage law.

current flow is shown by the arrow. First mark the polarity of the voltage drops across the resistors and trace the circuit in the direction of the current flow starting at Point A. Then write the voltage equation around the circuit:

$$-E_3 - E_B - E_2 - E_1 + E_A = 0$$

Solve for E_B:

$$E_B = E_A - E_3 - E_2 - E_1$$
$$E_B = 15V - 2V - 6V - 3V$$
$$E_B = 4V$$

Since E_B was found to be positive, the assumed direction of current is in fact the actual direction of current.

In its most general form, Kirchhoff's voltage law states that the algebraic sum of all the potential differences in a closed loop is equal to zero. A closed loop means any completely closed path consisting of wire, resistors, batteries, or other components. For series circuits, this implies that the sum of the voltage drops around the circuit is equal to the applied voltage. For parallel circuits, this implies that the voltage drops across all branches are equal.

Think About It

Putting It All Together

Draw four 60W lamps in parallel with a 120V power source. What is the amperage in the circuit? What would happen to the amperage if you doubled the voltage?

SUMMARY

The relationships among current, voltage, resistance, and power in Ohm's law are the same for both DC series and DC parallel circuits. Understanding and being able to apply these concepts is necessary for effective circuit analysis and troubleshooting. DC series-parallel circuits also have these fundamental relationships. Since DC series-parallel circuits are a combination of simple series and parallel circuits, Kirchhoff's voltage and current laws will apply. Calculating I, E, R, and P for series-parallel circuits is no more difficult than calculating these values for simple series or parallel circuits. However, for series-parallel circuits, these calculations require more careful circuit analysis in order to use Ohm's law correctly.

1. The formula for calculating the total resistance in a series circuit with three resistors is _____.

    a. $R_T = R_1 + R_2 + R_3$
    b. $R_T = R_1 - R_2 - R_3$
    c. $R_T = R_1 \, R_2 \, R_3$
    d.
    $$R_T = \cfrac{1}{\cfrac{1}{R_1} + \cfrac{1}{R_2} + \cfrac{1}{R_3}}$$

2. Find the total resistance in a series circuit with three resistances of 10Ω, 20Ω, and 30Ω.

    a. 1Ω
    b. 15Ω
    c. 20Ω
    d. 60Ω

3. The formula for calculating the total resistance in a parallel circuit with three resistors is _____.

    a. $R_T = R_1 + R_2 + R_3$
    b. $R_T = R_1 - R_2 - R_3$
    c. $R_T = R_1 \, R_2 \, R_3$
    d.
    $$R_T = \cfrac{1}{\cfrac{1}{R_1} + \cfrac{1}{R_2} + \cfrac{1}{R_3}}$$

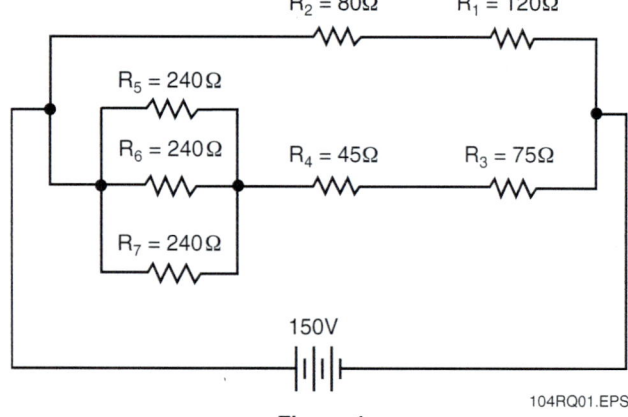

Figure 1

104RQ01.EPS

4. The total resistance in *Figure 1* is _____.

    a. 100Ω
    b. 129Ω
    c. 157Ω
    d. 1,040Ω

5. In a parallel circuit, the voltage across each path is equal to the _____.

    a. total circuit resistance times path current
    b. source voltage minus path voltage
    c. path resistance times total current
    d. applied voltage

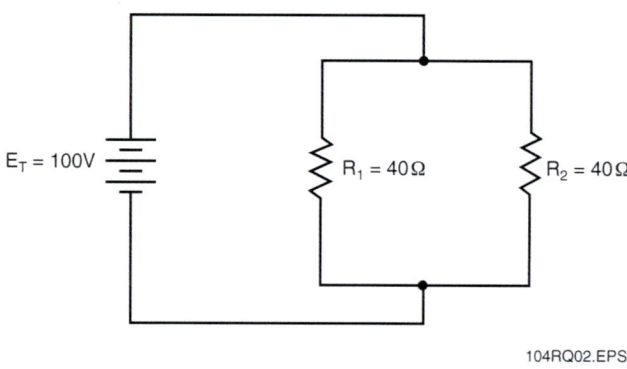

Figure 2

104RQ02.EPS

6. The value for total current in *Figure 2* is _____.

    a. 1.25A
    b. 2.50A
    c. 5A
    d. 10A

7. A resistor of 32Ω is in parallel with a resistor of 36Ω, and a 54Ω resistor is in series with the pair. When 350V is applied to the combination, the current through the 54Ω resistor is _____.

    a. 2.87A
    b. 3.26A
    c. 4.93A
    d. 5.86A

8. A 242Ω resistor is in parallel with a 180Ω resistor, and a 420Ω resistor is in series with the combination in a 27V circuit. A current of 22mA flows through the 242Ω resistor. The current through the 180Ω resistor is _____.

    a. 29.6mA
    b. 36.4mA
    c. 59.4mA
    d. 60.3mA

9. Kirchhoff's voltage law states that the algebraic sum of the voltages around any closed current path is _____.

  a. infinity
  b. zero
  c. twice the current
  d. always less than the individual voltages due to voltage drop

10. Two 24Ω resistors are in parallel, and a 42Ω resistor is in series with the combination. When 78V is applied to the three resistors, the voltage drop across the 42Ω resistor is about _____.

  a. 49.8V
  b. 55.8V
  c. 60.7V
  d. 65.3V

Trade Terms Quiz

Fill in the blank with the correct term that you learned from your study of this module.

1. _____ states that the total amount of current flowing through a parallel circuit is equal to the sum of the amounts of current flowing through each current path.

2. _____ states that the sum of all the voltage drops in a circuit is equal to the source voltage of the circuit.

3. _____ contain both series and parallel current paths.

4. _____ contain two or more parallel paths through which current can flow.

5. _____ contain only one path for current flow.

Trade Terms

Kirchhoff's current law
Kirchhoff's voltage law

Parallel circuits
Series circuits

Series-parallel circuits

1. Which formula does *not* apply to a parallel circuit?

   a. $R_T = \dfrac{R}{N}$

   b. $R_T = \dfrac{1}{\dfrac{1}{R_1} + \dfrac{1}{R_2} + \dfrac{1}{R_3} + \dfrac{1}{R_n}}$

   c. $R_T = R_1 + R_2 + R_3$

   d. $R_T = \dfrac{R_1 \times R_2}{R_1 + R_2}$

2. If two or more resistors are connected in parallel, the _____.
   a. total resistance is higher than any single resistor
   b. total resistance is lower than any single resistor
   c. current flow varies based on voltage fluctuation
   d. resistance depends on the power rating of the circuit

3. True or False? You can use Ohm's law to find the value of each branch current in a parallel circuit.

4. True or False? Parallel circuits are current dividers.

5. Which statement is *not* correct relative to Kirchhoff's current law?
   a. $I_A + I_B - I_C = 0$
   b. The total current entering a point must equal total current leaving a point.
   c. The total line current must equal the sum of the branch currents.
   d. Kirchhoff's laws apply to current only.

6. True or False? In a series circuit, E_T must equal the sum of the individual voltage drops.

7. When calculating series-parallel combination circuits, it is important to _____.
   a. reduce to series loads, then add the loads
   b. replace parallel combinations with one value
   c. reduce to simpler circuits where possible
   d. All of the above.

8. True or False? Kirchhoff's voltage law can be used to determine voltage drops.

9. Two resistors are connected in parallel; R_1 is 90 ohms, R_2 is 45 ohms. What is the total resistance of R_1 and R_2 in parallel? _____

10. Two resistors are connected in parallel; R_2 is 90 ohms, R_3 is 45 ohms. A third resistor (R_1, which is 20 ohms) is connected in series with R_2 and R_3.

    a. What is the total resistance of this circuit? _____

    b. If the voltage source of this circuit is 150V, what is the total current flow? _____

(Hint: sketch the circuit before solving.)

CIRCUIT

James Mitchem

TIC—The Industrial Company

Jim Mitchem serves as a troubleshooter for a large electrical contractor. During his career in the electrical industry, he worked his way up from apprentice to technical services manager.

How did you become an electrician?
Quite by accident. A couple of years after college, I was working as a relief operator in a plant when the lead electrician retired, creating a vacancy. I liked the idea that electricians were expected to use their knowledge and initiative to keep the place running. I applied and was accepted as a trainee.

How did you get your training?
I took an electrical apprenticeship course by correspondence, and I was fortunate enough to work with good people who helped me along. I worked in an environment that exposed me to a variety of equipment and applications, and just about everyone I've ever worked with has taught me something. Now I'm passing my knowledge on to others.

What kinds of work have you done in your career?
I've worked as an apprentice, journeyman, instrument and controls technician, instrument fitter, foreman, general foreman, superintendent, and startup engineer. Each of these positions required that I learn new skills, both technical and managerial. My experience in many disciplines and types of projects has given me a high level of credibility with my employer and our clients.

Now I act as a technical resource and trouble-shooter for job sites and in-house functions such as safety, quality assurance, and training. I visit job sites to help solve problems and help out with commissioning and startup.

What factor or factors have contributed the most to your success?
There are several factors. Two very important ones have been a desire to learn and a willingness to do whatever is asked of me. I also keep an eye on the big picture. When I'm on a job, I'm not just pulling wire, I'm building a power plant or whatever the project is. I also think it has helped me to remain with the same employer for 18 years.

Any advice for apprentices just beginning their careers?
Keep learning! And don't depend on others to train you. Take the initiative to buy or borrow books and trade journals. Take licensing tests and do whatever is necessary to keep your licenses current. Finally, make sure you know your own personal and professional values and work with a company that shares those values.

Trade Terms Introduced in This Module

Kirchhoff's current law: The statement that the total amount of current flowing through a parallel circuit is equal to the sum of the amounts of current flowing through each current path.

Kirchhoff's voltage law: The statement that the sum of all the voltage drops in a circuit is equal to the source voltage of the circuit.

Parallel circuits: Circuits containing two or more parallel paths through which current can flow.

Series circuits: Circuits with only one path for current flow.

Series-parallel circuits: Circuits that contain both series and parallel current paths.

Additional Resources

This module presents thorough resources for task training. The following resource material is suggested for further study.

Electronics Fundamentals: Circuits, Devices, and Applications, Thomas L. Floyd. New York: Prentice Hall.

Principles of Electric Circuits, Thomas L. Floyd. New York: Prentice Hall.

Figure Credits

AGC of America, Module opener
Topaz Publications, Inc., 104SA01

CONTREN® LEARNING SERIES — USER UPDATE

NCCER makes every effort to keep its textbooks up-to-date and free of technical errors. We appreciate your help in this process. If you find an error, a typographical mistake, or an inaccuracy in NCCER's Contren® materials, please fill out this form (or a photocopy), or complete the online form at www.nccer.org/olf. Be sure to include the exact module number, page number, a detailed description, and your recommended correction. Your input will be brought to the attention of the Authoring Team. Thank you for your assistance.

Instructors – If you have an idea for improving this textbook, or have found that additional materials were necessary to teach this module effectively, please let us know so that we may present your suggestions to the Authoring Team.

NCCER Product Development and Revision
3600 NW 43rd Street, Building G, Gainesville, FL 32606

Fax: 352-334-0932
Email: curriculum@nccer.org
Online: www.nccer.org/olf

☐ Trainee Guide ☐ AIG ☐ Exam ☐ PowerPoints Other _____

Craft / Level: _____ Copyright Date: _____

Module Number / Title: _____

Section Number(s): _____

Description: _____

_____

_____

_____

Recommended Correction: _____

_____

_____

_____

Your Name: _____

Address: _____

_____

Email: _____ Phone: _____

Electrical Test Equipment

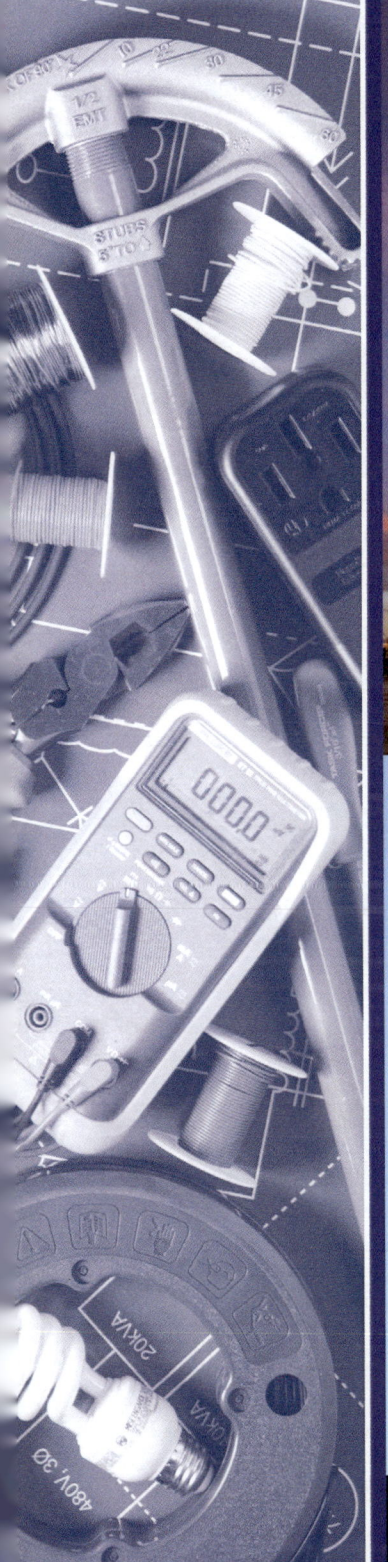

The Biodesign Institute at Arizona State University

The Biodesign Institute is the centerpiece of new growth at Arizona State University (ASU). This unique complex was built to meet the stringent demands posed by experimental programs in biotechnology and nanotechnology, and to enhance communication and collaboration between researchers with an open, shared lab design and a central atrium linking all floors.

Trainees with successful module completions may be eligible for credentialing through NCCER's National Registry. To learn more, go to **www.nccer.org** or contact us at **1.888.622.3720.** Our website has information on the latest product releases and training, as well as online versions of our *Cornerstone* newsletter and Pearson's Contren® product catalog.

Your feedback is welcome. You may email your comments to **curriculum@nccer.org,** send general comments and inquiries to **info@nccer.org,** or use the User Update form at the back of this module.

 V.1 6/11

Objectives

When you have completed this module, you will be able to do the following:

1. Explain the operation of and describe the following pieces of test equipment:
   - Voltmeter
   - Ohmmeter
   - Clamp-on ammeter
   - Multimeter
   - Megohmmeter
   - Motor and phase rotation testers
2. Select the appropriate meter for a given work environment based on category ratings.
3. Identify the safety hazards associated with various types of test equipment.

Performance Tasks

Under the supervision of the instructor, you should be able to do the following:

1. Measure the voltage in your classroom from line to neutral and neutral to ground.
2. Use an ohmmeter to measure the value of various resistors.

Trade Terms

Coil
Continuity

d'Arsonval meter
movement

Frequency

Required Trainee Materials

1. Pencil and paper
2. Copy of the latest edition of the *National Electrical Code*®
3. Appropriate personal protective equipment

Note: *NFPA 70*®, *National Electrical Code*®, and *NEC*® are registered trademarks of the National Fire Protection Association, Inc., Quincy, MA 02269. All *National Electrical Code*® and *NEC*® references in this module refer to the 2011 edition of the *National Electrical Code*®.

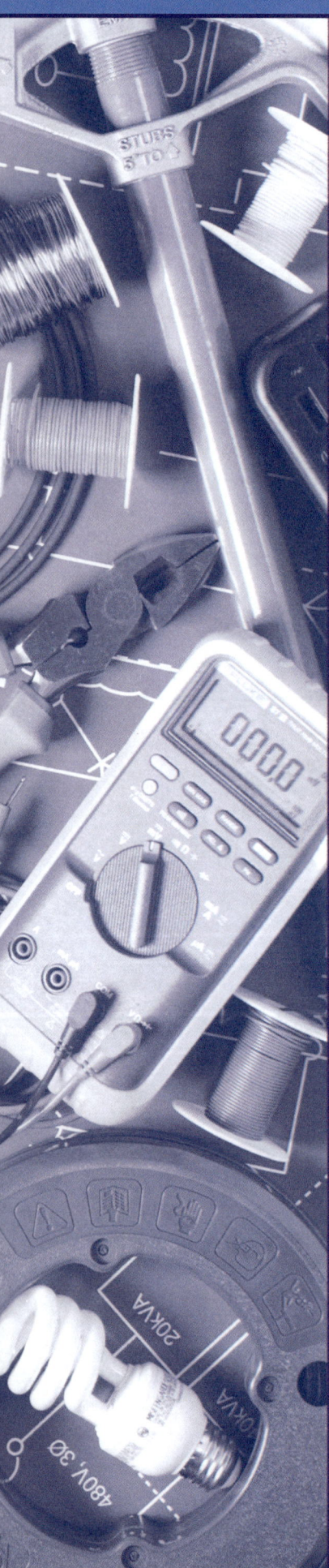

Contents ———————————————————————

Topics to be presented in this module include:

1.0.0 Introduction 1
2.0.0 Meters 1
   2.1.0 Voltmeter 2
   2.2.0 Ohmmeter 3
   2.3.0 Ammeter 5
   2.4.0 Multimeter 5
   2.5.0 Megohmmeter 7
   2.6.0 Motor and Phase Rotation Testers 7
   2.7.0 Recording Instruments 9
3.0.0 Category Ratings 9
4.0.0 Safety 10

Figures and Tables ——————————————————

Figure 1   d'Arsonval meter movement 1
Figure 2   Voltage tester 2
Figure 3   Multi-function voltage tester 3
Figure 4   Ohmmeter schematic 4
Figure 5   Clamp-on ammeter 5
Figure 6   Analog VOM 6
Figure 7   Digital VOM 6
Figure 8   Clamp-on multimeter 6
Figure 9   Battery-operated megohmmeter 7
Figure 10 Hand-crank megohmmeter 7
Figure 11 Motor rotation tester 8
Figure 12 Phase rotation tester 8
Figure 13 Data recording system 9
Figure 14 Category rating on a typical meter 10

Table 1 Overvoltage Installation Categories 10

1.0.0 INTRODUCTION

Electronic test instruments and meters are generally used for the following tasks:

- Troubleshooting electrical/electronic circuits and equipment
- Verifying proper operation of instruments and associated equipment

The test equipment selected for a specific task depends on the type of measurement and the level of accuracy required. This module will focus on some of the test equipment used by electricians. Upon completion of this module, you should be able to select the appropriate test equipment for a specific application and identify the applicable safety hazards.

2.0.0 METERS

In 1882, a Frenchman named Arsene d'Arsonval invented the galvanometer. This meter used a stationary permanent magnet and a moving coil to indicate current flow on a calibrated scale. The early galvanometer was very accurate but could only measure very small currents. Over the following years, many improvements were made

that extended the range of the meter and increased its ruggedness. The d'Arsonval meter movement (*Figure 1*) is the basis for analog meters.

A moving-coil meter movement operates on the electromagnetic principle. In its simplest form, the moving-coil meter uses a coil of very fine wire wound on a light aluminum frame. A permanent magnet surrounds the coil. The aluminum frame is mounted on pivots to allow it and the coil to rotate freely between the poles of the permanent magnet. When current flows through the coil, it becomes magnetized, and the polarity of the coil is repelled by the field of the permanent magnet. This causes the coil frame to rotate on its pivots, and the distance it rotates is determined by the amount of current that flows through the coil. By attaching a pointer to the coil frame and adding a calibrated scale, the amount of current flowing through the meter can be measured. Multiplier resistors are used to extend the range of the meter movement for voltage measurements, while shunt resistors are used to extend the range of the meter movement for current measurements.

Today, most meters are solid-state digital systems; they are easier to read than mechanical (analog) meters and have no meter movement or moving parts.

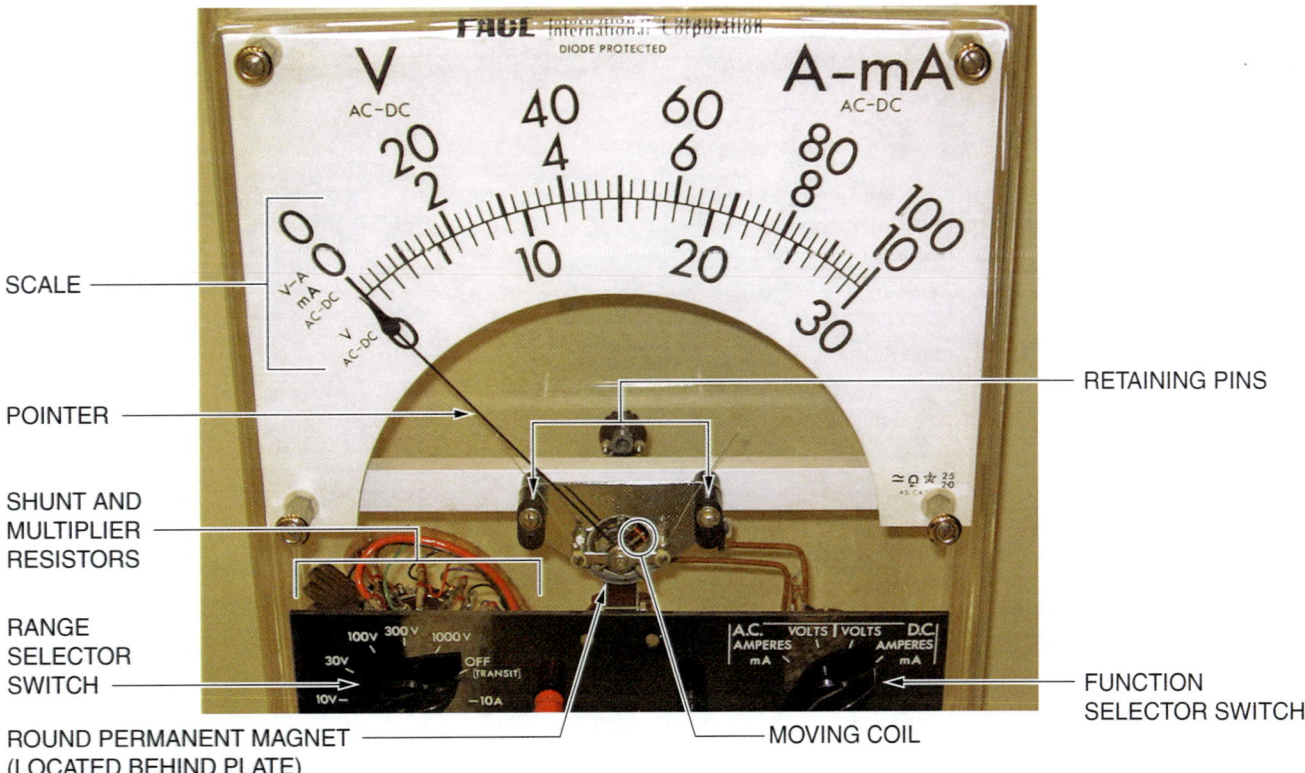

Figure 1 d'Arsonval meter movement.

Phantom Readings

The sensitivity of a digital meter can sometimes produce a low reading known as a phantom or ghost reading. This is due to the induction from the electrical field around the energized conductors in close proximity to the meter.

2.1.0 Voltmeter

A voltmeter is used to measure voltage, also known as potential difference or electromotive force (emf). It is connected in parallel with the circuit or component being measured. An analog meter uses the basic d'Arsonval meter movement with internally switched resistors to measure different voltage ranges. A digital meter uses an analog-to-digital converter chip to convert the sensed values into a digital or graphic display.

Many digital voltmeters are autoranging, which means that the meter will automatically search for the correct scale. When using a voltmeter that is not autoranging, always start with the highest voltage range and work down until the indication reads somewhere between half and three-quarter scale. This will provide a more accurate reading and prevent damage to the meter. On many meters, a DC value is indicated by a straight line with three dashes beneath it, while an AC value is indicated by a sine wave.

> **WARNING!**
>
> Measuring voltages above 50V exposes the technician to potentially life-threatening hazards. Follow all applicable safety procedures as found in *NFPA 70E*, OSHA standards, and company/institutional policies and standards.

A voltmeter is used when the exact value of the voltage is required. However, electricians are often concerned with identifying only whether voltage is present, and if so, the general range of the voltage. In other words, is it energized, and if so, is it at 120V, 240V, or 480V? In these cases, a voltage tester is used. The range of voltage and the type of current (AC and/or DC) that a voltage tester is capable of measuring are usually indicated on the scales that display the reading. See *Figure 2*.

Advanced voltage testers offer additional features, such as a digital readout, GFCI test capability, and even the ability to switch between use as a contact and noncontact detector. See *Figure 3*.

A voltage tester must be checked before each use to make sure that it is in good condition and is operating correctly. The external check of the tester should include a careful inspection of the insulation on the leads for cracks or frayed areas. Faulty leads constitute a safety hazard, so they must be replaced. As a check to make sure that the voltage tester is operating correctly, the probes of the tester are first connected to a known energized source. The voltage indicated on the tester should match the voltage of the source. If there is no indication, the voltage tester is not operating correctly, and it must be repaired or replaced. It must also be repaired or replaced if it indicates a voltage different from the known voltage of the source.

112F02.EPS

Figure 2 Voltage tester.

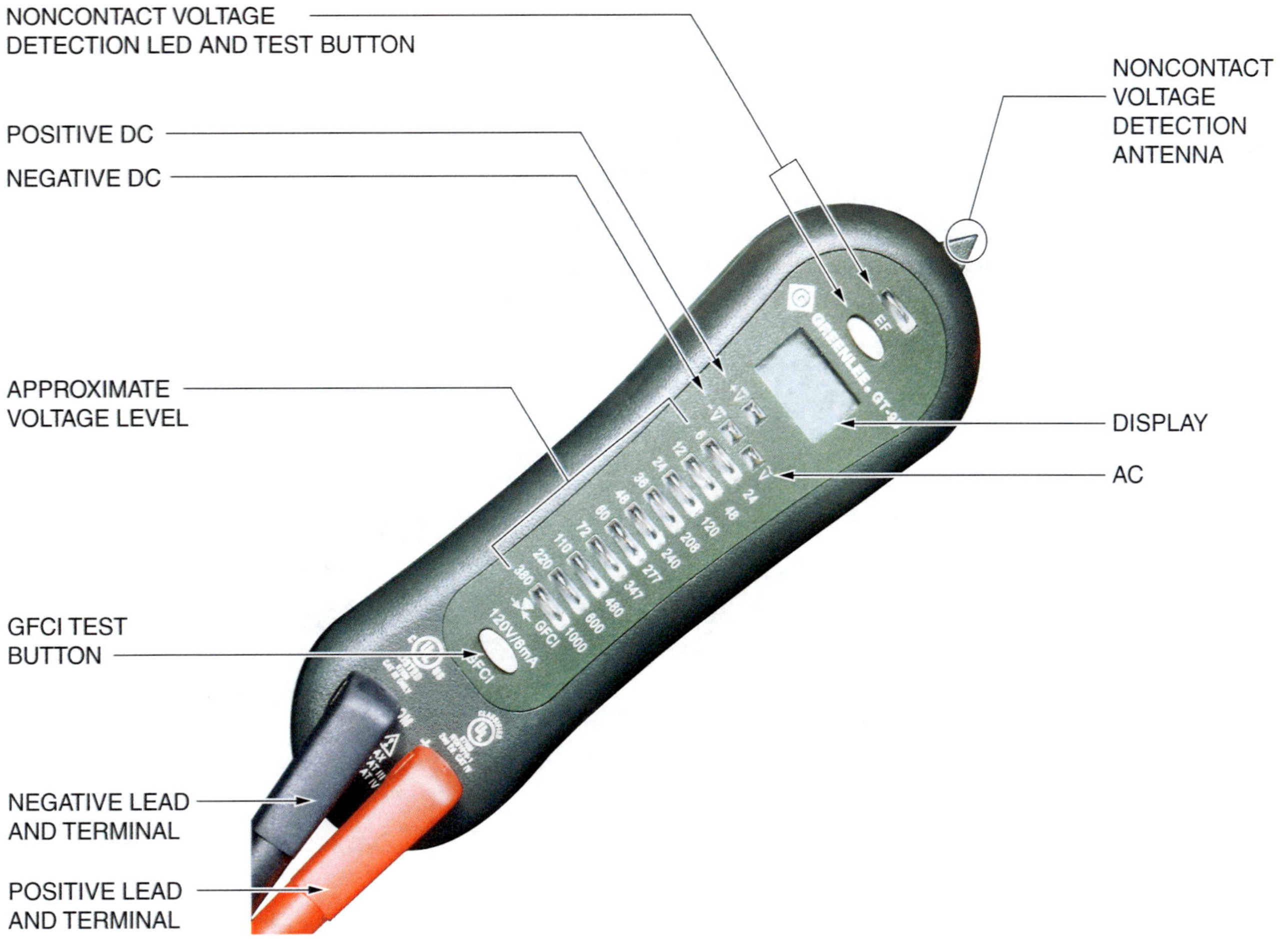

NONCONTACT VOLTAGE DETECTION LED AND TEST BUTTON

POSITIVE DC

NEGATIVE DC

NONCONTACT VOLTAGE DETECTION ANTENNA

APPROXIMATE VOLTAGE LEVEL

DISPLAY

AC

GFCI TEST BUTTON

NEGATIVE LEAD AND TERMINAL

POSITIVE LEAD AND TERMINAL

112F03.EPS

Figure 3 Multi-function voltage tester.

> **CAUTION**
>
> Care should be taken when placing the probes of the tester across the voltage source. Some voltage testers are designed to take quick readings and may be damaged if left in contact with the voltage source for too long.

Voltage testers are used to make sure that voltage is available when it is needed and to ensure that power has been cut off when it should have been. In a troubleshooting situation, it might be necessary to verify that power is available in order to be sure that lack of power is not the problem. For example, if there were a problem with a power tool, such as a drill, a voltage tester might be used to make sure that power is available to run the drill. A voltage tester might also be used to verify that there is power available to a three-phase motor that will not start.

> **WARNING!**
>
> When testing for voltage to verify that a circuit is de-energized as part of an electrical lockout, always perform a live-dead-live test. This involves first verifying operation of the test equipment on a known energized (live) source, then de-energizing and testing the target circuit to ensure it is dead, then again testing the known energized (live) source before making contact with the de-energized circuit. This is an OSHA requirement for systems above 600V but is a good practice at all voltage levels.

2.2.0 Ohmmeter

An ohmmeter measures the resistance of a circuit or component. It can also be used to locate open circuits or shorted circuits. An ohmmeter consists of a DC current meter movement, a low-voltage DC power source (usually a battery), and current-limiting resistors, all of which are connected in series with the meter (*Figure 4*).

Voltage Detectors

Simple noncontact (proximity) voltage detectors can also be used to indicate the presence of voltage within their specified range rating, but do not discriminate between ranges of values in the same way as a voltage tester. They are handy for quickly scanning for the presence of voltage in junction boxes or termination cabinets, and can even be used to trace circuits through walls. The voltage detector shown here glows in the presence of voltages between 50V and 1,000V.

112SA01.EPS

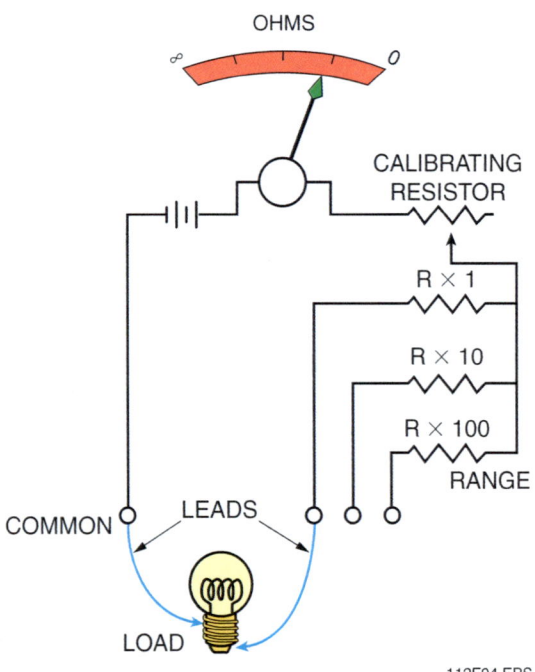

Figure 4 Ohmmeter schematic.

112F04.EPS

Before measuring the resistance of an unknown resistor or electrical circuit, connect the test leads together. This zeroes out or nulls the resistance of the leads. Some analog ohmmeters have a zero adjustment knob. With the leads connected together, turn the adjustment knob until the meter registers zero ohms. This adjustment must be made each time a different range is selected.

Many digital ohmmeters are autoranging. The correct scale is internally selected and the reading will indicate the range (ohms, K-ohms, or M-ohms). Analog ohmmeters require that you select the desired range. Most digital meters also have an audible tone when the measured value is very low or at zero ohms. This indicates a closed circuit and is useful when using the meter as a continuity tester. A continuity test is used to determine if a circuit is complete.

> **WARNING!**
>
> Prior to taking a reading with an ohmmeter, verify that both sides of the circuit are de-energized by using a voltmeter. If the circuit were energized, its voltage could cause a current to flow through the meter. This can damage the meter and/or circuit and cause personal injury.

Resistance

Why does the resistance vary when holding a resistor by pinching the meter leads against the resistor with your fingers when measuring it, versus measuring it while holding it in clips?

When making resistance measurements in circuits, each component in the circuit can be tested individually by removing the component from the circuit and connecting the ohmmeter leads across it. However, the component does not have to be totally removed from the circuit. Usually, the part can be effectively isolated by disconnecting one of its leads from the circuit. Note that this method can still be somewhat time consuming.

2.3.0 Ammeter

A clamp-on ammeter, also known as a clamp meter, can measure current without having to make contact with uninsulated wires (*Figure 5*). This type of meter operates by sensing the strength of the electromagnetic field around the wire(s).

Clamp-on ammeters measure current by using simple transformer principles. The conductor(s) being measured would be the primary and the jaws (clamp) of the meter would be the secondary. The current in the primary winding induces a current in the secondary winding. If the ratio of the primary winding to the secondary winding is 1,000, then the secondary current is $\frac{1}{1000}$ of the current flowing in the primary. The smaller secondary current is connected to the meter's input. For example, a 1A current in the conductor will produce 0.001A (1mA) in the meter.

To measure the current, open the jaws of the meter and close them around the conductor(s) that are to be measured. Make sure that the jaws are clean and close tightly. Then read the magnitude of the current on the meter display.

Many meters have a Hold function. This is useful in tight locations when it is hard to

Current Measurements

What happens if you loop the conductor so that the meter measures two turns instead of one?

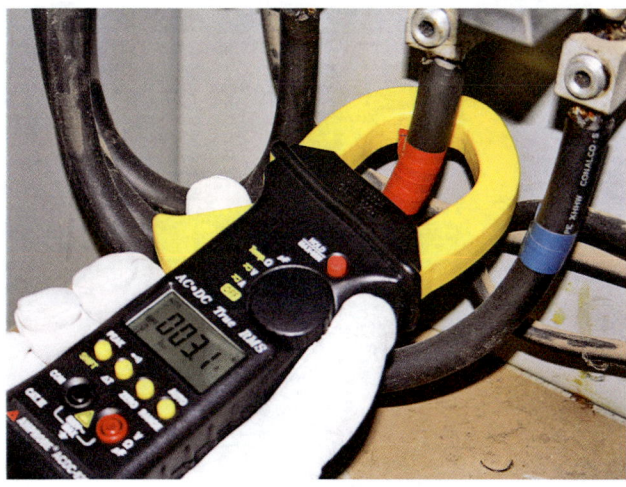

112F05.EPS

Figure 5   Clamp-on ammeter.

read the meter while it is clamped around the conductor(s). Just press the Hold button when the value is measured, then remove and read the meter.

CAUTION

When using a clamp-on ammeter, make sure that the range of the meter is at least as high as the current to be measured. If the meter is digital and the current is too high, the display will read OL (overload). This means that the meter has been overloaded. If the meter is an analog meter, the indicating needle will peg (move) above the maximum limit on the scale, which might damage the meter.

Some meters have a Min/Max or peak function. This allows the technician to record the maximum inrush, as with a motor start.

WARNING!

Using a clamp-on ammeter may expose you to energized systems and equipment. Never use this type of meter unless you are qualified and are following NFPA, OSHA, and company/institutional safety procedures.

2.4.0 Multimeter

The multimeter is also known as a volt-ohm-milliammeter (VOM). An analog VOM is shown in *Figure 6*. It is a multi-purpose instrument that combines the three previous meters discussed. When using an analog meter, you must select the proper voltage (DC or AC) and range

Figure 6   Analog VOM.

(in volts, amps, or ohms). When using a digital VOM (*Figure 7*), you must select the proper voltage. Most have an autoranging feature for the magnitude.

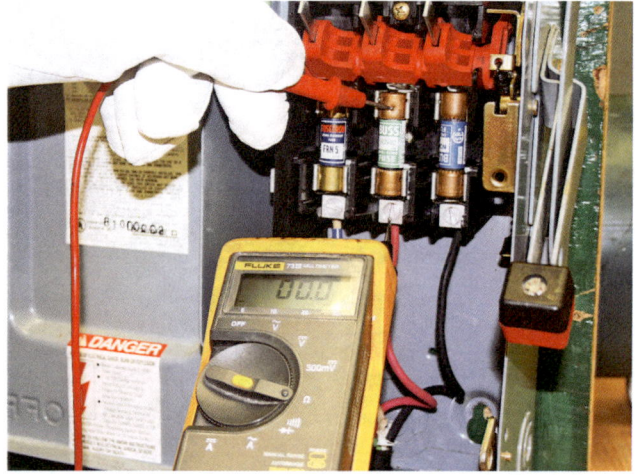

Figure 7   Digital VOM.

Some multimeters can also measure the frequency of an AC waveform—this function is useful for diagnosing harmonic problems in an electrical distribution system.

Another additional feature is the Min/Max memory function. It will record the minimum and maximum readings over the time period selected. Other common multimeter functions include capacitance measurement, diode and transistor testers, temperature measurement, and true RMS measurement for accurate voltage and current readings at different frequencies. Refer to the manufacturer's instructions for the meter in use. In addition to the current clamps used with standard multimeters, clamp-on multimeters are also available (*Figure 8*). They are used to measure AC current, AC and DC voltage, resistance, and other values.

> **WARNING!**
>
> Before use, a VOM must be checked on a known power source.

Use of a VOM is the same as using the individual voltmeter, ohmmeter, and clamp-on ammeter. Current clamps can be used with most multimeters for measuring AC and DC currents above the milliamp level. These can be plugged into either the amp or voltage test lead connections, depending on the current clamp. Always refer to the manufacturer's instructions for the device in use. Current clamps are available in various current ranges, from 50A to several thousand amps. The jaws are available in different shapes and sizes to suit various applications, from round to rectangular, and even flexible.

Figure 8   Clamp-on multimeter.

2.5.0 Megohmmeter

An ordinary ohmmeter cannot be used for measuring resistances of several million ohms, such as those found in conductor insulation or between motor or transformer windings. The instrument used to measure very high resistances is known as a megohmmeter. Megohmmeters are also called Meggers®, or insulation resistance testers. They can be powered by alternating current, battery (*Figure 9*), or hand cranking (*Figure 10*).

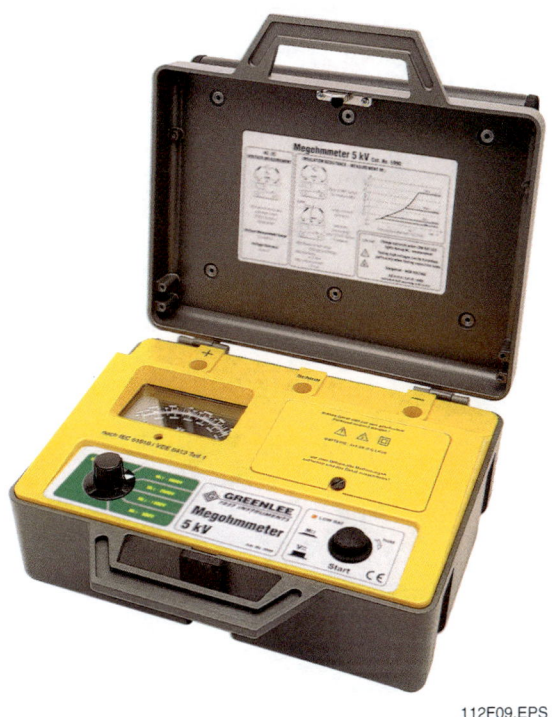

112F09.EPS

Figure 9   Battery-operated megohmmeter.

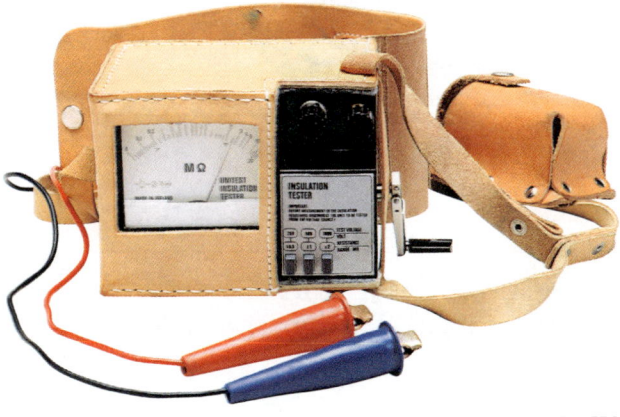

112F10.EPS

Figure 10   Hand-crank megohmmeter.

When using a megohmmeter, you could be injured or cause damage to the equipment you are working on if the following minimum safety precautions are not followed:

- High voltages are present when using a megohmmeter. For example, in 600V class systems, applied megohmmeter voltages are typically 500V and 1,000V. Only qualified individuals may use this equipment. Always wear appropriate personal protective equipment when approaching energized parts.
- De-energize and verify the de-energization of the circuit before connecting the meter. Make sure all capacitors are discharged.
- If possible, disconnect the item being checked from the other circuit components before using the meter.
- Do not exceed the manufacturer's recommended voltage test levels for the cable or equipment under test. Many manufacturers have different test levels based on the age of the cable or equipment being tested.
- Never touch the test leads when the meter is energized or powered. Meggers generate high voltage, and touching the leads could result in injury or electrical shock.
- After the test, discharge any energy that may be left in the circuit by grounding the conductor or equipment for a period of time equal to the duration of the test.
- When megging cables or busducts where you have exposed parts that are remote from your testing position, safely secure or barricade the exposed end to protect others from inadvertent contact with the test voltage.

> **CAUTION**
>
> If a megohmmeter is used to test switchgear, all of the electronics must be disconnected prior to testing the switchgear. The voltage produced by the megohmmeter may damage electronic equipment.

Meter manufacturers supply detailed manuals for testing various devices and equipment. Always follow these instructions.

2.6.0 Motor and Phase Rotation Testers

Before connecting a three-phase motor to a circuit, you must first match the legs or windings of the motor (T1, T2, and T3) to the phases of the circuit (L1, L2, and L3). This ensures that the motor rotates in the proper direction. A motor

Meter Care

Like all meters, a megohmmeter is a sensitive instrument. Treat it with care and keep it in its case when not in use.

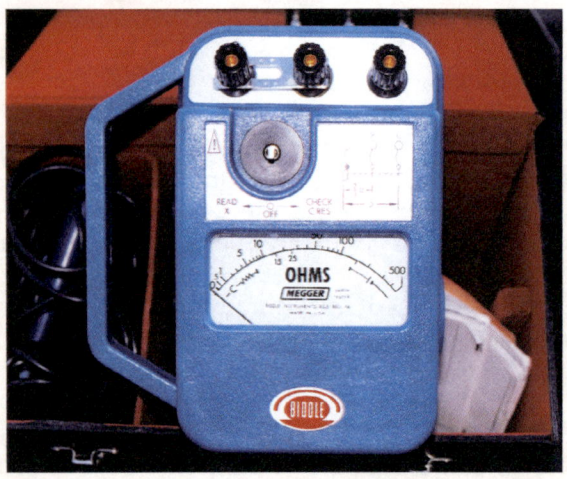

112SA02.EPS

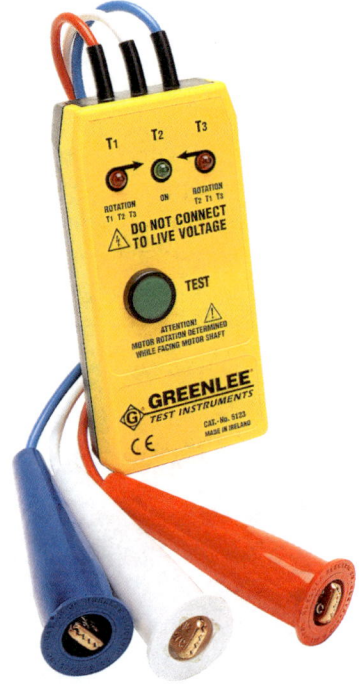

112F11.EPS

Figure 11   Motor rotation tester.

rotation tester can be used to identify the legs of the motor (*Figure 11*), while a phase rotation tester can be used to identify the phases of the circuit (*Figure 12*).

A motor rotation tester is a passive device; it operates on residual magnetism present in the motor after it has been run or tested by the manufacturer prior to shipping. To use a motor rotation tester, connect the three motor wires to the T1, T2, and T3 leads on the tester, then rotate the motor shaft a half-turn while pressing the Test button (the direction of rotation depends on the tester in use; always follow the manufacturer's instructions). Either the clockwise or counterclockwise LED will light up. If the required rotation is clockwise and the clockwise LED lights up, tag the motor wires to correspond to the motor rotation leads. If the required rotation is clockwise and the counterclockwise LED lights up, switch a pair of leads and retest.

112F12.EPS

Figure 12   Phase rotation tester.

A phase rotation tester, also called a phase sequence indicator, is used on three-phase electrical systems to indicate the phase sequence rotation of the voltages. These testers typically have LEDs to indicate the phase rotation. A phase sequence is measured as clockwise or counterclockwise rotation.

A phase rotation tester is used when it is necessary to ensure the same phase rotation throughout a facility. To test phase rotation, de-energize and lock out power to the circuit, then connect the three leads of the tester to the phase conductors in the circuit. Next, safely energize the circuit and observe the meter. Make note of the color scheme of the connected leads to the system, along with the phase sequences as indicated on the meter. This is necessary to ensure that the added equipment follows the same phase rotation. De-energize and lock out the circuit before disconnecting the leads.

2.7.0 Recording Instruments

The term *recording instrument* describes many instruments that make a permanent record of measured quantities over a period of time. Recording instruments use either a paper strip or electronic accessible memory. Those using electronic memory are usually called data loggers. These instruments record electrical quantities, including potential difference, current, power, resistance, and frequency. They can also record nonelectrical quantities by electrical means, such as a temperature recorder that uses a potentiometer system to record thermocouple output.

It is often necessary to know the conditions that exist in an electrical circuit over a period of time to determine such things as peak loads, voltage fluctuations, and so on. An automatic recording instrument can be connected to take readings at specified intervals for later review and analysis. Some meters can upload data to a PC for real-time data logging and graphing. See *Figure 13*.

3.0.0 CATEGORY RATINGS

Distribution systems and loads are becoming more complex, increasing the risk of transient power spikes. Lightning strikes on outdoor transmission lines and switching surges from normal switching operations can also produce dangerous high-energy transients. Motors, capacitors, variable speed drives, and power conversion equipment can also generate power spikes.

Safety systems are built into test equipment to protect electricians from transient power spikes. The International Electrotechnical Commission (IEC) developed a safety standard, *IEC 1010*, for test equipment that was adapted as *UL Standard UL3111-1*. These standards define four overvoltage installation categories, often abbreviated as CAT I, CAT II, CAT III, and CAT IV (*Table 1*). These categories identify the hazards posed by transients; the higher the category number, the greater the risk to the electrician. A higher category number refers to an installation with higher power available and higher-energy transients.

When selecting a meter, choose a meter rated for the highest category you will be working in. Then select the appropriate voltage level. In addition, make sure that your test leads are rated as high as your meter. Choose meters that are independently tested and certified by UL, CSA, or another recognized testing organization. Certified meters are marked with the category rating on the meter housing (*Figure 14*).

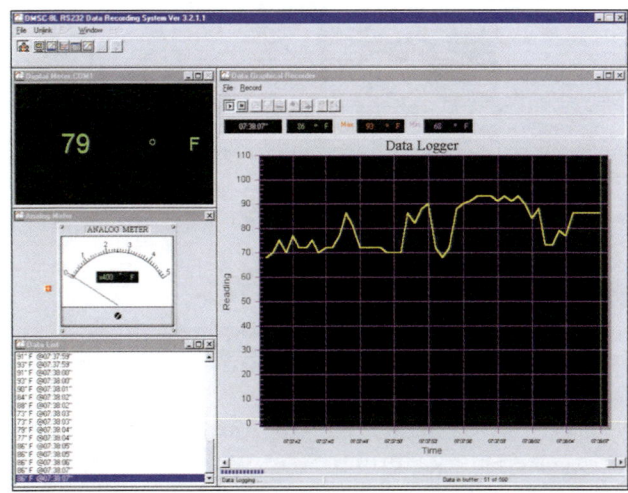

112F13.EPS

Figure 13 Data recording system.

Table 1 Overvoltage Installation Categories

Overvoltage Category	Installation Examples
CAT I	Electronic equipment and circuitry
CAT II	Single-phase loads such as small appliances and tools, outlets at more than 30 feet from a CAT III source or 60 feet from a CAT IV source
CAT III	Three-phase motors, single-phase commercial or industrial lighting, switchgear, busduct and feeders in industrial plants
CAT IV	Three-phase power at meter, service-entrance, or utility connection, any outdoor conductors

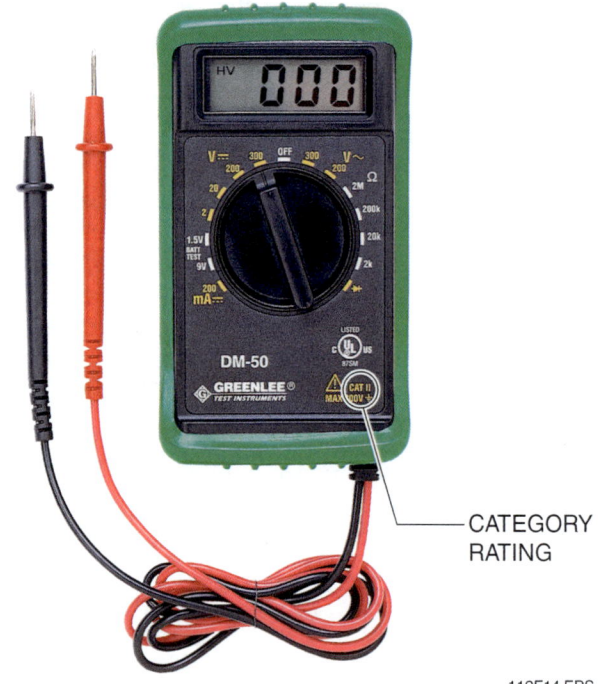

CATEGORY RATING

112F14.EPS

Figure 14 Category rating on a typical meter.

4.0.0 SAFETY

Safety must be the primary responsibility of all personnel on a job site. The safe installation, maintenance, and operation of electrical equipment requires strict adherence to local and national codes and safety standards, as well as facility and company safety policies. Carelessness can result in serious injury or death due to electrical shock, burns, falls, flying objects, etc. After an accident has occurred, investigation almost invariably shows that it could have been prevented by the exercise of simple safety precautions and procedures. It is your personal responsibility to identify and eliminate unsafe conditions and unsafe acts that cause accidents.

You must bear in mind that de-energizing main supply circuits by opening supply switches will not necessarily de-energize all circuits in a given piece of equipment. A source of danger that has often been neglected or ignored, sometimes with tragic results, is the input to electrical equipment from other sources, such as backfeeds. The rescue of a victim shocked by the power input from a backfeed is often hampered because of the time required to determine the source of power and isolate it. Always turn off all power inputs before working on equipment and lock out and tag, then check with an operating voltage tester to be sure that the equipment is safe to work on.

> **WARNING!**
> When performing lockout/tagout procedures, remember that other forms of energy may be present and must also be locked out and tagged. These include water pressure, steam, springs, gravity, and other forms of energy.

Remember that the common 120V power supply voltage is not a low, relatively harmless voltage but is a voltage that has caused more deaths than any other.

Safety can never be stressed enough. There are times when your life literally depends on it. Always observe the following precautions:

- Thoroughly inspect all test equipment before each use. Check for broken leads or knobs, damaged plugs, or frayed cords. Do not use equipment that is wet or damaged.
- Make sure the rating of any leads or accessories meets or exceeds the rating of the meter.
- Do not work with energized equipment unless you are both qualified and approved by your supervisor.
- Never shortcut safety; strictly adhere to all energized work policies and procedures.
- When testing circuits, test at higher ranges first, then work your way down to lower ranges.
- Always have a standby person present during hot work; he or she should know whom to contact in case of emergency and how to disconnect the power.

Putting It All Together

What kind of measuring device would you select or need for the following tasks, and how would you apply the device?

- Identify a short circuit in house wiring
- Measure the secondary voltage of an AC transformer
- Identify a blown fuse in a circuit
- Identify the contact configuration of a three-way switch or multi-pole relay

SUMMARY

Meters and other devices are used to test and troubleshoot circuits and electrical equipment. One of the most important tests you will perform is verifying the absence of voltage before working on a device or circuit. This is often done using a voltage tester. Other common test equipment includes multimeters, clamp-on ammeters, megohmmeters, and motor/phase rotation testers.

You must understand the operation of and safety precautions for each piece of test equipment. In addition, you must be able to select the appropriate test equipment based on the task and category rating of the environment in which the work is to be performed. Always inspect and verify the operation of all test equipment before using it. Your life may depend on it.

1. Digital meters do *not* have a(n) _____.
   a. red lead wire
   b. battery
   c. meter movement
   d. autoranging setting

2. A voltmeter is used to test _____.
   a. exact voltages
   b. voltage ranges
   c. power
   d. sine waves

3. In order to ensure safety, before measuring low voltages, you should first test for _____.
   a. resistance
   b. current
   c. vibration
   d. higher voltages

4. An ammeter is used to measure _____.
   a. current
   b. voltage
   c. resistance
   d. insulation value

5. Clamp-on ammeters operate by _____.
   a. using d'Arsonval meter movement
   b. sensing the strength of the electromagnetic field around the wire
   c. measuring the high resistance end of a power transformer
   d. using a resistive shunt

6. An insulation tester is another name for a(n) _____.
   a. megohmmeter
   b. ammeter
   c. multimeter
   d. continuity tester

7. A motor rotation tester _____.
   a. tests an energized motor
   b. gets connected to the motor supply conductors
   c. works on residual magnetism
   d. works only on clockwise rotation

8. The highest level of protection is provided by instruments rated as _____.
   a. CAT I
   b. CAT II
   c. CAT III
   d. CAT IV

9. The International Electrotechnical Commission (IEC) developed a safety standard for overvoltage installation categories of CAT I, CAT II, CAT III, and CAT IV for _____.
   a. wiring
   b. electrical equipment
   c. test equipment
   d. signaling circuits

10. De-energizing the main supply circuit will always de-energize all circuits in a given piece of equipment.
    a. True
    b. False

Trade Terms Quiz

Fill in the blank with the correct term that you learned from your study of this module.

1. Used for electromagnetic effects or for providing electrical resistance, a _____ is a number of turns of wire.

2. A _____ uses a permanent magnet and moving coil arrangement to move a pointer across a scale.

3. Usually expressed in hertz, _____ is the number of cycles completed each second by a given AC voltage.

4. _____ is an uninterrupted electrical path for current flow.

Trade Terms

Coil

Continuity

d'Arsonval meter movement

Frequency

Module 26112-11
Supplemental Exercises

1. The measurement of the electromotive force of a circuit is accomplished using a(n) _____.
   a. ammeter
   b. wattmeter
   c. voltmeter
   d. ohmmeter

2. When using a voltmeter that is not autoranging, start with the highest setting and work down until the meter reads somewhere between _____.

3. A(n) _____ is used to extend the range of a meter movement for current measurements.

4. True or False? Always connect an ohmmeter in parallel with a load.

5. The voltage range of a meter movement can be extended by adding a(n) _____ in series.

6. Short circuits can be detected by using a(n) _____.

7. What type of test equipment would you use to check the resistance between motor windings?

   _____

8. The phase sequence of a circuit is identified using a(n) _____.

9. True or False? A voltage tester is used for precise voltage measurements.

10. A(n) _____ is used to take electrical readings at specified intervals.

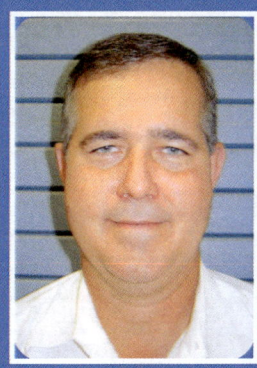

Clarence "Ed" Cockrell

HR/Safety Manager
Vector Electric & Controls, Inc.

How did you get started in the construction industry?
I worked as a summer electrical helper during high school and college and found it very rewarding. I was looking for a job that would hold my interest for more than a year. I studied electrical engineering at Louisiana State University for two years.

Who inspired you to enter the industry? Why?
My brother-in-law and father-in-law inspired me to enter the industry.

What do you enjoy most about your job?
I started off as an electrician's apprentice, which offered many potential job opportunities. I enjoyed seeing the work progress from dirt to a functional building and finding new challenges as new technology changes the way we install the electrical components. My electrical training also opened the door to the possibility of being a field superintendent, project manager, senior office manager, human resource/training manager, and then human resource/safety/training manager.

Do you think training and education are important in construction? If so, why?
Training gives an apprentice the opportunity to become a great electrician and not an electrical laborer, by that I mean not just a conduit or cable tray installer or wire puller, but a well-rounded electrician. It also allows the apprentice to advance beyond the limits of being an electrician. Almost all of our supervisors, estimators, and project managers have been trained by an apprenticeship program and followed it up with more NCCER training.

How important are NCCER credentials to your career?
I went through a state certified apprenticeship and then completed the CSST training. Without these certifications, I believe I would be no more than a second rate electrician.

How has training/construction impacted your life and you career?
It has allowed me to raise a family, buy a house and raise a child in a comfortable fashion. I have advanced many times at work, which has led to increased wages. Through my job, I have met and had dealings with many interesting people from all walks of life.

Would you suggest construction as a career to others? If so, why?
Yes. It offers a rewarding career opportunity to anyone willing to take pride and ownership in their learning and work.

How do you define craftsmanship?
I believe that craftsmanship is always delivering a great quality job in whatever you do. It takes pride and self-esteem to deliver work deemed to meet this definition of craftsmanship. This pride in their work helps build life qualities that become the building blocks of a truly honorable life. They walk through their community as a positive contributor as well as helping to build a secure and prosperous business. They teach their children through their actions how those who build contribute to the well-being of their community and country.

Trade Terms Introduced in This Module

Coil: A number of turns of wire, especially in spiral form, used for electromagnetic effects or for providing electrical resistance.

Continuity: An electrical term used to describe a complete (unbroken) circuit that is capable of conducting current. Such a circuit is also said to be closed.

d'Arsonval meter movement: A meter movement that uses a permanent magnet and moving coil arrangement to move a pointer across a scale.

Frequency: The number of cycles completed each second by a given AC voltage; usually expressed in hertz. One hertz equals one cycle per second.

Additional Resources

This module presents thorough resources for task training. The following resource material is suggested for further study.

ABCs of Multimeter Safety, Everett, WA: Fluke Corporation.

ABCs of DMMs, Multimeter Features and Functions Explained, Everett, WA: Fluke Corporation.

Clamp Meter ABCs, Everett, WA: Fluke Corporation.

Electronics Fundamentals: Circuits, Devices, and Applications, Thomas L. Floyd. New York: Prentice Hall.

Power Quality Analyzer Uses for Electricians, Everett, WA: Fluke Corporation.

Principles of Electric Circuits, Thomas L. Floyd. New York: Prentice Hall.

Figure Credits

AGC of America, Module opener

Tim Dean, Figure 1, Figure 5, Figure 7, Figure 8

Greenlee Textron, Inc., a subsidiary of Textron Inc., Figure 2, Figure 3, Figure 6, Figures 9–14, 112SA01

Topaz Publications, Inc., 112SA02

CONTREN® LEARNING SERIES — USER UPDATE

NCCER makes every effort to keep its textbooks up-to-date and free of technical errors. We appreciate your help in this process. If you find an error, a typographical mistake, or an inaccuracy in NCCER's Contren® materials, please fill out this form (or a photocopy), or complete the online form at www.nccer.org/olf. Be sure to include the exact module number, page number, a detailed description, and your recommended correction. Your input will be brought to the attention of the Authoring Team. Thank you for your assistance.

Instructors – If you have an idea for improving this textbook, or have found that additional materials were necessary to teach this module effectively, please let us know so that we may present your suggestions to the Authoring Team.

NCCER Product Development and Revision
3600 NW 43rd Street, Building G, Gainesville, FL 32606

Fax: 352-334-0932
Email: curriculum@nccer.org
Online: www.nccer.org/olf

☐ Trainee Guide ☐ AIG ☐ Exam ☐ PowerPoints Other _____

Craft / Level: _____ Copyright Date: _____

Module Number / Title: _____

Section Number(s): _____

Description: _____

_____

_____

_____

_____

Recommended Correction: _____

_____

_____

_____

_____

Your Name: _____

Address: _____

_____

Email: _____ Phone: _____

58104-11

Electrical Wiring

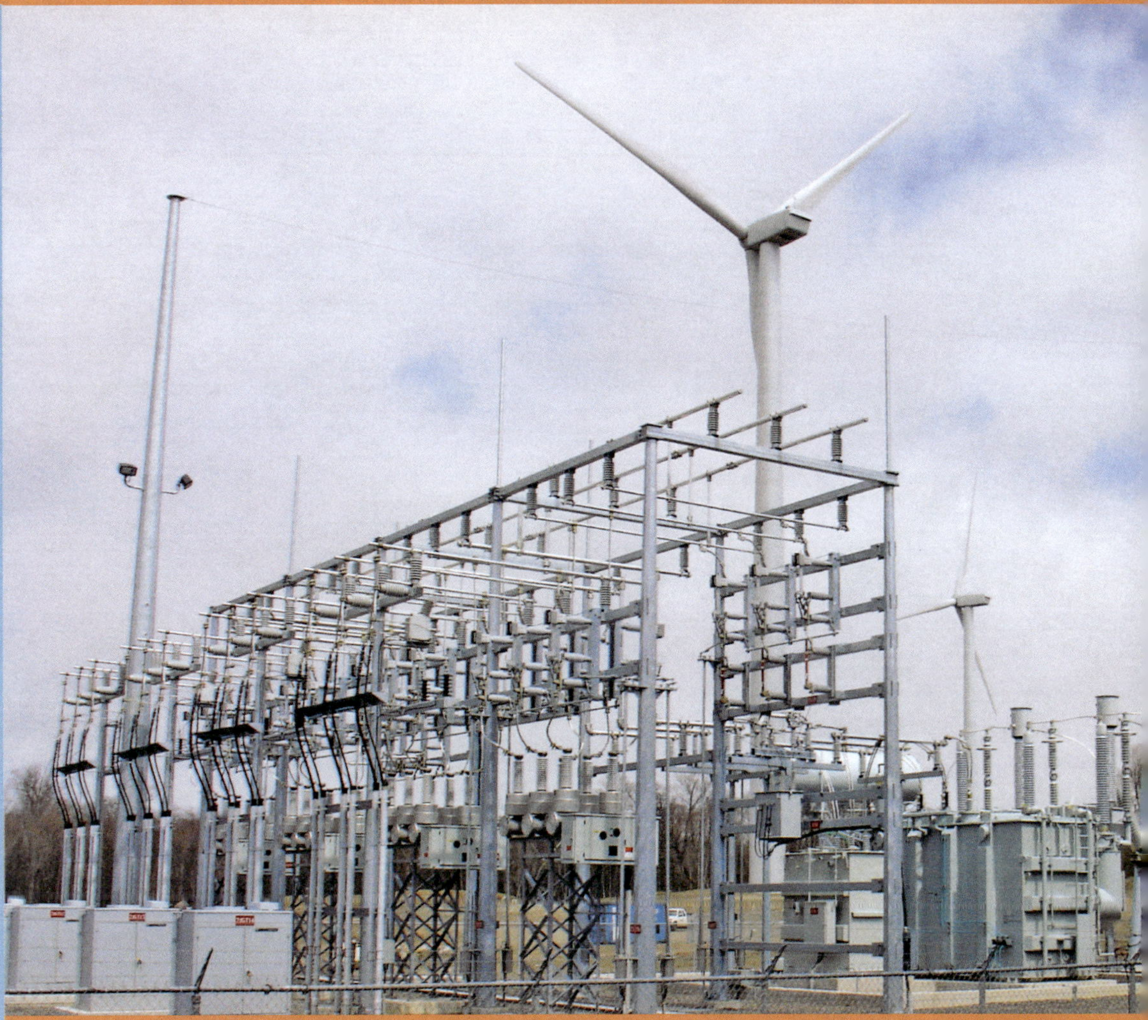

Trainees with successful module completions may be eligible for credentialing through NCCER's National Registry. To learn more, go to **www.nccer.org** or contact us at **1.888.622.3720**. Our website has information on the latest product releases and training, as well as online versions of our *Cornerstone* newsletter and Pearson's Contren® product catalog.

Your feedback is welcome. You may email your comments to **curriculum@nccer.org,** send general comments and inquiries to **info@nccer.org**, or use the User Update form at the back of this module.

 V.1 6/11

Objectives

When you have completed this module, you will be able to do the following:

1. Interpret cable and wire markings to determine the physical and functional properties of wire and cable.
2. Select the appropriate conductor based on the ampacity, insulation type, and temperature rating using the *National Electrical Code®* tables.
3. Describe types of conductors and cables used in wind turbine installations based on the application.
4. Identify the various types of data cable used in wind turbine systems.
5. Prepare and terminate various low-voltage conductors and cables.

Performance Tasks

Under the supervision of the instructor, you should be able to do the following:

1. Prepare and terminate various low-voltage conductors and cables.
2. Select the appropriate conductor based on the ampacity, insulation type, and temperature rating using the *NEC®* tables.

Trade Terms

Ambient temperature	Distribution conductors	Nominal output voltage	Solid-state relay
Collection system	Load	Power-limited cable	Stator windings
Commercial power grid system	Local area network (LAN)	Programmable logic controller (PLC)	Thermoplastic insulation
	Messenger wire	Site substation	Thermoset insulation
			Turbine transformer

Required Trainee Materials

Copy of the latest edition of the *National Electrical Code®*

Industry Recognized Credentials

If you're training through an NCCER-accredited sponsor you may be eligible for credentials from NCCER's Registry. The module ID number for this module is 58104-11. Note that this module may have been used in other NCCER curricula and may apply to other level completions. Contact NCCER's Registry at 888.622.3720 or go to nccer.org for more information.

Note: *NFPA 70®*, *National Electrical Code®*, and *NEC®* are registered trademarks of the National Fire Protection Association, Inc., Quincy, MA 02269. All *National Electrical Code®* and *NEC®* references in this module refer to the 2011 edition of the *National Electrical Code®*.

Contents

Topics to be presented in this module include:

1.0.0 Introduction 1
2.0.0 Properties of Conductors and Cables.. 2
  2.1.0 Physical Size... 2
    2.1.1 AWG System.. 2
    2.1.2 Circular Mils.. 3
    2.1.3 Stranding 4
  2.2.0 Conductor Material.. 4
    2.2.1 Conductivity.. 4
    2.2.2 Workability 4
  2.3.0 Insulating Material... 5
    2.3.1 Thermoplastic- and Thermoset-Insulated Conductors 5
    2.3.2 High-Temperature Insulated Conductors 6
    2.3.3 Insulation Color-Coding 7
  2.4.0 Conductor Temperature Rating... 7
  2.5.0 Ampacity Rating.. 8
    2.5.1 Determining Ampacity 9
    2.5.2 Conductor Ampacity Derating and Correction.............. 9
3.0.0 Applications of Conductors and Cables 10
  3.1.0 Data and Communications Cables 10
  3.2.0 Low-Voltage Power Conductors or Cables 12
    3.2.1 Generator Stator (Output) Conductors 12
  3.3.0 Medium-Voltage Cables 13
    3.3.1 Three-Phase Medium-Voltage Cables......................... 13
    3.3.2 Medium-Voltage, Single-Phase Collection System Cables............ 14
4.0.0 Cable Connectors and Conductor Terminations 14
  4.1.0 Stripping and Cleaning Conductors.................................... 15
    4.1.1 Stripping Small Conductors 15
    4.1.2 Stripping Power Cables and Large Conductors........... 16
    4.1.3 Stripping Copper Communications
           Conductors and Data Cables 18
  4.2.0 Wire Connections Under 600 Volts..................................... 19
    4.2.1 Dissimilar Metal Conductors and Connections 20
    4.2.2 Heat-Shrink Insulators 21
    4.2.3 Control and Sensor Cables 21
    4.2.4 Low-Voltage Connectors and Terminals 22
  4.3.0 Installing Compression Connectors..................................... 23
    4.3.1 Crimping Tools 24
    4.3.2 Making a Crimped Connection................................... 26
Appendix A *NEC Table 310.15(B)(16)* 33
Appendix B *NEC Table 310.15(B)(17)* 34
Appendix C *NEC Table 310.15(B)(2)(a)* 35

Figures and Tables

Figure 1 Wind turbine site substation 1

Figure 2 Solid-state relay 1

Figure 3 Hydraulic pump motor 2

Figure 4 Electronics cooling fan 2

Figure 5 Cross-sectional area of a conductor 3

Figure 6 Wire size markings 3

Figure 7 Strand configurations 4

Figure 8 Confined area within a nacelle 5

Figure 9 Fiberglass nacelle 5

Figure 10 Cables powering a nacelle space heater 5

Figure 11 Racked thermoplastic-insulated conductors 6

Figure 12 Typical power cable insulation color codes 7

Figure 13 Nacelle-mounted PLC 11

Figure 14 Shielded power-limited cables 11

Figure 15 Simplex (single) optical fiber cable 12

Figure 16 Low-voltage wiring to a cabinet fan 13

Figure 17 Parallel generator stator output cables 13

Figure 18 Stator cables terminated on switchgear busbars 13

Figure 19 Cross-section of medium-voltage,
          three-phase distribution cable 14

Figure 20 Cable ready to be pulled up to the nacelle 14

Figure 21 Turbine power collection system 14

Figure 22 Interconnection of output cables from a string of turbines 15

Figure 23 Wire stripper/crimper 15

Figure 24 Wire strippers 16

Figure 25 Ratchet-type cable cutter 16

Figure 26 Heavy-duty cable stripper 17

Figure 27 Types of cable stripping 17

Figure 28 Round cable slitting and ringing tool 17

Figure 29 Chisel point left on a conductor 18

Figure 30 Cable and wire stripping tools 18

Figure 31 Proper stripping length 19

Figure 32 Crimp-on wire lugs 19

Figure 33 Various mechanical compression connectors 20

Figure 34 Installing heat-shrink insulators 21

Figure 35 Multi-pair control or sensor cable 22

Figure 36 Basic crimp connector structure 22

Figure 37 Tongue styles of crimped connectors 23

Figure 38 Mechanical strength versus electrical
          performance of a crimped connection 24

Figure 39 Hand crimpers 24

Figure 40 Leveraged crimping tool 24

Figure 41 Crimping tools used to crimp large connectors 25

Figures and Tables (*continued*)

Figure 42 Battery-operated crimping tool..26
Figure 43 Corded crimping tool...26
Figure 44 Universal crimping tool..26
Figure 45 Multiple crimps 27

Table 1 AWG and Physical Size of
         Stranded Conductors...3
Table 2 Circular Mil Wire Sizes
         and Physical Dimensions 3
Table 3 Typical Color Codes...23
Table 4 Recommended Tightening Torques for Various Bolt Sizes................27

1.0.0 INTRODUCTION

This module covers electrical wiring in four basic areas of a wind turbine site typically maintained by a wind turbine maintenance technician. The four areas include the nacelle, tower, turbine transformer, and collection system. The collection system collects power from multiple turbine transformers and, therefore, is not part of the wind turbine itself. Site substation wiring beyond the collection system is not normally the responsibility of the wind technician, due to the higher voltage and extreme short-circuit energy potential. *Figure 1* shows a typical wind farm substation.

Conductors and cables carrying voltages of less than 1,000 volts AC (VAC) are referred to as low-voltage cables and conductors. Conductors and cables carrying voltages greater than 1,000VAC and as high as 69,000VAC are referred to as medium-voltage conductors and cables. The operating voltage range on most wind turbine electrical systems ranges from 24VAC to a maximum of 34,500VAC leaving the turbine transformer.

Sensors and other control or monitoring devices, including solid-state relays (*Figure 2*), operate at voltage levels as low as 24VAC. Lighting fixtures and utilization equipment, such as hydraulic pump motors (*Figure 3*), cooling fans (*Figure 4*), and case heaters within nacelles, typically operate at voltage levels of 120, 230, 575, or 690VAC.

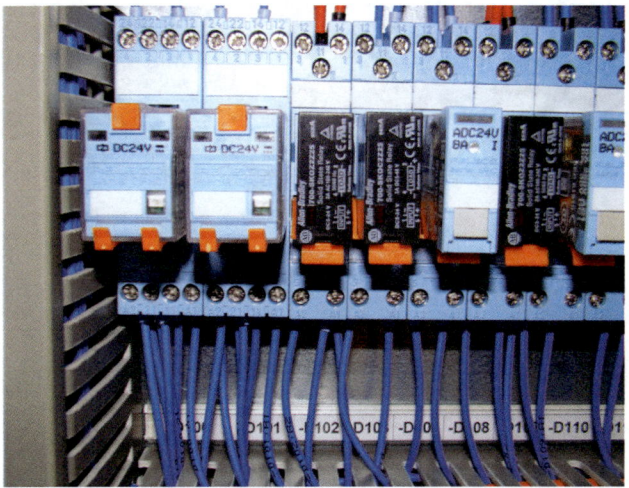

58104-11_F02.EPS

Figure 2 Solid-state relay.

58104-11_F01.EPS

Figure 1 Wind turbine site substation.

Figure 3 Hydraulic pump motor.

58104-11_F03.EPS

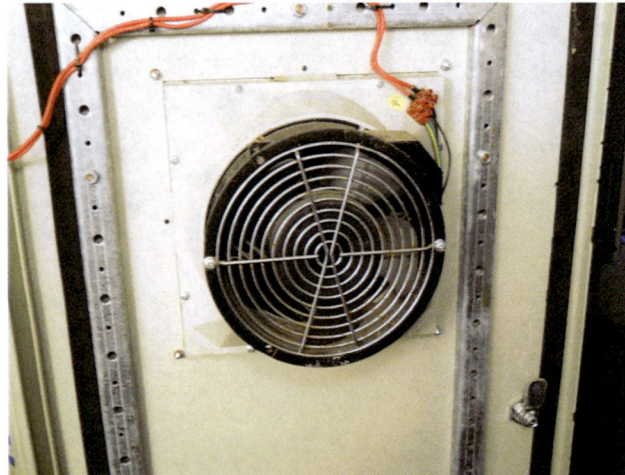

Figure 4 Electronics cooling fan.

58104-11_F04.EPS

The nominal output voltage level from a turbine generator can be 575, 690, or 1,200VAC. An output of 690VAC is the most common from existing turbine generators. This voltage places generator output wiring in the low-voltage category.

It is essential to remember that wind turbine models and brands are different in terms of construction methods, material specifications, and design. Since they are constructed across the world, no single set of standards is applied to every turbine. Although the practices outlined in this module often apply, the technician must be intimately familiar with the specific turbine being serviced and the standards that were applied to its construction. Acquiring and reading the manufacturer's specifications and guidelines for each turbine model is the only way to ensure that electrical wiring repairs and installations are completed properly. Do not proceed with electrical installations or repairs without consulting the applicable literature.

2.0.0 PROPERTIES OF CONDUCTORS AND CABLES

The term *conductor* can mean two different things in an electrical context. It can be used to describe only the current-carrying portion of a wiring or cable assembly. However, it can also be used to describe the complete assembly composed of a current-carrying inner core and its outer covering.

Conductors are identified by their physical size, conductive material, insulating material, temperature rating, and ampacity.

2.1.0 Physical Size

Wire sizes are expressed in gauge numbers. The standard system of determining wire size in the United States is the American Wire Gauge (AWG) system. Insulation or any other jacket added to a wire has no impact on the wire gauge size. The AWG size refers to the conductive current-carrying part of the conductor only. Even though two wires may share the same AWG wire size, various types of insulation and outer jackets often make their outside diameter different.

2.1.1 AWG System

The AWG system uses numbers to identify the different sizes of wire and cable as shown for stranded wire in *Table 1*. The larger the AWG number, the smaller the cross-sectional area of the wire will be, as illustrated in *Figure 5*. For example, a No. 18 AWG wire size cross section is smaller than a No. 12 AWG wire size. The AWG numbering system ranges from No. 50 down to No. 1, followed by No. 1/0, No. 2/0, No. 3/0, and No. 4/0. The latter sizes are generally spoken as one aught, two aught, and so on. They may also be written as No. 0, No. 00, No. 000, and No. 0000. The AWG size markings are printed on the outer jacket of conductors larger than No. 16 AWG (*Figure 6*).

Table 1 AWG and Physical Size of Stranded Conductors

AWG Size	Physical Size (in²)
#50 to #18	0.002
#16	0.003
#14	0.004
#12	0.006
#10	0.011
#8	0.017
#6	0.027
#4	0.042
#3	0.053
#2	0.067
#1	0.087
#0 or #1/0	0.109
#00 or #2/0	0.137
#000 or #3/0	0.173
#0000 or #4/0	0.219

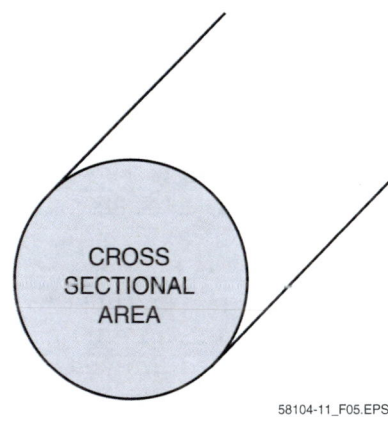

58104-11_F05.EPS

Figure 5 Cross-sectional area of a conductor.

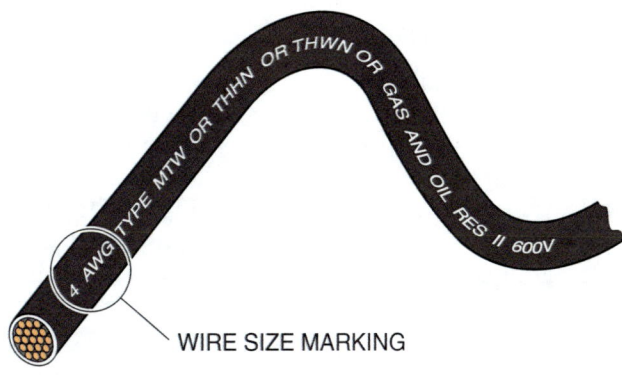

58104-11_F06.EPS

Figure 6 Wire size markings.

Think About It

Wire Size

Why is wire size a critical factor in a wiring system? Other than load, what other factors may dictate wire size? What can happen when a wire is not properly sized for the load?

2.1.2 Circular Mils

A circular mil is a circle that has a diameter of 1 mil. A mil is 0.001 inch, or $\frac{1}{1000}$ of an inch. When a wire size is 250 kcmil, the cross-sectional area of the current-carrying portion of the wire is the same as 250,000 circles having a diameter of 0.001 inch. This may seem to be a rather clumsy way of sizing wire at first. However, the alternative would be to size the wire as a function of its cross-sectional area expressed in square inches.

According to *NEC Chapter 9, Table 8*, the overall cross-sectional area of a 250-kcmil conductor is 0.260 square inch. However, if a conductor is to be sized by cross-sectional area, it is much easier to express the wire size in circular mils (or thousandths of circular mils) than in square inches. *Table 2* lists circular mil wire sizes and their equivalent physical dimensions.

Table 2 Circular Mil Wire Sizes and Physical Dimensions

kcmil Wire Size	Physical Size (in²)
250	0.575
300	0.630
350	0.681
400	0.728
500	0.813
600	0.893
700	0.964
750	0.998
800	1.030
900	1.094
1,000	1.152
1,250	1.289
1,500	1.412
1,750	1.526
2,000	1.632

2.1.3 Stranding

NEC Chapter 9, Table 8 contains descriptive information on wire sizes and other conductor properties. The number of strands for a given wire size is listed, along with other physical and functional data. A strand quantity of one indicates that it is actually a single solid wire. All insulated wire sizes larger than No. 6 AWG are stranded. Wires sizes No. 6 AWG or smaller are available with either solid or stranded conductive material. Solid wire larger than No. 6 AWG is available by special order if desired. However, *NEC Section 310.106(C)* only permits the use of stranded wire in a raceway for wire sizes No. 8 AWG and larger. As a result, the need for solid wire larger than No. 6 AWG is rare.

Per this section of the *NEC®*, stranded wire contains the following number of strands for each wire size:

- Wire sizes No. 18 up to No. 2 have 7 strands
- Wire sizes No. 1 up to No. 4/0 contain 19 strands
- Wire sizes between 250 kcmil and 500 kcmil have 37 strands
- Wire sizes between 600 kcmil and 100 kcmil have 61 strands
- Wire sizes 1,250 kcmil and 1,500 kcmil contain 91 strands
- Wire sizes 1,750 kcmil and 2,000 kcmil have 127 strands

The purpose of stranding is to increase the flexibility of the wire. Terminating solid wire larger than No. 8 AWG in pull boxes, disconnect switches, and panels would be very difficult and might result in damage to both equipment and the wire insulation. Pulling solid wire of this size through conduit and around bends would also be challenging, if not impossible.

The reason for specifying the precise strand count for these wire sizes is to provide a flexible, nearly round conductor. In order for a conductor to be flexible, the individual strands must not be too large. *Figure 7* shows how these conductors are configured.

2.2.0 Conductor Material

The most common and preferred conductor material is copper. Copper is used because of its excellent conductivity (low resistance), ease of use, and overall value. The value of a material as an electrical conductor is determined by several factors, including conductivity, cost, availability, and workability.

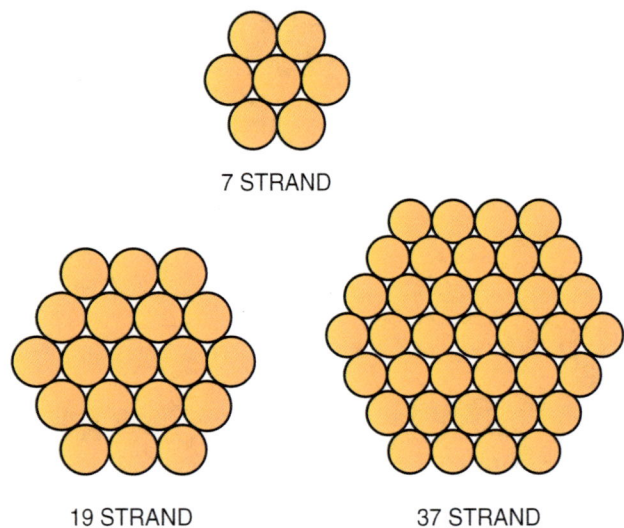

7 STRAND

19 STRAND   37 STRAND

58104-11_F07.EPS

Figure 7 Strand configurations.

2.2.1 Conductivity

Conductivity is a word that describes the ease or difficulty of travel presented to the flow of electric current by a conductor. If a conductor has a low resistance to current flow, it has a high conductivity. Silver and gold are both very good conductors since they have very low resistance and high conductivity. Copper has high conductivity, although not as high as silver or gold. The significant difference in cost makes copper the overwhelming choice over silver and gold.

Aluminum is also a conductor with good conductivity and is available at a low cost. However, aluminum is seldom used in nacelle wiring because of its undesirable characteristics. Aluminum has a tendency to oxidize, expands and contracts significantly as temperature changes, and is relatively fragile when subjected to stress such as repeated bending. Aluminum conductors are often used in the medium-voltage distribution systems of wind turbine sites in spite of these characteristics due to the extreme cost of copper for large wire sizes.

2.2.2 Workability

Conductor workability is a priority when selecting conductors and cables for wind turbine and tower installations due to the limited space within nacelles and towers. Cable and cable fittings are often the choice for wiring within the nacelle and tower. The alternative of installing individual conductors in raceways would require complex raceway routing and multiple bends in the cramped nacelle area, as shown in *Figure 8*.

Figure 8 Confined area within a nacelle.

58104-11_F08.EPS

In addition, because most nacelles (*Figure 9*) are constructed of fiberglass or other non-conductive material, securing raceway systems to these structures is difficult. The flexibility of cable can be a significant advantage in areas where some natural structural movement is expected.

2.3.0 Insulating Material

Conductor and cable insulation serves to protect the conductor inside as it shields it from unintentional contact by humans, grounded objects, or other conductors. Low-voltage conductor and cable insulation is much less complex than that applied to medium-voltage conductors used in a wind turbine system.

Most accessible cables within a nacelle carry voltages less than or equal to 690VAC, usually in the range of 120 to 240VAC, to operate pumps, lighting, heaters, and sensors (*Figure 10*).

Wiring within a nacelle is normally protected from wet weather and, therefore, an insulation that is rated for wet locations is not required. However, the heat generated in an enclosed nacelle, coupled with high ambient temperatures, does require that conductor and cable insulation be rated to handle potentially extreme temperature conditions. High temperatures also must be considered during wire sizing.

2.3.1 Thermoplastic- and Thermoset-Insulated Conductors

The two most common classes of conductor and cable insulation are thermoplastic insulation and thermoset insulation.

Thermoplastic insulation initially becomes hard but remains somewhat pliable when first compounded. If it is exposed to high tempera-

58104-11_F09.EPS

Figure 9 Fiberglass nacelle.

58104-11_F10.EPS

Figure 10 Cables powering a nacelle space heater.

tures, it tends to soften or even melt like most plastics. Thermoplastics are one of the most common insulating materials since they are very effective and inexpensive. Some types of thermoplastic insulation have an outer jacket of nylon to reduce friction when pulling the wire into raceways and conduits. *Figure 11* shows a rack of thermoplastic-insulated conductors prepared for installation.

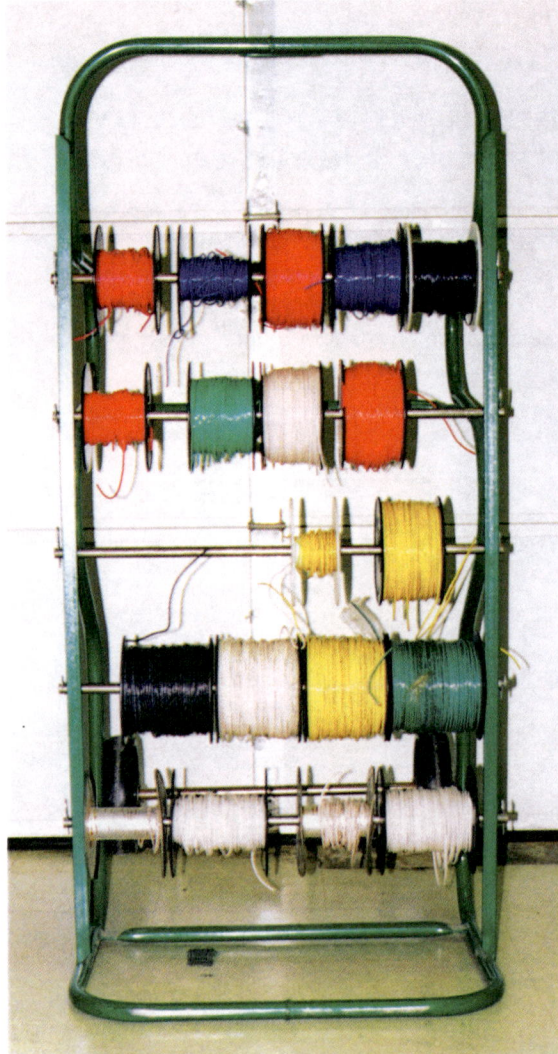

58104-11_F11.EPS

Figure 11 Racked thermoplastic-insulated conductors.

Examples of common thermoplastic-insulation materials include:

- *THHN* – A flame-retardant, heat-resistant thermoplastic with an outer covering of nylon and an overall insulation rating of 90°C (194°F) when installed in dry and damp locations.
- *THHW* – A flame-retardant, moisture-, and heat-resistant thermoplastic with no outer covering and an overall insulation rating of 75°C (167°F) when installed in wet locations, or 90°C (194°F) for dry locations.
- *THW* – A flame-retardant, moisture-, and heat-resistant thermoplastic with no outer covering and an overall insulation rating of 75°C (167°F) when installed in dry or wet locations, and 90°C (194°F) when installed in special applications within electric discharge lighting equipment.

- *THWN* – A flame-retardant, moisture-resistant thermoplastic with an outer covering of nylon and an overall insulation rating of 75°C (167°F) whether installed in dry or wet locations.
- *TW* – A flame-retardant, moisture-resistant thermoplastic with no outer covering and an overall insulation rating of 60°C (140°F) whether installed in dry or wet locations.

Thermoset insulation is rubbery when compounded and maintains its shape and form when heated, unlike thermoplastic materials. Conductors and cables coated with thermoset insulating materials are more commonly used in high-temperature environments as a result. Some common examples of thermoset-insulated products include:

- *RHH* – A flame-retardant, moisture-resistant conductor with a non-metallic covering and an overall insulation rating of 90°C (194°F) when installed in dry or damp locations.
- *RHW* – A flame-retardant, moisture-resistant conductor with a non-metallic covering and an overall insulation rating of 75°C (167°F) when installed in dry or damp locations.
- *XHH* – A flame-retardant conductor with no additional outer covering and an overall insulation rating of 90°C (194°F) when installed in dry or damp locations.
- *XHHW* – A flame-retardant, moisture-resistant thermoset with no outer covering and an overall insulation rating of 90°C (194°F) when installed in dry or damp locations; rated at 75°C (167°F) when installed in wet locations.

Note that all of the wire types listed that can be used in wet outdoor applications end with the letter W.

2.3.2 High-Temperature Insulated Conductors

Since conductors may be installed in above-normal temperature environments, insulations with higher temperature ratings must be available. Although common thermoplastics do not perform well at high temperatures, several special compounds are used extensively in high-temperature environments. There are several special thermoplastic-insulated types available for high-temperature environments, including:

- *FEP or FEPB* – A thermoplastic-insulated (fluorinated ethylene propylene) conductor with a rating of 200°C (392°F) when installed in dry locations, or 90°C (194°F) when installed in wet locations. FEPB incorporates a glass braid in the insulation that is not found in FEP.

- *PFA* – A thermoplastic-insulated (perfluoro-alkoxy) conductor with an overall insulation rating of 90°C (194°F) when installed in damp locations, or 200°C (392°F) when installed in dry locations.
- *PFAH* – Much like PFA, but with a nickel or nickel-coated copper conductor and an overall insulation rating of 250°C (482°F) when installed in dry locations. PFAH is specifically for power leads within an apparatus or within raceways connected to an apparatus.
- *SA* – A silicone rubber-insulated conductor with a glass (or other suitable material) braid and an overall insulation rating of 90°C (194°F) (No. 14 through No. 4/0 AWG) when installed in dry or damp locations. For 250 kcmil through 2,000 kcmil conductors, where design conditions require maximum conductor operating temperatures, it is rated for 200°C (392°F) in dry locations.
- *TFE* – A thermoplastic-insulated (polytetrafluoroethylene) conductor with an overall insulation rating of 250°C (392°F) when installed in dry locations and specifically for leads within apparatus or within raceways connected to apparatus. Like PFA, the conductor itself is nickel or nickel-coated copper.
- *Z* – A thermoplastic-insulated (ethylene tetra-fluoroethylene) conductor with an insulation rating of 90°C (194°F) for dry locations, or 150°C (302°F) for special applications in dry locations.
- *ZW* – Same as Z, but usable in wet locations, with a temperature rating of 75°C (167°F).

2.3.3 Insulation Color-Coding

An insulation color code is used to identify some wires by color. This makes it easier to install and properly terminate the wires. *Figure 12* is an example of typical color-coding in multiple conductor cables.

The grounding conductor may be bare, green, or green with a yellow stripe. The ungrounded conductors may be any color with the exception of white, gray, or green. However, it is a wise practice to follow a standardized color-coding system for all conductors installed at a given site or project. Wiring that is internal to manufactured equipment is often one color, with the wires numbered to help identify them during troubleshooting and repair.

2.4.0 Conductor Temperature Rating

The temperature ratings of conductors listed in this module are considered as the maximum operating temperature at which the conductor can safely carry its maximum allowed load current under the design conditions. It is essential that the conductor's temperature rating is not confused with the design ambient temperature. The ambient temperature is related to the air temperature of the area, while the conductor's temperature rating refers to its own operating temperature while carrying an electrical load. A wire's resistance causes it to heat up to a higher temperature than its surroundings.

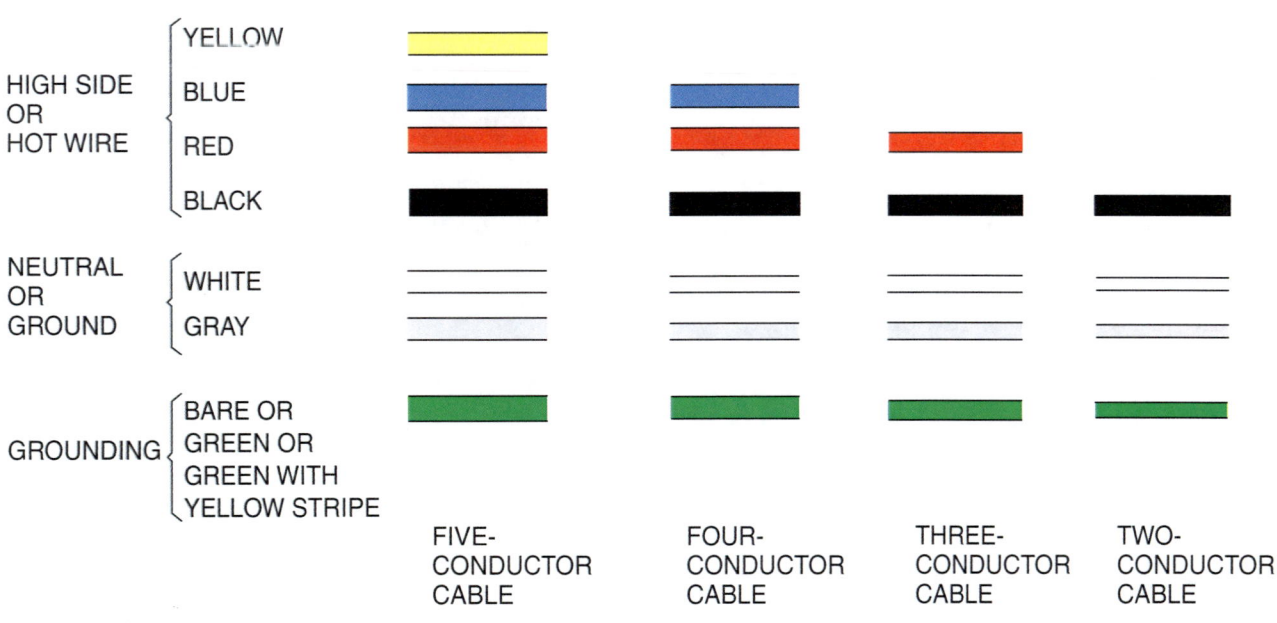

Figure 12 Typical power cable insulation color codes.

Since current flowing through a conductor creates heat, the level of current flowing, the ambient temperature in which the conductor and load are installed, and the number of conductors routed together factor into the overall operating temperature of the conductor. For example, ampacity limitations exist for bundled conductors and cables with four or more conductors in use. For these situations, at less than 2,000V, with conductors or a cable bundled together tightly for a distance greater than 600mm (24 inches), you must consult *NEC Table 310.15(B)(3)(a)* for an ampacity adjustment factor. Other derating tables exist for higher voltages. The speed at which heat can be dissipated may also be a factor, which is generally determined by the type of insulation used.

There is no perfect method to select conductors based on their insulation temperature rating. Many tables in *NEC Article 310* are devoted to determining the ampacity of a conductor based on their temperature ratings. Those tables include *NEC Tables 310.15(B)(16), 310.15(B)(17), 310.15(B)(18), 310.15(B)(19), 310.15(B)(20), and 310.15(B)(21)*. However, few electricians seriously consider the ambient temperature when selecting conductors unless it is to be used in an extremely hot environment, such as near a furnace or kiln. With typical temperature ratings of 90°C (194°F), few applications truly require careful attention to temperature. Conservative wire sizing by ampacity also helps avoid problems related to temperature.

2.5.0 Ampacity Rating

By *NEC*® definition, a conductor's ampacity rating is the maximum current in amperes that a conductor can carry continuously under the conditions of use and without exceeding its temperature rating. Ampacity is the primary factor in selecting a conductor.

NEC Table 310.15(B)(16) is the most frequently applied *NEC*® table when selecting conductors for raceway installation, cabling, or earth burial. It lists the ampacity rating of common thermoplastic- and thermoset-insulated conductors according to their AWG size and temperature ratings.

The ampacity values in this table only apply if there are no more than three current-carrying conductors contained in a raceway, cable, or earth-buried installation. Further, the ambient

temperature surrounding the installation must not exceed 30°C (86°F). For ambient temperatures greater than 30°C, refer to *NEC Table 310.15(B)(2)(a)* for a correction factor that is applied to the value shown in *Table 310.15(B)(16)*.

Conductor ampacity ratings are higher for single-insulated conductors installed in free air than those contained in a raceway, cable, or buried in the earth. This is because the conductors remain cooler in open air. *NEC Table 310.15(B)(17)* lists the ampacity for thermoplastic- or thermoset-insulated single conductors installed in free-air based on an ambient temperature of 30°C (86°F).

NEC Table 310.15(B)(18) gives the ampacity for specialty high-temperature-rated insulated conductors including Type Z, FEP, FEPB, PFA, and SA with no more than three current-carrying conductors inside the raceway, conduit, or cable. In this table, the ambient temperature allowed is 40°C (104°F). *NEC Table 310.15(B)(19)* lists the ampacity ratings for these same conductors installed in free air.

Ampacity ratings for larger insulated conductors supported by a messenger wire, including overhead electrical service installations, can be found using *NEC Table 310.15(B)(20)*. The AWG range included in this table is No. 8 AWG – 1,000 kcmil.

Finally, *NEC Table 310.15(B)(21)* is typically referenced by electrical utility companies to determine the ampacity for bare or covered conductors installed in free air, based on an ambient temperature of 40°C with a wind velocity factor included.

It is important to note that the primary table for determining the ampacity of wire in a given situation provides the related table for ambient temperature correction in its footnotes. Whenever a table is consulted for wire sizing, always examine the footnotes to see if any are related to your specific application.

2.5.1 Determining Ampacity

The ampacity value of a given wire size and type can be determined using the *NEC*® tables and other information related to the installation. The tables can be used to determine the ampacity for a selected wire size, or they can help you determine which wire size is sufficient to carry a known load. Both approaches are best demonstrated through the following example problems. Copies of *NEC Tables 310.15 (B)(16) and (17)* have been provided in *Appendix A* and *B* of this module.

Problem 1:

Determine the ampacity of a No. 12 AWG THW copper conductor when installed in a raceway with two other current-carrying conductors in an ambient temperature of 86°F.

Solution:

1. Reference *NEC Table 310.15(B)(16)*, which covers wiring installed in raceways.

2. Find the column in which THW copper conductors are located, which is the 75°C column (second column from the left under the Copper heading). The 75°C value represents the conductor's temperature rating with these insulation types. Ensure that you do not select the column related to aluminum or copper-clad aluminum wire on the right side of the table.

3. Locate the row that contains the No. 12 AWG wire size in the far left column of the table.

4. Go right across the row to the 75°C column.

5. The value found at the intersection of the No. 12 AWG row and the 75°C copper conductor column contains the ampacity for a single THW copper No. 12 AWG wire when installed in a raceway with two other current-carrying conductors in an ambient temperature of 86°F. Under the specified conditions, No. 12 AWG THW wire has an ampacity of 25 amps.

Problem 2:

Determine what AWG wire size you will need to support a load of 48 amperes when the conductors are installed in free air. The design ambient temperature is 30°C and the wire type is specified as copper THHN.

Solution:

1. Reference *NEC Table 310.15(B)(17)*. This table, per its description, covers wiring in free air.

2. Find the column in which THHN copper conductors are located, which is the 90°C column (third column from the left under the COPPER heading).

3. Read down that column until you reach an ampacity value that is equal to or greater than 48 amperes. The appropriate value found is 55 amps.

4. Go left across the row to the AWG column on the far left.

5. The matching wire size contained in this column, No. 10 AWG, is the minimum copper THHN wire size needed to carry the 48-amp load when installed in free air.

2.5.2 Conductor Ampacity Derating and Correction

The ampacity ratings provided by the *NEC*® tables only apply when the specific number of conductors and ambient temperature conditions are met. If either or both conditions are not met, the actual conductor ampacity must be recalculated by applying factors shown in additional tables within *NEC Article 310*.

NEC Article 310.15 provides direction for determining ampacities using *NEC Tables 310.15(B)(16) and 310.15(B)(17)*. A review of this article provides the designer with guidance when the conditions of the job do not precisely meet the conditions of the tables. The article points out that *NEC Table 310.15(B)(3)(a)* lists adjustment factors that are applied to ampacity values found in *NEC Tables 310.15(B)(16)* and *310.15(B)(18)* if the maximum number of three current-carrying conductors is exceeded. The number of conductors is shown in the left column of the table, and the factor to be applied is shown in the right column. Note that this factor is a percentage. To apply the adjustment factor, first determine the ampacity of the conductor using either *Table 310.15(B)(16)* or *310.15(B)(18)*. If the number of current-carrying conductors is 4 to 6, for example, multiply the ampacity of the selected wire by 80 percent, or its decimal equivalent of 0.80.

In Problem 1, it was determined that the ampacity of the No. 12 AWG wire size was 25 amps. If the actual number of conductors to be installed together in the raceway is eight, a correction factor of 70 percent must be applied. The actual ampacity for the wire would then be calculated as follows:

25 amps × 0.70 correction factor = 17.5 amps

There are also correction factors to adjust ampacity values for temperatures outside of normal table conditions. *NEC Table 310.15(B)(2)(a)* contains the correction factors that must be applied to conductor ampacity values found in *NEC Tables 310.15(B)(16)* or *310.15(B)(17)* for higher or lower temperatures. These two tables are based on a maximum ambient temperature of 30°C (86°F). A copy of *NEC Table 310.15(B)(2)(a)* is provided in *Appendix C*.

NEC Table 310.15(B)(2)(b) contains the correction factors that must be applied to conductor ampacity values selected from the ampacity *NEC Tables 310.15(B)(18)*, *310.15(B)(19)*, *310.15(B)(20)*, *or 310.15(B)(21)*, which are the four tables based on a maximum ambient temperature of 40°C (104°F).

In order to select the appropriate correction factor to apply to the base ampacity, you must know the expected ambient temperature of the wire location. In addition, the temperature rating of the selected wire must be known. This is found at the top of the column headings in the ampacity *NEC Tables 310.15(B)(16) and 310.15(B)(17)*.

The left column of *NEC Tables 310.15(B)(2)(a) and 310.15(B)(2)(b)* shows the temperature range in °C, while the right column shows the temperature range in °F. Find the correct temperature range for your application, and then move across the row toward the center of the table to find the required correction factor. Choose the factor from one of the three columns that represents the temperature rating of the wire. The value shown is the factor that must be applied to the base ampacity.

Example 1:

Use *Table 310.15(B)(16)* to determine the ampacity of one No. 8 AWG THHN copper conductor installed in a conduit with two additional current-carrying conductors. The design ambient operating temperature is 60°C (140°F). The table shows that the ampacity of the wire, at an ambient operating temperature of 30°C, is 55 amps.

Using *NEC Table 310.15(B)(2)(a)*, determine the correction factor to apply for the 60°C design condition. For the THHN wire, use the 90°C column; this corresponds to the 90°C operating temperature rating (as opposed to the ambient temperature rating) shown at the top of the column in *NEC Table 310.15(B)(16)*. Using the table, find the ambient temperature range that includes

60°C. Then read across to the 90°C column to find the value. The correction factor shown here is 0.71.

The ampacity is then corrected by applying the 0.71 correction factor to the ampacity value of 55 amps as shown here:

55 amps × 0.71 correction factor = 39.05 amps

As demonstrated here, the higher ambient temperature caused a 30 percent reduction in the ampacity rating of the wire. If the correction factor is not applied and the wire is installed under a load exceeding 39 amps, it could overheat and fail.

Note that the ambient temperature correction factors are not just for temperatures higher than those stated in the base ampacity tables. In some cases, the ambient operating temperature may be cooler than 30°C (86°F). In these cases, the correction factor actually works in favor of the designer. The correction factors in the tables that are greater than 1.00 represent an increase to the wire's ampacity value, rather than a decrease. As a result, a smaller wire size may be suitable for the application.

Some situations require a correction factor to be applied for more than one abnormal condition. For example, a project may require 12 conductors together in a conduit, installed in an area where the expected ambient temperature is 60°C (140°F). In these cases, both correction factors must be applied. You may simply multiply the basic ampacity value by one correction factor, then multiply the result by the second factor. The order in which they are applied has no effect on the final result.

3.0.0 APPLICATIONS OF CONDUCTORS AND CABLES

The installation application for a conductor or cable is a major factor in determining what type of conductor to use. Conductor or cable applications at a wind turbine site include communications and data transfer, low-voltage (less than 1,000VAC) power, and medium-voltage (1,000 to 69,000VAC) power. Medium-voltage cables that carry power from the turbine transformer all the way to the substation are collectively known as **distribution conductors** or cables.

3.1.0 Data and Communications Cables

Most wind turbines contain two **programmable logic controllers (PLCs)**. One is typically located in the nacelle while the other is located at the base of the tower. The nacelle-mounted PLC is equipped with input and output (I/O) modules that receive sensor input from variable-output devices within the nacelle, such as gear-case temperature sensors, wind speed sensors, and shaft rotation speed sensors. A typical PLC with I/O modules is shown in *Figure 13*. Most wiring material used for control and sensor signals at the I/O modules of the PLC is referred to as power-limited cable. Per *NEC Article 725*, a power-limited circuit is one where the voltage is less than 30V and has a current flow of less than 1,000 volt-amps. Some sensors transport their signals to the I/O modules through fiber-optic cable instead.

The control power inputs transmitted on the copper power-limited cables to the nacelle PLC are converted into an optical signal, and then transmitted by bundled optical fiber cable down to the second PLC located in the base of the tower. This PLC is connected to a local area network (LAN) by means of a standard Ethernet cable.

The LAN communicates with a remote supervisory control and data acquisition (SCADA) system, sometimes located thousands of miles away. The SCADA system uses the data to adjust turbine operation and position, or to alert on-site operations personnel when local attention is required.

A power-limited cable may contain a single pair of conductors routed to an individual device such as a sensor, or it may contain multiple pairs of conductors serving a number of devices. Each pair of conductors is often surrounded, or shielded, by a thin film made from aluminum foil

bonded to a plastic film. This shielding material prevents undesirable signals or electrical noise from penetrating the enclosed signal wires, causing inaccurate data transmission from the sensing device to the I/O module. One end of a grounding wire, also known as a drain, is attached to the shield within the cable. The other end must be connected to earth ground to complete the shielding process. Any undesirable signal power that is encountered simply goes to ground through the drain. *Figure 14* shows both a single twisted pair

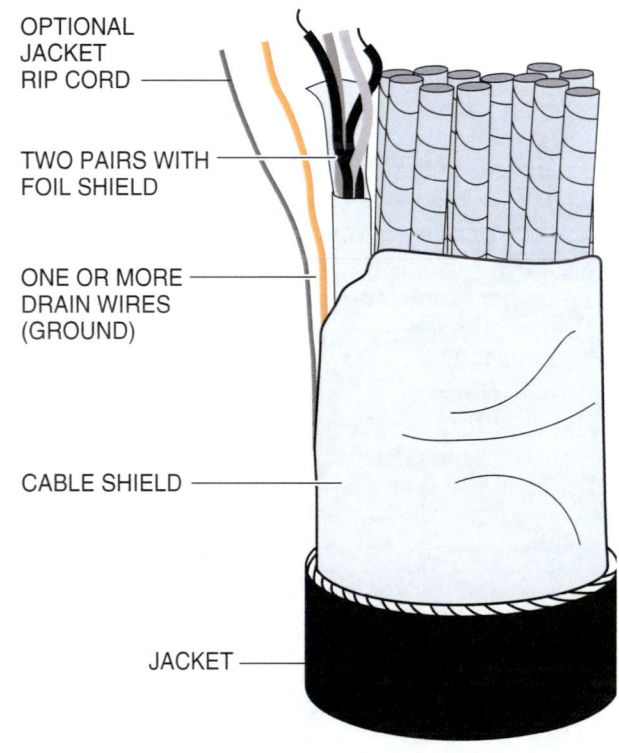

58104-11_F13.EPS

Figure 13 Nacelle-mounted PLC.

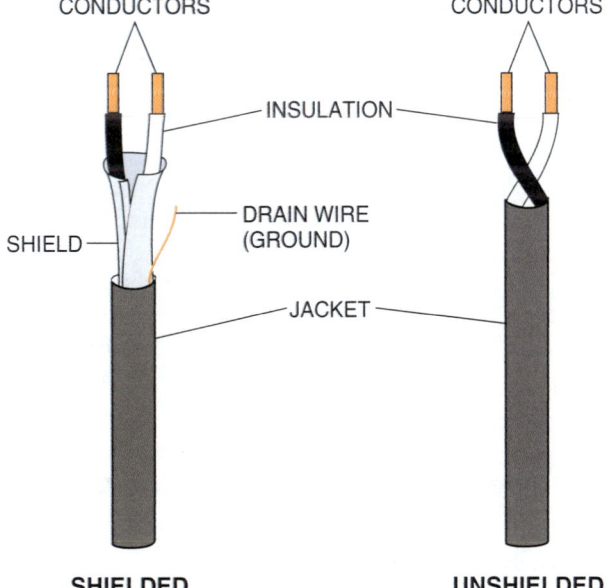

Figure 14 Shielded power-limited cables.

58104-11_F14.EPS

cable and a shielded, multi-conductor power-limited cable.

Optical fiber cable is used to conduct a light signal instead of an electrical current. *Figure 15* shows a typical simplex (single) optical fiber cable and its optical fiber light conductor and sheathing. The components shown are described as follows:

- *Jacket* – An outer covering made from various materials, from flexible metal to soft PVC. The jacket color usually identifies one of two types of fiber used inside. Yellow typically indicates single-mode, while orange often signifies multi-mode fiber. However, other color codes are used.
- *Strength member* – Non-metallic fiber strands under the jacket that provide strength and flexibility to the cable. Strength members are usually made of aramid (such as Kevlar®) yarn to which certain types of connectors can be attached. They can also be used to pull the cable into position at installation.
- *Buffer layer* – An intermediate layer that is usually 900 microns in diameter. A micron is one millionth of a meter or 1/25,000 of one inch. It is also known a micrometer (μm). This layer can be either tight or loose. For indoor cable, the layer is a soft plastic that is tight against the layer below it. This is known as a tight buffer. For outdoor cable, the layer consists of a gel-filled plastic tube that protects the optical fiber from damage due to impact, temperature extremes, or water. This type of buffer layer is called a loose-tube buffer. Both types of buffers contribute to the strength and flexibility of the cable.
- *Acrylate coating* – The acrylate coating, or acrylate buffer, provides strength and flexibility to the optical fiber and improves the fiber's handling characteristics.
- *Cladding* – The cladding and core make up the optical fiber that transmits the light signal. The cladding surrounds the core. It is made of glass or plastic that is purer than the core glass. The difference in purity between the core and

cladding glass (or plastic) creates a reflective boundary where the two meet. This reflective surface prevents the light signals from escaping the core.

- *Core* – The core, which is made of very pure glass or plastic, carries the modulated light signals through the fiber. Light travels through the core as the result of total internal reflection. As the light travels through the core, it is reflected off the boundary between the core and cladding at a shallow angle, allowing the light to continue through the core. Light travels in the core on a path called a mode. The number of modes depends on the core diameter. Optical fiber is either single-mode or multi-mode. In any installation in which optical fiber cables are joined together, it is essential that the cables have the same size core and cladding. Joining cables with different cores or cladding results in unacceptable signal losses and poor performance.

3.2.0 Low-Voltage Power Conductors or Cables

Low-voltage power conductors and cables carry voltages less than 1,000V and are commonly thermoplastic- or thermoset-insulated. These conductors supply power to lighting fixtures, hydraulic-pump motors, relay coils, ventilation fans, case heaters, sensing devices, and other equipment within a nacelle or tower, as shown in *Figure 16*. They may be bundled in a cable assembly or installed in a raceway or conduit.

3.2.1 Generator Stator (Output) Conductors

The power conductors or cables connected to the stator windings of the wind turbine generator carry the output current to the main switchgear within the nacelle. *Figure 17* shows stator output cables connected to the junction box at the generator. In this particular turbine, there are six stator

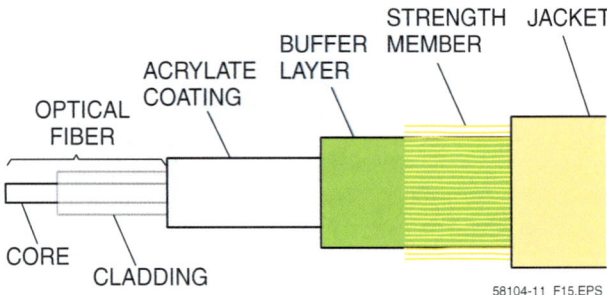

Figure 15 Simplex (single) optical fiber cable.

58104-11_F15.EPS

Figure 16 Low-voltage wiring to a cabinet fan.

Figure 17 Parallel generator stator output cables.

output cables. Two cables are dedicated to each phase of the three-phase stator output, meaning that each phase is paralleled. From the stator junction box, the parallel-phase cables are routed to the generator switchgear cabinet. The wires are terminated on busbars, with two conductors per phase as shown in *Figure 18*.

The generator output voltage level is dependent on the manufacturing design of the turbine. Some manufacturers generate at voltage levels of 575V, while others generate 1,200V. However, the majority of wind turbine systems generate power at 690V.

Stator output conductor ampacity ratings are based on the amount of current available from the generator, which concurs with the kilowatt (kW) or megawatt (MW) capacity rating of the turbine. Most new commercial and utility-scale wind turbines located on wind sites generate in a megawatt capacity range.

Figure 18 Stator cables terminated on switchgear busbars.

3.3.0 Medium-Voltage Cables

Medium-voltage cables, carrying power at voltages in excess of 1,000V, are specially designed cables that require a more complex system of insulation. This is required to protect the conductors inside and to prevent unintentional insulation penetration or current leakage due to cracks or insulation fatigue.

3.3.1 Three-Phase Medium-Voltage Cables

Most turbine manufacturers locate the turbine transformer on a slab at the foot of the tower. This is done primarily because of the excessive weight of the transformer and the lack of space in the nacelle. Power output is then routed from the generator in the nacelle down to the transformer. Since most generators operate at less than 1,000V, the cabling down to the transformer would be classified as low-voltage cabling.

Figure 19 shows a cross-section of a medium-voltage three-phase cable used in a turbine designed with the power transformer in the nacelle. This cable will conduct the 34,500-volt output from a transformer to switchgear located in the base of the tower. *Figure 20* shows a reel of this same cable being prepared for installation up to the transformer in the nacelle.

On Site

Turbine Transformer

The function of the turbine transformer is to increase the turbine output voltage for distribution to the site substation. Increasing the voltage significantly reduces the required wire size.

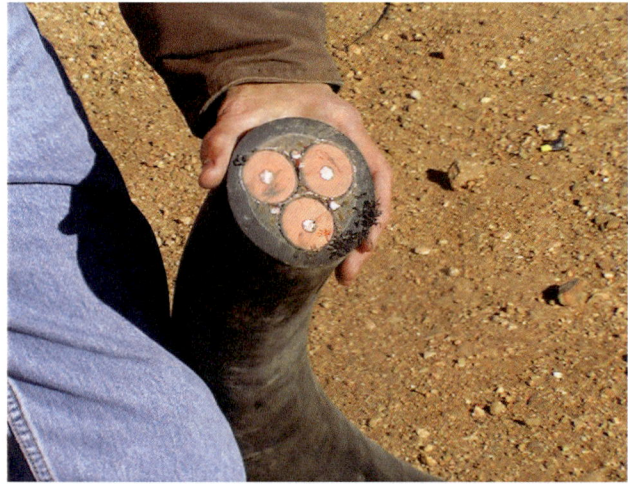
Figure 19 Cross-section of medium-voltage, three-phase distribution cable.

Figure 20 Cable ready to be pulled up to the nacelle.

3.3.2 Medium-Voltage, Single-Phase Collection System Cables

A universal wind site design involves connecting several turbine power outputs together, forming a circuit or string. The outputs of individual turbine transformers within a string are connected together at the base of each turbine in a daisy-chain fashion. Single-phase, medium-voltage conductors are the appropriate choice, since the transformer outputs are in the middle of the medium-voltage range. The interconnection of all turbine outputs in a circuit or string is referred to as the collection system, as shown in *Figure 21*.

The collection system carries the combined turbine outputs to the site substation where the cables terminate at the load side of the circuit switchgear. Each circuit or string of turbines is designed in a similar fashion. The site substation

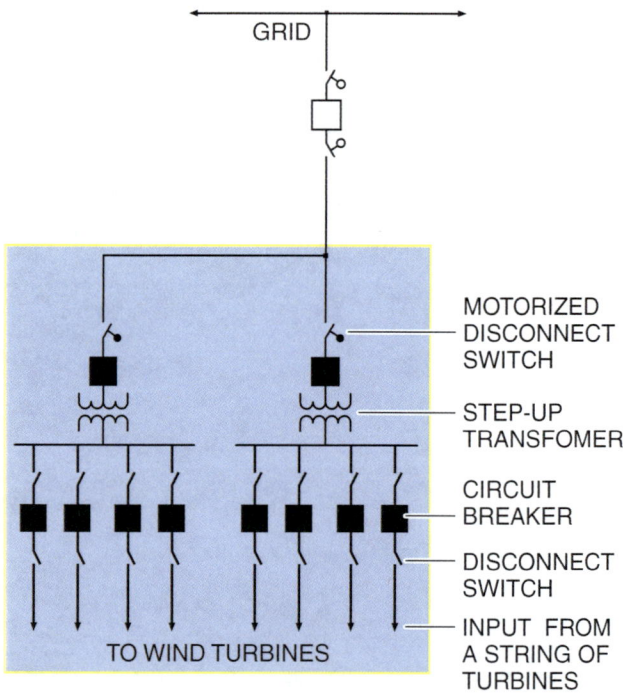
Figure 21 Turbine power collection system.

transforms the collected power to an even higher voltage, matching that of the commercial power grid system nearby. The commercial grid system provides the means for generated power to be distributed to consumers.

The interconnection point of one turbine to an adjacent turbine is dependent on the design of the turbine. For those turbines equipped with a transformer in the nacelle, the interconnection between turbine outputs is made at an isolating switch in the base of the tower (*Figure 22*). In turbine systems that are designed with both the transformer and isolating switch located at the tower base, the collection system connections are made inside the transformer housing.

4.0.0 CABLE CONNECTORS AND CONDUCTOR TERMINATIONS

Wind turbine technicians must be familiar with wire connectors and splicing, as the related materials and skills are both required to complete an electrical installation on a wind turbine site.

A poorly made electrical connection will always be a source of trouble. Joints will overheat under load, operate intermittently, and eventually fail or create a fire hazard. Good connections must be both mechanically and electrically secure, and be insulated as well as, or better than, the existing conductors. A properly made splice or connection should last as long as the insulation on the wire itself.

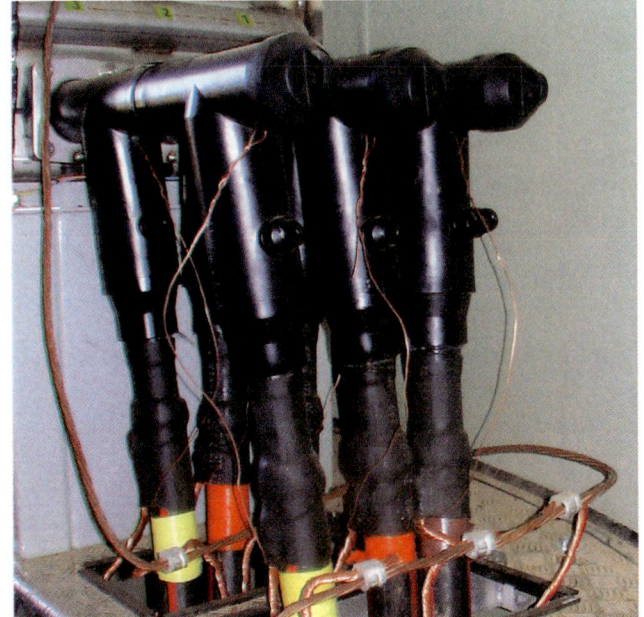

Figure 22 Interconnection of output cables from a string of turbines.

Every splice made in a wire represents a point of possible failure. Wire splicing should never be used as a solution to shorten the distance of wire or cable pulls. In some areas and applications, the use of splices may be prohibited.

There are many different types of electrical joints. The selection of the proper type for a given application often depends on how and where the splice or connection is to be used. Electrical joints are often made with solderless pressure connectors or lugs to save time.

It is important to note that appropriate training, and possibly specific certification, is required to make splices and terminations in medium-voltage applications. Before proceeding with this type of work, ensure that you are aware of the requirements and have the proper credentials.

4.1.0 Stripping and Cleaning Conductors

Before any connection or splice can be made, the ends of the conductors must be properly cleaned and stripped. To ensure a low-resistance connection and avoid contamination, clean the areas of the cable where it is to be cut and stripped. Remove any dirt, oil, grease, or water to avoid contaminating the exposed conductor.

Stripping is the removal of insulation from the end of the conductor or at the location of the splice. Conductors should only be stripped with the appropriate stripping tool. This helps prevent cuts and nicks in the wire, which can reduce the conductor area and weaken the conductor.

Conductor Terminations and Splices

Poor electrical connections are responsible for a large percentage of equipment burnouts and fires. Many of these failures are the direct result of improper terminating hardware, unsuitable splicing devices, or poor electrical workmanship.

Poorly stripped conductors can result in nicks, scrapes, or burnishes. Any of these can lead to a stress concentration in the damaged area. Heat, rapid temperature changes, mechanical vibration, and oscillatory motion in the wiring can aggravate the damage. Lost strands are a problem in splice or crimp-type terminals. Using the wrong stripping tool can result in the accidental removal of strands, especially in smaller wire sizes. Leaving any strands exposed at a termination or splice can also be a safety hazard.

Faulty stripping can pierce, scuff, or split the insulation. This can cause changes in dielectric strength and lower the conductor's resistance to moisture and abrasion. Insulation particles or pieces can be trapped in solder and crimp joints. A variety of factors determine how precisely a conductor can be stripped, including the wire size and type of insulation.

The conductive cores of stranded and solid conductors of the same AWG size do not have the same diameter. This is a very important consideration in selecting the proper blades for strippers. The specific stripping tool used depends on both the size and type of wire being stripped.

4.1.1 Stripping Small Conductors

There are many kinds of wire strippers available. *Figure 23* shows a common wire stripper for small conductors. It can be used to cut and strip wire, and crimp terminations or splices from wire size No. 22 AWG through No. 10 AWG. To use this tool, insert the conductor into the proper size knife groove, then squeeze the tool handles. The

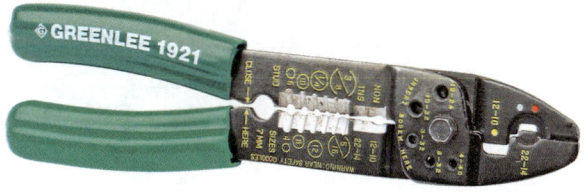

Figure 23 Wire stripper/crimper.

tool cuts through the conductor insulation without cutting the wire beneath it when the correct knife groove is chosen. A quick pull then slides the cut insulation from the end of the wire. The length of the strip is regulated by the amount of wire extending beyond the blades when it is inserted in the knife groove.

Figure 24 shows two production-grade stripping tools. *Figure 24(A)* can be used to strip conductors from No. 20 AWG to No. 10 AWG, while *Figure 24(B)* strips wires from No. 18 AWG to No. 6 AWG. Note that this tool has front entry jaws for use in tight spaces. These two strippers do not have a crimping mechanism; they are for stripping wire only.

4.1.2 Stripping Power Cables and Large Conductors

Larger conductors can be cut using a ratchet-type cable cutter. The cable cutter shown in *Figure 25* can be used to cut cables up to 1,000 kcmil. This unit is for cutting only. Heavy-duty strippers

58104-11_F25.EPS

Figure 25 Ratchet-type cable cutter.

are used to strip large power cables, such as stator output and medium-voltage cables. *Figure 26* shows a heavy-duty stripper used to strip power cables with outside diameters ranging from No. 1/0 AWG through 1,000 kcmil.

Strippers can be used to remove insulation from the end of a cable or to make window cuts (*Figure 27*). All stripping tools should be operated according to the manufacturer's instructions. The procedures for using the tool shown in *Figure 26* to strip insulation from the end of a cable and make a window cut are described here.

> **WARNING!**
>
> Keep fingers away from the blade when using any stripping tool.

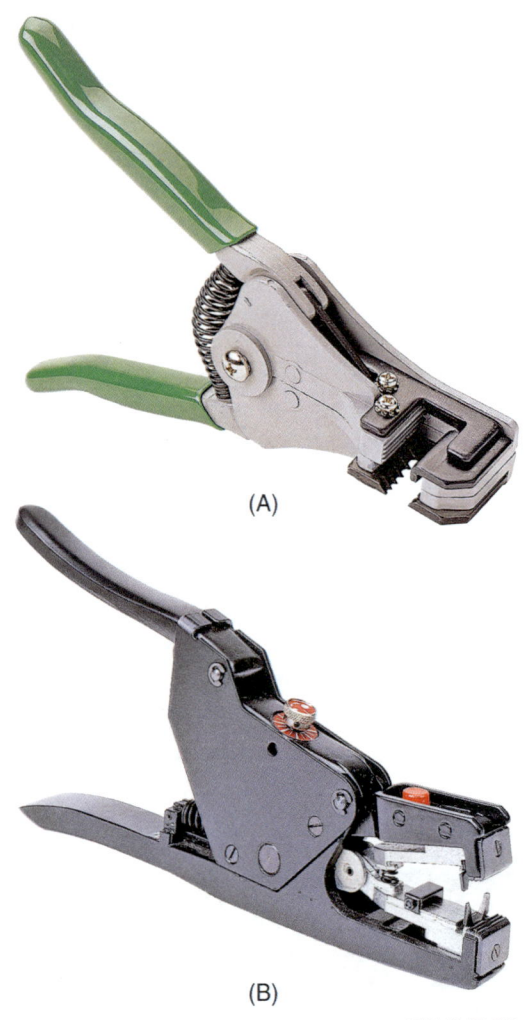

(A)

(B)

58104-11_F24.EPS

Figure 24 Wire strippers.

On Site

Cable Strippers

Open-blade (knife) stripping can damage the conductors and often results in minor or major injuries to workers. As a result, the practice is prohibited on many job sites. Adjustable strippers such as the tool shown here are used instead.

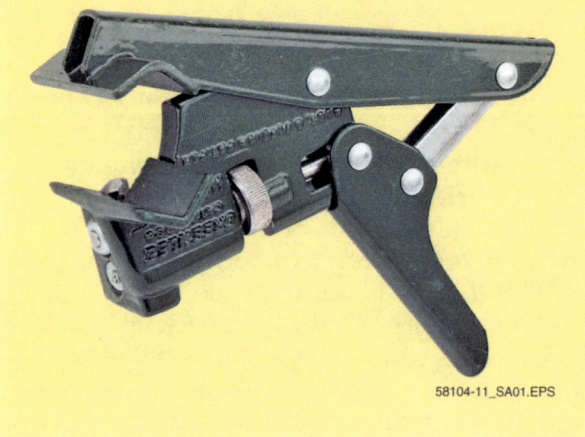

58104-11_SA01.EPS

To strip insulation from the end of a cable, proceed as follows:

Step 1 Loosen the locking knob to open the tool to its maximum position.

Step 2 Place the cable in the V-groove and close the tool firmly around the cable.

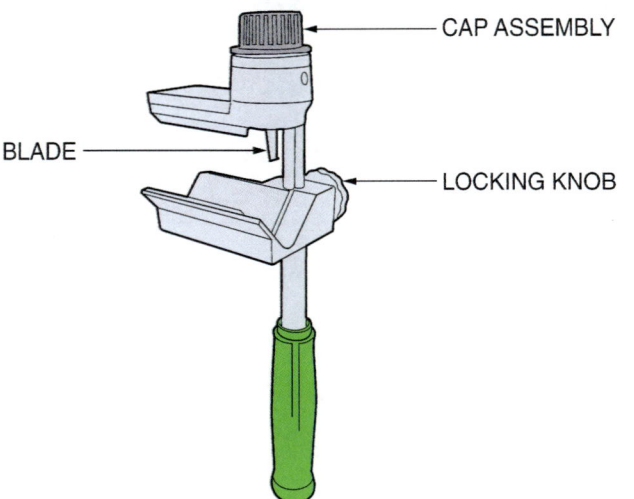

CAP ASSEMBLY

BLADE

LOCKING KNOB

58104-11_F26.EPS

Figure 26 Heavy-duty cable stripper.

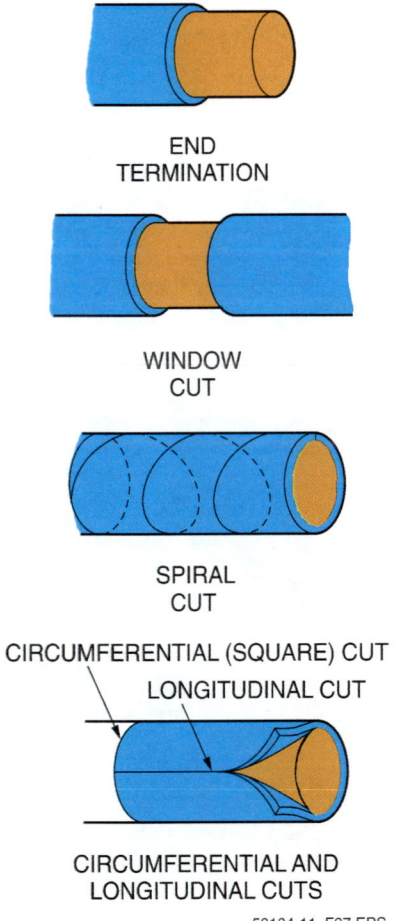

END
TERMINATION

WINDOW
CUT

SPIRAL
CUT

CIRCUMFERENTIAL (SQUARE) CUT
LONGITUDINAL CUT

CIRCUMFERENTIAL AND
LONGITUDINAL CUTS

58104-11_F27.EPS

Figure 27 Types of cable stripping.

Step 3 Tighten the locking knob.

Step 4 Turn the cap assembly until the blade reaches the required depth.

Step 5 Rotate the tool around the cable, advancing to the required strip length.

Step 6 Rotate the tool in the reverse direction to produce a square end cut.

Step 7 Loosen the locking knob to release the tool and remove it from the cable.

Step 8 Peel off the insulation.

> **CAUTION**
>
> Do not allow the blade to contact the conductor because damage to the conductor and/or the blade can result.

To make a window cut, proceed as follows:

Step 1 With the tool opened to the maximum position, place the cable in the V-groove and close the tool firmly around the cable. Tighten the locking knob.

Step 2 Turn the cap assembly until the blade reaches the required depth.

Step 3 Rotate the tool to produce the first square cut.

Step 4 Rotate the tool in the reverse direction to cut the required window strip length.

Step 5 Rotate the tool in the original direction to produce the second square cut.

Step 6 Loosen the locking knob assembly to release the tool and remove it from the cable.

Step 7 Peel off the insulation.

Figure 28 shows a round cable slitting and ringing tool that can be used to strip single- or multi-conductor cables. This tool can be used to cut around the cable (square cut) or slit the length of the cable jacket (longitudinal cut) for easy removal. The tool blade is adjustable to accommodate different jacket thicknesses.

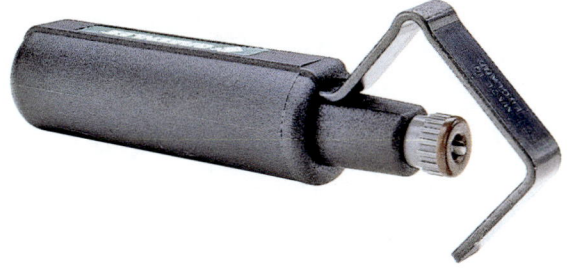

58104-11_F28.EPS

Figure 28 Round cable slitting and ringing tool.

4.1.3 Stripping Copper Communications Conductors and Data Cables

A scissor-action type of cutting tool is preferable to cutting tools with jaws that butt against each other when stripping copper communications conductors and data cables. Jaws that butt have a tendency to produce a flattened chisel end on the conductor, especially when the cutting edges become dull, as shown in *Figure 29*. This makes it difficult to insert the conductor into the barrel of a terminal.

Observe the following points when stripping conductors:

- Remove the cable jacket using strippers with an adjustable blade or a die designed for the particular wire size (*Figure 30*). Terminal manufacturers recommend a stripping length. Be careful to avoid nicking or stretching the conductor or insulation in a multi-conductor cable. If this type of damage occurs, it is likely that the wrong groove of the stripping tool was used, the tool was improperly set or damaged, the wrong type of stripping blade was used, or the tool was used incorrectly.
- Cut the insulation so that no frayed pieces or threads extend past the point of cutoff. Frayed pieces or threads of insulation indicate the use of an improper tool or dull cutter blades.
- If possible, do not twist, spread, or disturb the wire strands from their original position in the cable. Retwisting or tightening the twist of the strands eventually results in damage.
- Terminate stripped wire as soon as possible. The exposed strands invariably become bent and spread during handling, making termination difficult. A minimum amount of handling and storage after stripping results in better terminations.

Figure 30 Cable and wire stripping tools.

Figure 31 shows the positioning of the wire

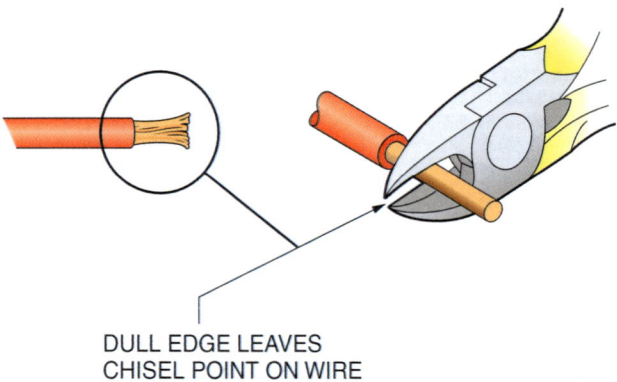

DULL EDGE LEAVES
CHISEL POINT ON WIRE

58104-11_F29.EPS

Figure 29 Chisel point left on a conductor.

in the crimp barrel when stripped to the proper length. The conductor insulation must be in the belled mouth of the terminal. This relieves stress on the wire strands and increases the strength of the connection. Allowing conductor strands to protrude out of the inspection hole more than $\frac{1}{32}$" interferes with the terminal screw. Cutting the strands too short reduces the contact surface area.

4.2.0 Wire Connections Under 600 Volts

NEC Section 110.14 governs electrical connections, including terminations and splices. Wire connections are used to connect a wire or cable to electrically operated devices, such as fan coil units, duct heaters, oil burners, motors, pumps, and control circuits of all types.

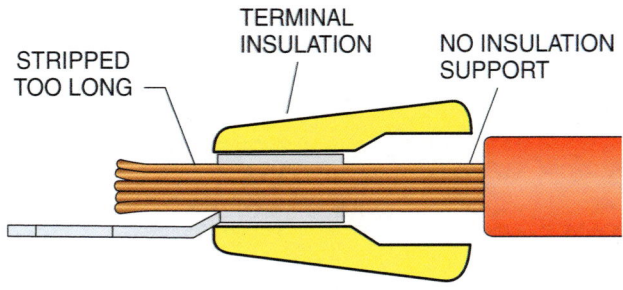

STRIPPING THAT IS TOO LONG WILL INTERFERE WITH THE TERMINAL SCREW

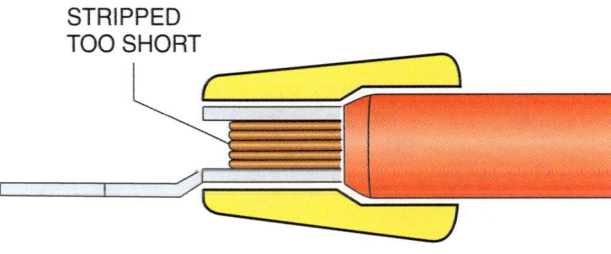

STRIPPING THAT IS TOO SHORT DOES NOT PROVIDE ENOUGH CONTACT SURFACE

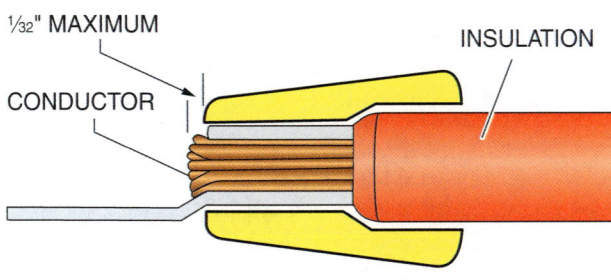

PROPER STRIPPING LENGTH WITH INSULATION INSIDE THE TERMINAL

58104-11_F31.EPS

Figure 31 Proper stripping length.

Several sizes of crimp-on wire lugs for stranded wire are shown in *Figure 32*. These connectors are available in various sizes to accommodate wire sizes No. 22 AWG and larger. They can be installed with crimping tools having a single or double indenter. The wire-size range is normally stamped on the tongue of each terminal.

Crimp-type connectors for wires smaller than No. 8 AWG are normally made to accept at least two wire sizes and are often color-coded. For example, one manufacturer's color code is red for connectors fitting No. 18 AWG or No. 20 AWG wire, blue for No. 16 AWG or No. 14 AWG wire, and yellow for No. 12 AWG or No. 10 AWG wire. Crimp-type connectors for wire sizes No. 8 AWG and larger, commonly called lugs, are made to accept one specific conductor size. Reducing connectors are used to connect two different size wires.

Mechanical compression-type terminators are also available to accommodate wires from No. 8 AWG through 1,000 kcmil. One-hole lugs, two-hole lugs, split-bolt connectors, and other types are shown in *Figure 33*. Mechanical compression-type connectors and lugs are available to accommodate a range of different wire sizes.

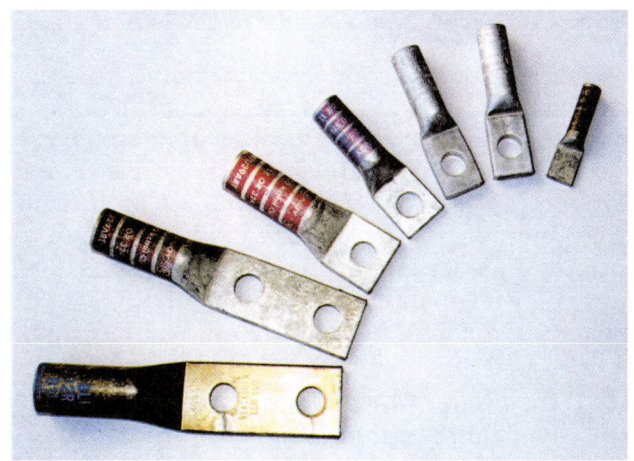

58104-11_F32.EPS

Figure 32 Crimp-on wire lugs.

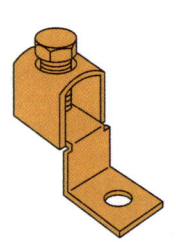

ONE BARREL, OFFSET TONGUE
ONE HOLE
NO. 14 AWG THROUGH 1,000 KCMIL

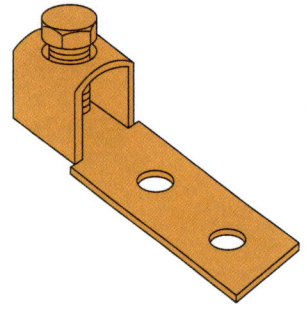

ONE BARREL, STRAIGHT TONGUE
TWO HOLE
NO. 14 AWG THROUGH 1,000 KCMIL

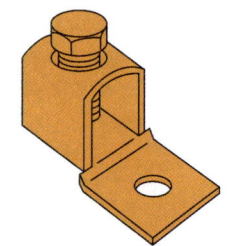

ONE BARREL, FIXED TONGUE
ONE HOLE
NO. 14 AWG THROUGH 500 KCMIL

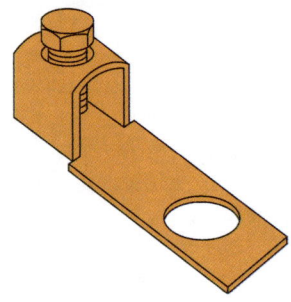

ONE BARREL, STRAIGHT TONGUE
ONE HOLE
NO. 14 AWG THROUGH 1,000 KCMIL

SINGLE HOLE
NO. 14 AWG THROUGH 4/0

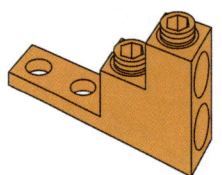

TWO HOLE, PANELBOARD CONNECTOR
NO. 2 AWG THROUGH 750 KCMIL

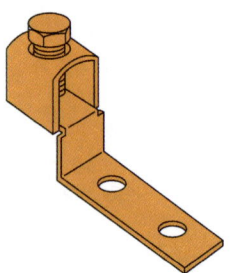

ONE BARREL, OFFSET TONGUE
TWO HOLE
NO. 14 AWG THROUGH 1,000 KCMIL

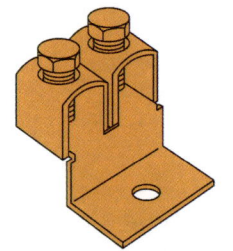

TWO BARRELS, OFFSET TONGUE
ONE HOLE
NO. 6 AWG THROUGH 500 KCMIL

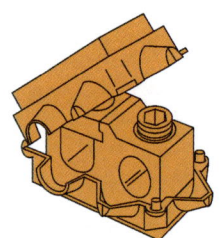

PARALLEL-TAP CONNECTOR
WITH INSULATED COVER
(VARIOUS WIRE SIZE COMBINATIONS)

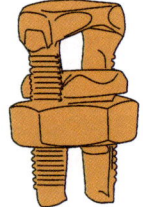

SPLIT BOLT CONNECTOR
(2) NO. 14 AWG THROUGH (2) 1,000 KCMIL
RUN AND TAP COMBINATIONS

58104-11_F33.EPS

Figure 33 Various mechanical compression connectors.

The parallel-tap connector with an insulated cover shown in *Figure 33* is one example of a pre-insulated, molded, mechanical compression connector. There are several kinds available. They come in setscrew with pressure plate and insulation-piercing configurations made for use in a variety of feeder tap and splice applications. Because they are equipped with an insulating cover, the requirement for taping the joint is eliminated.

4.2.1 Dissimilar Metal Conductors and Connections

NEC Section 110.14 prohibits conductors made of dissimilar metals (such as copper and aluminum,

copper and copper-clad aluminum, or aluminum and copper-clad aluminum) from being mixed in a terminal or splicing connector unless the device is identified for that purpose. Markings on connectors can help identify the proper use, as noted here:

- Connectors marked with only the wire size should only be used with copper conductors.
- Connectors marked with AL or ALR should only be used with aluminum conductors.
- Connectors marked with AL-CU or the newer designation of CO/ALR may be safely used with either copper or aluminum.

Tightening Compression Connector Screws and Bolts

Mechanical compression connectors are often tightened to a specified torque using a torque screwdriver or torque wrench. Over-tightening can cut the wires or break the fitting, while under-tightening may lead to loose connections, resulting in overheating and eventual failure.

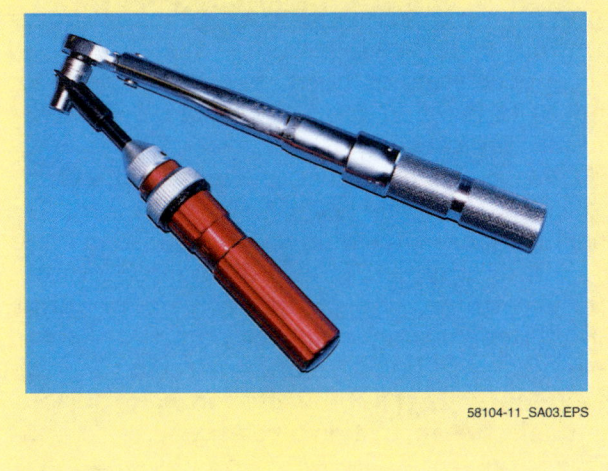

58104-11_SA03.EPS

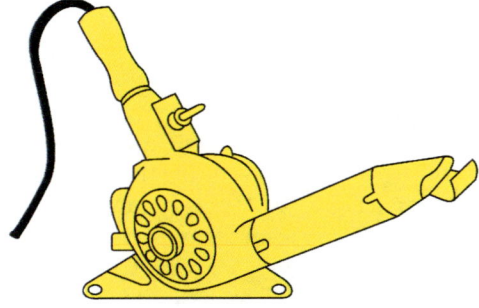

HEAT GUN

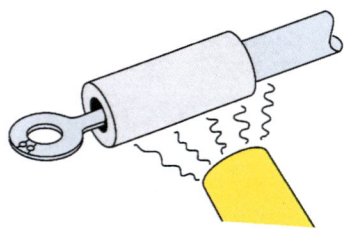

SLIP INSULATOR OVER
OBJECT TO BE INSULATED, THEN
APPLY HEAT FOR A FEW SECONDS

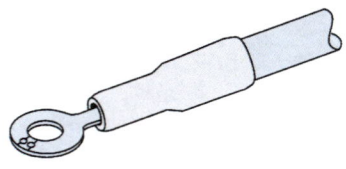

WHEN FINISHED, IT PROVIDES
PERMANENT INSULATION
PROTECTION

58104-11_F34.EPS

Figure 34 Installing heat-shrink insulators.

4.2.2 Heat-Shrink Insulators

Heat-shrink insulators for connectors provide skin-tight insulation protection and are fast and easy to use. They are designed to slip over wires, taper pins, connectors, terminals, and splices. When heat is applied, the insulation shrinks significantly, becoming semi-rigid and providing some positive strain relief at the flex point of the conductor. A vapor-proof band seals and protects the conductor against abrasion, chemicals, dust, gasoline, oil, and moisture. Extreme temperatures, both hot and cold, will not affect the performance of these insulators. The source of heat used varies, but most manufacturers of these insulators produce or recommend a heat gun specially designed for use on heat-shrink insulators. It is similar to a conventional hair dryer, in appearance only, as shown in *Figure 34*.

In general, a heat-shrink insulator may be thought of as tubing with a memory. After it is initially manufactured, it is heated and expanded to a predetermined diameter and then cooled. Upon a second application of heat, the tubing compound remembers its original size and shrinks to that smaller diameter. This property enables it to conform to the contours of any object. Heat shrink insulators are made from a variety of materials to suit different applications. Some are even transparent to make visual inspection of the connection possible.

4.2.3 Control and Sensor Cables

Wind turbine service technicians do not typically install control and sensor cables in wind turbines, but on occasion they must repair or replace these cables.

Control cables carry low-level signals that require less current than power cables. Conductor sizes range from No. 12 AWG to No. 24 AWG, with fiber-optic products also used as needed. Copper conductors within the cables can be tinned or bare, solid or stranded, and twisted or parallel. Shielding, if used, is usually an aluminum-coated Mylar® film with one or more drain wires inside a jacket. The jacket and conductor insulation are rated for the applicable usage. *Figure 35* shows a typical control or sensor cable containing multiple pairs of conductors.

To be effective, the shielding drain wire(s) must be earth-grounded. The installation loop diagrams or instructions for the system equipment indicate how and where the shields are connected and grounded. Typically, only one end is grounded, and the drain wire at the other end of the cable is simply isolated by folding it back and taping over it. This prevents the electrical noise collected by the shield from recirculating through the system and interfering with the control signals.

4.2.4 Low-Voltage Connectors and Terminals

Low-voltage control circuits typically use compression-type crimp connectors. These connectors are often color coded by wire size.

Compression-type connectors for connecting conductors to screw terminals on low-voltage circuits include those in which hand tools indent or crimp tube-like sleeves onto a conductor. Proper crimping action changes the size and shape of the connector and deforms the conductor strands enough to provide good electrical conductivity and mechanical strength.

Figure 36 shows the basic structure of a crimp connector. The crimp barrel receives the wire. The barrel is then crimped to secure the wire in place. The Vs, or dimples, inside the barrel improve wire-to-terminal contact, and increase the termination's tensile strength. Most crimp connectors have nylon or vinyl insulation covering the barrel to reduce the possibility of shorting to adjacent terminals. The insulation is color coded according to the connector's wire range to reduce the problem of wire-to-connector mismatch. An inspec-

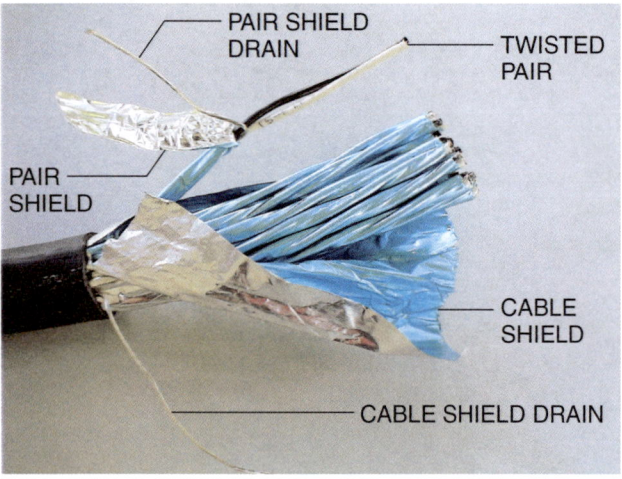

58104-11_F35.EPS

Figure 35 Multi-pair control or sensor cable.

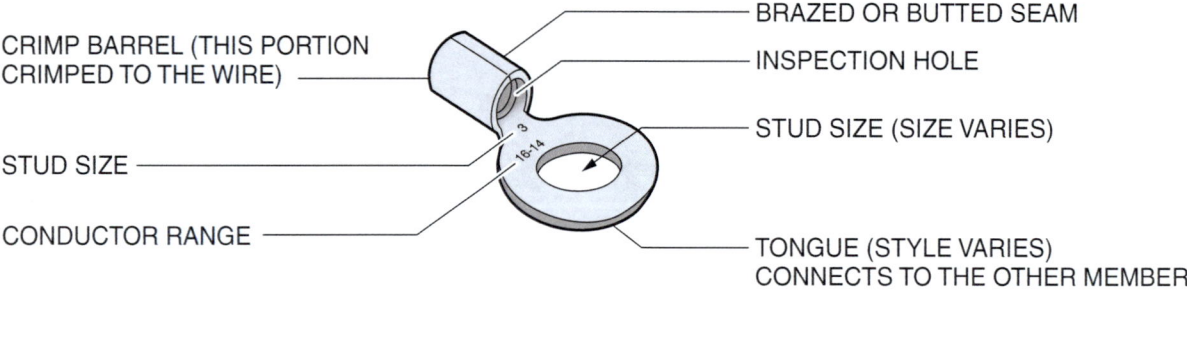

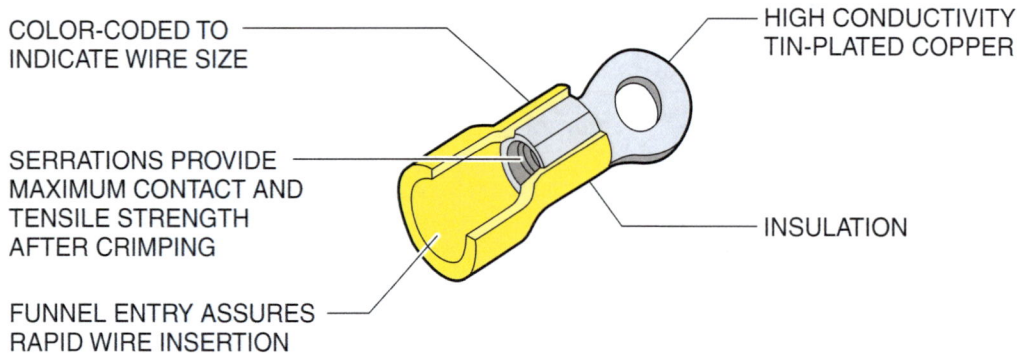

58104-11_F36.EPS

Figure 36 Basic crimp connector structure.

tion hole is provided at the end of the barrel to allow visual inspection of the wire position. For the smaller wire sizes, a sleeve is crimped over the conductor insulation in the process of crimping the barrel, providing strain relief for the conductor.

The barrel is connected to the terminal tongue, which physically connects the wire to the termination point, such as a terminal screw. Information about the connector size and conductor range is usually stamped on the tongue by the manufacturer. Tongue styles vary depending on termination requirements. *Figure 37* shows various tongue styles. The styles most frequently used are the ring tongue, and the flanged or locking fork. These types are preferred because the terminals do not slip off the terminal screw as the screw is tightened. They are also compatible with most vendor-supplied termination points.

Most manufacturers color code the barrel insulation to provide quick identification for installation and as an aid to inspection. Different colors, or a combination of colors, have special meanings. Although manufacturers do vary somewhat,

common or standard colors have been accepted. *Table 3* lists typical color codes.

Color combinations are sometimes varied to indicate the class or grade rating of an individual lug or splice. Some manufacturers use a clear plastic or other suitable insulation on the crimp barrel with a colored line to indicate wire-size range.

4.3.0 Installing Compression Connectors

The task of fastening a compression connector to a wire requires the use of the proper connector, crimping tool, and installation procedure. Always use the correct tool for the connector and follow the manufacturer's instructions.

Table 3 Typical Color Codes

AWG Wire Size	Color Code
22–16	Red
16–14	Blue
12–10	Yellow

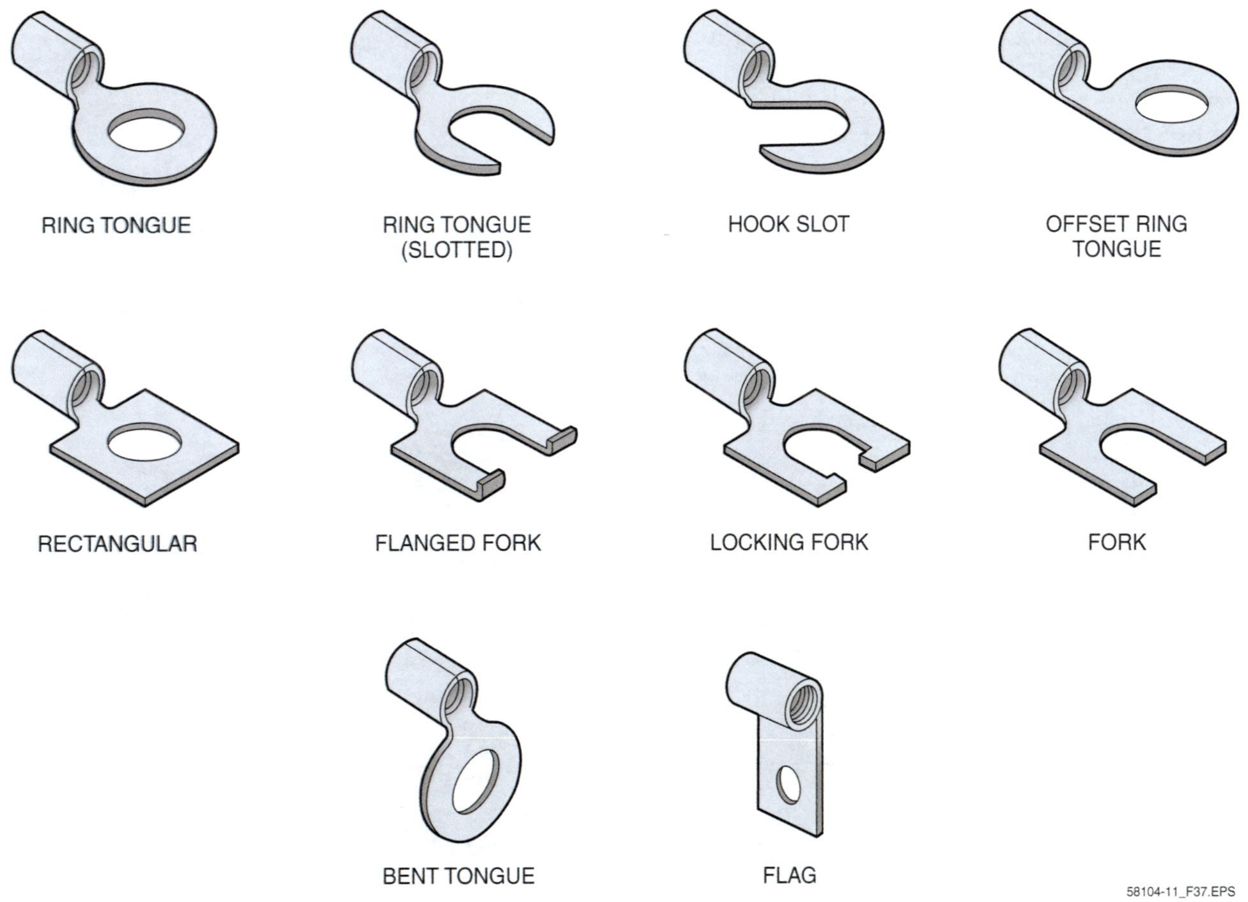

RING TONGUE	RING TONGUE (SLOTTED)	HOOK SLOT	OFFSET RING TONGUE
RECTANGULAR	FLANGED FORK	LOCKING FORK	FORK

BENT TONGUE	FLAG

58104-11_F37.EPS

Figure 37 Tongue styles of crimped connectors.

4.3.1 Crimping Tools

With a compression connector, an electrical connection between a wire and a terminal can be made by tightly compressing the crimp barrel with an ordinary pair of pliers. However, such a connection would not necessarily be made to the required pressure or in the correct location to ensure a good connection. A crimping tool is required to produce consistently good connections.

Figure 38 shows the relationship between the amount of crimping force and the mechanical and electrical performance. The maximum mechanical strength (A) occurs at a lower crimping force than the maximum electrical performance (B). The point of intersection (C) represents the ideal crimping force. Using a crimping die that is too large results in poor electrical performance, and using a die that is too small produces a weak mechanical connection.

A simple pliers-type crimping tool was the earliest crimping tool developed and continues to be used for repair operations or where only a few installations are being made. These tools are similar in construction to ordinary mechanics' pliers except the jaws are specially shaped, as shown in *Figure 39*.

It takes a tremendous amount of force at the lug to make a good crimp. Ratchet tools provide a means of increasing the human output force. The typical force capability of a hand for repetitive operations is 75 pounds for adult men and 50 pounds for adult women. The amount the tool multiplies the hand force is termed the mechanical advantage (MA) of the tool.

Simple pliers are basically constant-MA tools; the MA is the same whether the crimp is being started or finished. By adjusting linkage or cam mechanisms connecting the handles to the crimp dies, the MA can be varied so that it is low at the start and high at the finish of the crimp stroke. Thus, when the handles start to close from the open position and little or no crimping is being done, the MA is low. As the handles are closed farther, the crimp dies begin to compress the terminal, and the MA increases. In this manner, the MA is patterned to the crimp force requirements and distributed so that a high MA is achieved over the portion of the cycle where required.

Figure 40 shows a high MA tool equipped with a ratchet control. The ratchet mechanism prevents opening the tool and removing the crimped terminal before the handle has been closed all the way and the crimp completed. These provide a consistent, reliable crimp that meets the terminal manufacturer's requirements.

The crimp dies of this tool are interchangeable and may contain two or three positions for crimping different size terminals. These dies may be color coded to be used with a color-coded terminal lug for easy identification and to ensure proper crimping force. The crimp die of the tool determines the completed crimp configuration.

58104-11_F39.EPS

Figure 39 Hand crimpers.

58104-11 F40.EPS

Figure 40 Leveraged crimping tool.

INCREASING CRIMPING FORCE ➡

ELECTRICAL – – –
MECHANICAL ———

58104-11_F38.EPS

Figure 38 Mechanical strength versus electrical performance of a crimped connection.

There are a variety of configurations in use, such as the simple nest and indenter type of die or the more complicated four-indent die.

Figure 41 shows hand-operated and hydraulic crimping tools typical of those used to crimp connectors for stranded wires ranging from No. 8 AWG to 750 kcmil. These tools normally develop about 12 tons of compression force at 10,000 pounds per square inch (psi). Guidelines for the use of these tools are described here. Many other tools operate in a similar manner.

> **NOTE**
>
> Some manufacturers design a crimping tool that must be used with their connectors. Use the crimping tool recommended by the manufacturer.

DIE SET

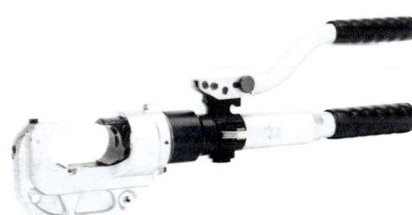

HAND-OPERATED

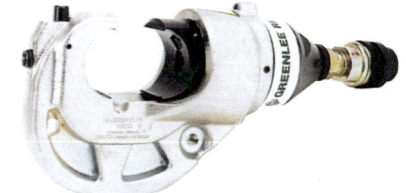

HYDRAULIC

58104-11_F41.EPS

Figure 41 Crimping tools used to crimp large connectors.

Several different configurations may work equally well for some applications, while for others, a certain shape is superior. Many considerations affect the determination of crimp die configuration, including the type of terminal (size, shape, material, and function), as well as the type and size of wires to be accommodated.

To use a hand-operated crimping tool, proceed as follows:

Step 1 Select the proper die for use with the connector to be crimped. Do not operate the tool without the die.

Step 2 Push the die release button on the C-head and slide one of the die halves into position until the retainer snaps. Insert the other die half in the piston body by pushing the die release button and sliding the die in until the retainer snaps.

Step 3 Place the tool C-head in position over the connector to be crimped. Pump the handle until compression is complete, as indicated by the dies touching at their flat surfaces nearest the throat of the C-head.

Step 4 Retract the ram and remove the connector after completion of the crimp. This is done by raising the pump handle slightly, rotating it clockwise until it stops, and then pushing the handle down in a pumping motion until the pressure release snaps.

> **WARNING!**
>
> Always read and follow the manufacturer's instructions when using power tools.

To operate a hydraulic crimping tool, proceed as follows:

Step 1 Connect the hydraulic pump to the crimping tool using a suitable hydraulic hose.

Step 2 Select the proper dies for use with the connector to be crimped. Do not operate the tool without the dies.

Step 3 Push the die release button on the C-head and slide one of the die halves into position until the retainer pin snaps. In a similar manner, install the other die half in the piston body.

Step 4 Place the tool C-head in position over the connector to be crimped. Operate the remote pump until compression is complete, as indicated by the dies touching on the frame side.

Step 5 Release the pressure at the hydraulic pump to retract the lower die half, and then remove the connector from the tool.

In addition to hand-operated and hydraulic crimpers, battery-operated and corded crimping tools are also available (*Figures 42* and *43*). Battery-operated tools offer the advantage of freedom of movement, while corded tools allow extensive use without worrying about battery changes. Universal crimping tools are also available (*Figure 44*). They operate in the same way as other crimpers but do not require separate dies.

58104-11_F42.EPS

Figure 42 Battery-operated crimping tool.

4.3.2 Making a Crimped Connection

To make a crimp connection, proceed as follows:

Step 1 Select a crimp connector of proper size and appropriate material for the wire size you are using. Copper connectors should be used with copper wires and aluminum connectors with aluminum wires. Dual-rated connectors may be used with both copper and aluminum wires.

Step 2 Use a suitable wire stripper to remove the insulation from the end of the wire. Be careful not to nick the wire. Strip the insulation back far enough so that the bare conductor will goes fully into the connector. Make sure not to strip off too much insulation; it should fit close to the connector when the wire is fully inserted into the connector.

Step 3 Clean the stripped portion of the wire. Use a wire brush for large wire sizes. Also clean the related unplated terminal pad and the surface to which the connector will be attached.

Step 4 Obtain the crimping tool and dies made for the type and size of connector to be crimped.

Step 5 Insert the stripped end of the wire completely into the connector. Position the crimping tool in place over the connector, and then operate the tool to fully crimp the connector, as directed in the tool manufacturer's instructions. Make sure that the crimping tool jaws are fully closed, indicating that a full-compression crimp has been made. If multiple crimps are required (*Figure 45*), crimp from the lug back to the barrel base, rotating the crimper as necessary to avoid deforming the connector.

58104-11 F43.EPS

Figure 43 Corded crimping tool.

Step 6 Using a bolt or screw and washers (if required), secure the crimped connector and attached wire to the correct terminal in the equipment. Tighten the terminal bolt and torque to the level specified by the equipment manufacturer. Too little or too much torque can adversely affect the performance of the connection.

Table 4 lists some torque values typical of those used for tightening common sizes of steel and aluminum terminal bolts.

Figure 44 Universal crimping tool.

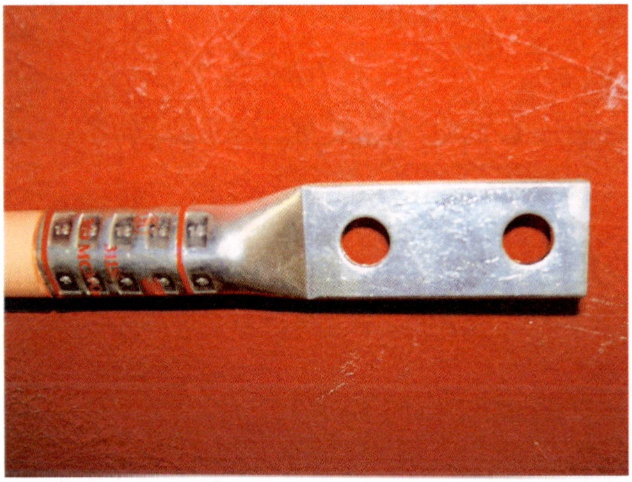

Figure 45 Multiple crimps.

Table 4 Recommended Tightening Torques for Various Bolt Sizes

Steel Hardware		Aluminum Hardware	
Bolt Size	Recommended Torque (Inch-Pounds)	Bolt Size	Recommended Torque (Inch-Pounds)
¼–20	80	½–13	300
⁵⁄₁₆–18	180	¾–11	480
⅜–16	240	¾–10	650
½–13	480	—	—
⅝–11	660	—	—
¾–10	1,900	—	—

On Site

Tightening Torques

Many terminations and types of equipment are marked with tightening torque information. For items of equipment, torque information is often marked on the equipment and/or listed in the manufacturer's installation instructions. When specific torque requirements are given, always follow the manufacturer's recommendations. Good sources of tightening information are manufacturers' catalogs and product literature.

SUMMARY

Although wind turbine nacelles are constructed and wired as a complete unit including all wiring, wind turbine technicians must maintain the systems and wiring, which often requires repair or replacement of conductors or cables. In order to perform the tasks associated with this maintenance, the technician must understand and recognize the difference between the conductors and cables as well as their applications.

Recognizing differences includes selecting and sizing conductors and cables based on their usage, along with preparing and terminating the conductor ends as necessary. Additional training may be necessary when it comes to preparing and terminating large complex medium-voltage cables, low-voltage sensor cables, and fiber-optic cables.

1. Site substation wiring beyond the collection system is not normally included in the duties of the wind technician because of its high voltage levels and _____.

   a. elevation from ground level
   b. distance from the turbines
   c. extreme short-circuit energy potential
   d. lack of electrical equipment

2. The most common nominal output voltage from existing wind turbine generators is _____.

   a. 230VAC
   b. 690VAC
   c. 1,000VAC
   d. 34,500VAC

3. AWG is an abbreviation for _____.

   a. Aluminum Wire Gauge
   b. Ampacity Wire Gauge
   c. Accepted Wire Gauge
   d. American Wire Gauge

4. The larger the AWG number, the smaller the cross-sectional area of the wire.

   a. True
   b. False

5. A circular mil is a circle that has a diameter of 1 mil, which is also equal to what part of an inch?

   a. 0.0001
   b. 0.001
   c. 0.01
   d. 1

6. Most accessible cables within a wind turbine nacelle carry AC voltages equal to or less than _____.

   a. 120V
   b. 230V
   c. 240V
   d. 690V

7. The maximum operating temperature rating of a conductor is the same as the design ambient temperature where the wire will be used.

   a. True
   b. False

8. The maximum current in amps that a conductor can carry, under the *NEC*®-specified conditions of use, without exceeding its temperature rating is known as its _____.

   a. workability rating
   b. ampacity rating
   c. conductivity rating
   d. ambient temperature rating

9. Assume that THHW copper wire has been specified for a circuit feeding a nacelle heater, with a load ampacity of 25.6 amps at 230 volts. The design ambient temperature is 30°C, and only two conductors are in the cable. What would the appropriate AWG wire size be for the installation? (Refer to *NEC Table 310.15(B)(16)*, in *Appendix A.*)

   a. No. 14 AWG
   b. No. 12 AWG
   c. No. 10 AWG
   d. No. 8 AWG

10. Assume that a 120 volt power supply was installed for a ventilation fan using No. 16 AWG THHN copper wire. There are three conductors together in the cable, and the ambient temperature is 30°C. What is the maximum ampacity of the individual conductors? (Refer to *NEC Table 310.15(B)(16)*, in *Appendix A.*)

    a. 18 amps
    b. 20 amps
    c. 25 amps
    d. 30 amps

11. An appropriate wire size is needed to power a load of 162 amps at 460 volts, 3 phase. The specified wire type is THHW copper-clad aluminum, and the ambient temperature is 30°C. What is the appropriate AWG wire size for each of the three conductors? (Refer to *NEC Table 310.15(B)(17)* in *Appendix B.*)

a. No. 2 AWG
b. No. 1 AWG
c. No. 1/0 AWG
d. No. 2/0 AWG

12. Which component of an optical fiber cable carries the light signal?

a. Jacket
b. Core
c. Cladding
d. Buffer layer

13. The power conductors or cables connected to the stator windings of the wind turbine generator carry the output current of the generator to the _____.

a. nacelle
b. up-tower PLC
c. main switchgear
d. down-tower PLC

14. The interconnection of all turbine outputs that are connected in a circuit or string is referred to as the _____.

a. stator output
b. site substation
c. isolating switch
d. collection system

15. What installation requirement makes shielded control and sensor wire different from other types of cable?

a. The shield must be firmly attached to every cable conductor.
b. The shield must be connected to ground.
c. The shield wire should be the first conductor used.
d. Shield wire has a maximum voltage rating of 15 volts.

Trade Terms Quiz

Fill in the blank with the correct term that you learned from your study of this module.

1. A computer incorporating input and output modules that can be instructed to provide specific outputs based on various input values is called a(n) _____.

2. A wire or cable that provides physical support for other conductors, but may or may not be used as a conductor itself, is known as a(n) _____.

3. A switching device that contains no moving parts, operating only on integrated circuitry, is called a(n) _____.

4. The temperature that surrounds the installation area of conductors, cables, or equipment is referred to as the _____.

5. When wiring must maintain its flexibility and is to be exposed to higher temperatures, the best choice of insulating material is _____.

6. Some examples of electrical _____would be electric heaters, motors, or lights.

7. The _____ is where turbine-generated power is collected, transformed to a higher voltage, and then made available to the power grid for consumption.

8. Although the _____ of a generator may be 690 volts, its actual output voltage may vary slightly.

9. The system of high-voltage power lines that interconnects wind-site substations and other power-generating facilities with power consumers is known as the _____.

10. A commonly used conductor insulation that changes form when exposed to higher temperatures, becoming more elastic or melting, is known as _____.

11. The wound portion of a generator that does not move or rotate when in operation is the _____.

12. The _____ interconnects a number of turbine transformers in a string-like fashion for delivery to the site substation.

13. An electromagnetic device that increases the voltage of the turbine generator before it is sent to the site substation is called a(n) _____.

14. A(n) _____ is a system designed to interconnect computers and related devices within a limited area.

15. The various conductors involved in transporting power all the way from the turbine generator to the site substation are referred to as _____.

16. Cable that is used for control and communication applications where current flow is very low, making overcurrent protection unnecessary is called _____.

Trade Terms

Ambient temperature	Distribution conductors	Messenger wire	Programmable logic controller (PLC)	Thermoplastic insulation
Collection system	Load	Nominal output voltage	Site substation	Thermoset insulation
Commercial power grid system	Local area network (LAN)	Power-limited cable	Solid-state relay	Turbine transformer
			Stator windings	

Tim Dean

Electrician/ Electrical Trades Instructor
Central Ohio ABC/Madison Comprehensive High School
Mansfield, Ohio

How did you get started in the construction industry?
Upon exiting The University of Toledo, I took a job with my brother-in-law who worked as an electrician in Akron, Ohio. I knew little or nothing about electricity, but needed a job to support myself and my wife.

Who inspired you to enter the industry? Why?
I suppose my brother-in-law, Tom Argenio, was my inspiration. He possessed a work ethic and craftsmanship that is very rare in our society today. He taught me not only how to be a good tradesman but to understand the pride of quality workmanship.

What do you enjoy most about your job?
In the early days of my career, I simply appreciated having a job. But as time and experience passed I acquired a thirst for knowledge and understanding. Knowing the why, when, where, and how brought new meaning to the skills I was attaining.

Do you think training and education are important in construction? If so, why?
The electrical trade is one of the most diverse and challenging of all the construction trades. Training and education are not only important but mandatory to stay safe, efficient, qualified, and prepared for the challenges of new technology.

How important are NCCER credentials to your career?
Being a part of NCCER has been one of the most rewarding experiences of my life. Acquiring the credentials needed as an SME has benefited not only me personally, by recognition and professional development, but the organization I work for as well. I enjoy the challenges required to stay abreast of our evolving industry. NCCER provides the vehicle and resources for keeping up with cutting-edge technological changes.

How has training/construction impacted your life and your career?
When I entered the trade, I was clueless. I had no idea what I was getting myself into. But as I continued, I found a challenge in understanding how and why things worked the way they did. I was engaged in the process and discovered that the more I knew and learned, the more I was worth to my employer.

Would you suggest construction as a career to others? If so, why?
I would recommend a career in the electrical trade to anyone who has a desire to learn and a thirst for understanding new technologies. The electrical construction trade is both rewarding and challenging. Being a part of this great industry opens doors to new and exciting opportunities in many different ways and is much more than a job. Career opportunities abound and the sky is the limit for what you can attain.

How do you define craftsmanship?
To me craftsmanship is the result of acquiring knowledge, developing skills, infusing moral and ethical values, and blending personal pride to construct a product or process of recognizable and enduring quality.

Trade Terms Introduced in This Module

Ambient temperature: The temperature surrounding the installation of a conductor, cable, or equipment.

Collection system: The interconnection of a number of turbine transformer outputs in a circuit or string. The power collected is then transferred to the site substation.

Commercial power grid system: The high-voltage transmission lines that interconnect wind-site substations to commercial power distribution substations.

Distribution conductors: As it refers to a wind turbine, the term includes all conductors that transmit power from the point of the turbine generator output to the wind-site substation.

Load: Equipment and devices that are connected to an electrical circuit and consume electrical power.

Local area network (LAN): A network designed to interconnect computers, electronic storage devices, and other computer accessory equipment within a local area or limited distance.

Messenger wire: A metallic supporting member (such as a cable), either solid or stranded, that may or may not perform the function of a conductor.

Nominal output voltage: A universally-accepted voltage value used to describe an output voltage that may be slightly less or greater than the actual output voltage. The nominal output voltage of a device generally indicates its design output value, which can vary in actual use for a number of reasons.

Power-limited cable: Cable used for communication and control applications where the voltage is less than 30 volts and the current is less than 1,000 volt-amps. As a result, overcurrent protection for the wire is typically not required.

Programmable logic controller (PLC): A computer containing input and output modules that may be programmed to deliver specific output values, sequences, and other functions based on input values.

Site substation: A wind-site, outdoor electrical facility containing switchgear, transformers, and other electrical equipment designed to connect turbine output power to the commercial grid system. Also known as a collector substation.

Solid-state relay: Relays that contain no moving parts, only integrated circuitry.

Stator windings: The stationary, non-moving winding within a generator. Stator windings deliver the generated power through wiring connections. The rotor rotates inside the stator winding, generating electrical power.

Thermoplastic insulation: Conductor insulation made from plastic materials, that become elastic or melt when heated, and return to their rigid state at room temperature.

Thermoset insulation: A rubbery conductor insulation, made from plastic materials that maintains its form when exposed to high temperatures.

Turbine transformer: An electromagnetic device that increases the turbine generator output voltage to the value required by the site substation, typically located adjacent to the tower base.

NEC TABLE 310.15(B)(16)

ALLOWABLE AMPACITIES OF INSULATED CONDUCTORS RATED UP TO AND INCLUDING 2000 VOLTS, 60° THROUGH 90°C (140°F THROUGH 194°F), NOT MORE THAN THREE CURRENT-CARRYING CONDUCTORS IN RACEWAY, CABLE, OR EARTH (DIRECTLY BURIED), BASED ON AMBIENT TEMPERATURE OF 30°C (86°F)*

Size AWG or kcmil	Temperature Rating of Conductor [See Table 310.104(A).]						Size AWG or kcmil
	60°C (140°F)	75°C (167°F)	90°C (194°F)	60°C (140°F)	75°C (167°F)	90°C (194°F)	
	Types TW, UF	Types RHW, THHW, THW, THWN, XHHW, USE, ZW	Types TBS, SA, SIS, FEP, FEPB, MI, RHH, RHW-2, THHN, THHW, THW-2, THWN-2, USE-2, XHH, XHHW, XHHW-2, ZW-2	Types TW, UF	Types RHW, THHW, THW, THWN, XHHW, USE	Types TBS, SA, SIS, THHN, THHW, THW-2, THWN-2, RHH, RHW-2, USE-2, XHH, XHHW, XHHW-2, ZW-2	
	COPPER			ALUMINUM OR COPPER-CLAD ALUMINUM			
18**	—	—	14	—	—	—	—
16**	—	—	18	—	—	—	—
14**	15	20	25	—	—	—	—
12**	20	25	30	15	20	25	12**
10**	30	35	40	25	30	35	10**
8	40	50	55	35	40	45	8
6	55	65	75	40	50	55	6
4	70	85	95	55	65	75	4
3	85	100	115	65	75	85	3
2	95	115	130	75	90	100	2
1	110	130	145	85	100	115	1
1/0	125	150	170	100	120	135	1/0
2/0	145	175	195	115	135	150	2/0
3/0	165	200	225	130	155	175	3/0
4/0	195	230	260	150	180	205	4/0
250	215	255	290	170	205	230	250
300	240	285	320	195	230	260	300
350	260	310	350	210	250	280	350
400	280	335	380	225	270	305	400
500	320	380	430	260	310	350	500
600	350	420	475	285	340	385	600
700	385	460	520	315	375	425	700
750	400	475	535	320	385	435	750
800	410	490	555	330	395	445	800
900	435	520	585	355	425	480	900
1000	455	545	615	375	445	500	1000
1250	495	590	665	405	485	545	1250
1500	525	625	705	435	520	585	1500
1750	545	650	735	455	545	615	1750
2000	555	665	750	470	560	630	2000

*Refer to 310.15(B)(2) for the ampacity correction factors where the ambient temperature is other than 30°C (86°F).

**Refer to 240.4(D) for conductor overcurrent protection limitations.

NEC TABLE 310.15(B)(17)

ALLOWABLE AMPACITIES OF SINGLE-INSULATED CONDUCTORS RATED UP TO AND INCLUDING 2000 VOLTS IN FREE AIR, BASED ON AMBIENT TEMPERATURE OF 30°C (86°F)*

Size AWG or kcmil	Temperature Rating of Conductor [See Table 310.104(A).]						Size AWG or kcmil
	60°C (140°F)	75°C (167°F)	90°C (194°F)	60°C (140°F)	75°C (167°F)	90°C (194°F)	
	Types TW, UF	Types RHW, THHW, THW, THWN, XHHW, ZW	Types TBS, SA, SIS, FEP, FEPB, MI, RHH, RHW-2, THHN, THHW, THW-2, THWN-2, USE-2, XHH, XHHW, XHHW-2, ZW-2	Types TW, UF	Types RHW, THHW, THW, THWN, XHHW	Types TBS, SA, SIS, THHN, THHW, THW-2, THWN-2, RHH, RHW-2, USE-2, XHH, XHHW, XHHW-2, ZW-2	
	COPPER			ALUMINUM OR COPPER-CLAD ALUMINUM			
18	—	—	18	—	—	—	—
16	—	—	24	—	—	—	—
14**	25	30	35	—	—	—	—
12**	30	35	40	25	30	35	12**
10**	40	50	55	35	40	45	10**
8	60	70	80	45	55	60	8
6	80	95	105	60	75	85	6
4	105	125	140	80	100	115	4
3	120	145	165	95	115	130	3
2	140	170	190	110	135	150	2
1	165	195	220	130	155	175	1
1/0	195	230	260	150	180	205	1/0
2/0	225	265	300	175	210	235	2/0
3/0	260	310	350	200	240	270	3/0
4/0	300	360	405	235	280	315	4/0
250	340	405	455	265	315	355	250
300	375	445	500	290	350	395	300
350	420	505	570	330	395	445	350
400	455	545	615	355	425	480	400
500	515	620	700	405	485	545	500
600	575	690	780	455	545	615	600
700	630	755	850	500	595	670	700
750	655	785	885	515	620	700	750
800	680	815	920	535	645	725	800
900	730	870	980	580	700	790	900
1000	780	935	1055	625	750	845	1000
1250	890	1065	1200	710	855	965	1250
1500	980	1175	1325	795	950	1070	1500
1750	1070	1280	1445	875	1050	1185	1750
2000	1155	1385	1560	960	1150	1295	2000

*Refer to 310.15(B)(2) for the ampacity correction factors where the ambient temperature is other than 30°C (86°F).

**Refer to 240.4(D) for conductor overcurrent protection limitations.

58104-11_A02.EPS

NEC Table 310.15(B)(2)(A)

**AMBIENT TEMPERATURE CORRECTION
FACTORS BASED ON 30°C (86°F)**

For ambient temperatures other than 30°C (86°F), multiply the
allowable ampacities specified in the ampacity tables by the
appropriate correction factor shown below.

Ambient Temperature (°C)	Temperature Rating of Conductor			Ambient Temperature (°F)
	60°C	75°C	90°C	
10 or less	1.29	1.20	1.15	50 or less
11–15	1.22	1.15	1.12	51–59
16–20	1.15	1.11	1.08	60–68
21–25	1.08	1.05	1.04	69–77
26–30	1.00	1.00	1.00	78–86
31–35	0.91	0.94	0.96	87–95
36–40	0.82	0.88	0.91	96–104
41–45	0.71	0.82	0.87	105–113
46–50	0.58	0.75	0.82	114–122
51–55	0.41	0.67	0.76	123–131
56–60	—	0.58	0.71	132–140
61–65	—	0.47	0.65	141–149
66–70	—	0.33	0.58	150–158
71–75	—	—	0.50	159–167
76–80	—	—	0.41	168 176
81–85	—	—	0.29	177–185

58104-11_A03.EPS

Additional Resources

This module presents thorough resources for task training. The following resource material is suggested for further study.

NFPA 70®, *National Electrical Code®*, Copyright © 2010, National Fire Protection Association, Quincy, MA.

Figure Credits

Power Tel Utilities Contractor Limited, Module opener and Figure 1

Topaz Publications, Inc., Figures 2–4, 8–11, 13, 16–20, 22, 30, 32, 35, and SA03

Jim Mitchem, Figure 45

Greenlee Textron, Inc., a subsidiary of Textron Inc., Figures 23–25, 28, 39–44, SA01, and SA02

Reprinted with permission from *NFPA 70®*, *National Electrical Code®*, Copyright © 2010, National Fire Protection Association, Quincy, MA. This reprinted material is not the complete and official position of the NFPA on the referenced subject, which is represented only by the standard in its entirety., Appendixes A–C

NCCER CURRICULA — USER UPDATE

NCCER makes every effort to keep its textbooks up-to-date and free of technical errors. We appreciate your help in this process. If you find an error, a typographical mistake, or an inaccuracy in NCCER's Curricula, please fill out this form (or a photocopy), or complete the online form at **www.nccer.org/olf**. Be sure to include the exact module number, page number, a detailed description, and your recommended correction. Your input will be brought to the attention of the Authoring Team. Thank you for your assistance.

Instructors – If you have an idea for improving this textbook, or have found that additional materials were necessary to teach this module effectively, please let us know so that we may present your suggestions to the Authoring Team.

NCCER Product Development and Revision
13614 Progress Blvd., Alachua, FL 32615

Email: curriculum@nccer.org
Online: www.nccer.org/olf

❏ Trainee Guide   ❏ AIG   ❏ Exam   ❏ PowerPoints   Other _____

Craft / Level: _____ Copyright Date: _____

Module Number / Title: _____

Section Number(s): _____

Description: _____

_____

_____

_____

_____

Recommended Correction: _____

_____

_____

_____

_____

Your Name: _____

Address: _____

_____

Email: _____ Phone: _____

Glossary

Ambient temperature: The temperature surrounding the installation of a conductor, cable, or equipment.

Ammeter: An instrument for measuring electrical current.

Ampere (A): A unit of electrical current. For example, one volt across one ohm of resistance causes a current flow of one ampere.

Anchor point: The location of attachments on a structure for all types of climbing and/or rigging systems.

Anenometer: A device used to measure wind speed that often incorporates wind direction as well.

Arc: A visible, luminous bridge formed between two surfaces when electrical power jumps between them.

Arc blast: An explosion similar to the detonation of dynamite that occurs during an arc flash incident.

Arc flash: A dangerous condition caused by the enormous release of thermal energy when arcing occurs.

Atom: The smallest particle to which an element may be divided and still retain the properties of the element.

Battery: A DC voltage source consisting of two or more cells that convert chemical energy into electrical energy.

Betz limit: The theoretical limit of 59.3 percent of the available wind power that a rotor can capture. The theory is named for its German developer, Albert Betz.

Body harness: A system of straps and rings worn on the body with the intent of distributing weight and force applied evenly across the shoulders, chest, waist, thighs, and pelvic area. The body harness must not be confused with a body belt, which is simply worn around the waist and is not approved as a fall arrest device.

Cable grabs: Components that ride along the length of a cable smoothly when moved casually, but lock onto the cable sharply when downward movement becomes too rapid or violent.

Carabiner (Kare-uh-BEAN-er): A chain-link shaped device which can be opened on one side for the insertion of a line, then closed securely. It is usually rated in Newtons, since it is tested by impact to determine its strength.

Centrifugal release units: Devices that use centrifugal force to open or close hydraulic valves in response to rotor speed.

Circuit: A complete path for current flow.

Coil: A number of turns of wire, especially in spiral form, used for electromagnetic effects or for providing electrical resistance.

Collection system: The interconnection of a number of turbine transformer outputs in a circuit or string. The power collected is then transferred to the site substation.

Commercial power grid system: The high-voltage transmission lines that interconnect wind-site substations to commercial power distribution substations.

Conductor: A material through which it is relatively easy to maintain an electric current.

Connecting devices: Devices used to connect the PFAS and positioning belts to anchor points and positioning points.

Continuity: An electrical term used to describe a complete (unbroken) circuit that is capable of conducting current. Such a circuit is also said to be closed.

Coulomb: An electrical charge equal to 6.25 × 10^{18} electrons or 6,250,000,000,000,000,000 electrons. A coulomb is the common unit of quantity used for specifying the size of a given charge.

Current: The movement, or flow, of electrons in a circuit. Current (I) is measured in amperes.

d'Arsonval meter movement: A meter movement that uses a permanent magnet and moving coil arrangement to move a pointer across a scale.

Distribution conductors: As it refers to a wind turbine, the term includes all conductors that transmit power from the point of the turbine generator output to the wind-site substation.

Dynamo: An apparatus that converts mechanical energy into electrical energy, typically in the form of direct current.

Effective ground level: The actual surface that air movement is passing across, as opposed to actual ground level. For example, the effective ground level for a wind blowing over a dense forest would be the tops of the trees rather than the ground itself.

Egress: An escape or emergence from one area into another.

Electron: A negatively charged particle that orbits the nucleus of an atom.

Equipotential plane: An electrical condition when all points have the same potential.

Fall arrest: A means of stopping or controlling a fall in progress with minimal or no injury to the climber.

Fall restraint: A means of preventing a fall from occurring or aiding in worker positioning.

Flash hazard analysis: A study investigating a worker's potential exposure to arc flash energy, conducted for the purpose of injury prevention, the determination of safe work practices, and the identification of appropriate PPE.

Flash protection boundary: An approach limit at a distance from exposed energized electrical conductors or components within which a person could receive a second-degree burn if an arc flash incident occurred.

Frequency: The number of cycles completed each second by a given AC voltage; usually expressed in hertz. One hertz equals one cycle per second.

Furling: One method of preventing excessive wind turbine rotor speed through yaw control by turning the rotor blades away from a direct wind facing.

Horizontal-axis wind turbine (HAWT): A wind turbine that spins on an axis which is horizontal or nearly so, much like the early windmills of the western and midwestern U.S. Also referred to as a conventional turbine or propeller-style, they are directional by design, i.e. the rotor must face into the wind for maximum performance.

Insulator: A material through which it is difficult to conduct an electric current.

Jib: The projecting arm of a crane from which the load is suspended. Generally, a jib is considered an extension of the main arm of the crane, rather than the main arm itself.

Joule (J): A unit of measurement that represents one newton-meter (Nm), which is a unit of measure for doing work.

Kilo: A prefix used to indicate one thousand; for example, one kilowatt is equal to one thousand watts.

Kinetic energy: The energy contained in a mass or body caused by its motion.

Kirchhoff's current law: The statement that the total amount of current flowing through a parallel circuit is equal to the sum of the amounts of current flowing through each current path.

Kirchhoff's voltage law: The statement that the sum of all the voltage drops in a circuit is equal to the source voltage of the circuit.

Lanyards: Flexible ropes, woven straps, and wire ropes that attach PFAS components to structures or other people.

Limited approach boundary: An approach limit at a distance from exposed energized electrical conductors or components within which a shock hazard exists.

Load: Equipment and devices that are connected to an electrical circuit and consume electrical power.

Local area network (LAN): A network designed to interconnect computers, electronic storage devices, and other computer accessory equipment within a local area or limited distance.

Longitudinal: Running lengthwise, or extending along the length of an object. A line drawn the length of an object would indicate its longitudinal axis.

Low-voltage distribution module (LVDM): The assembly of switchgear and circuit breakers that controls the flow of turbine-generated power. The voltage is low relative to the voltage that leaves the pad-mounted transformer on its way to the substation.

Matter: Any substance that has mass and occupies space.

Mega: A prefix used to indicate one million; for example, one megawatt is equal to one million watts.

Messenger wire: A metallic supporting member (such as a cable), either solid or stranded, that may or may not perform the function of a conductor.

Minimum approach distance (MAD): The distance from energized electrical conductors or components that a qualified person may approach before donning insulating PPE made of rubber.

Nacelle: A streamlined housing or enclosure that contains the major working components of the turbine system at the top of the tower.

Neutrons: Electrically neutral particles (neither positive nor negative) that have the same mass as a proton and are found in the nucleus of an atom.

Newtons: A measure of force applied, equal to the amount of force required to accelerate a mass of one kilogram at a rate of one meter per second, per second. One pound of force = 4.45 Newtons.

Nominal output voltage: A universally-accepted voltage value used to describe an output voltage that may be slightly less or greater than the actual output voltage. The nominal output voltage of a device generally indicates its design output value, which can vary in actual use for a number of reasons.

Nucleus: The center of an atom. It contains the protons and neutrons of the atom.

Occupational Safety and Health Administration (OSHA): A government organization that works to ensure that employers provide a safe workplace for their employees.

Ohm (Ω): The basic unit of measurement for resistance.

Ohm's law: A statement of the relationships among current, voltage, and resistance in an electrical circuit: current (I) equals voltage (E) divided by resistance (R). Generally expressed as a mathematical formula: $I = E/R$.

Ohmmeter: An instrument used for measuring resistance.

Parallel circuits: Circuits containing two or more parallel paths through which current can flow.

Permit-required confined space: A confined space that has been evaluated and found to have specific actual or potential hazards, such as a toxic atmosphere or other serious safety or health hazard. Workers need written authorization to enter.

Personal fall arrest system (PFAS): System consisting of anchor points, full body harness, and connecting devices.

Pitch control: Management and turning of a turbine blade's position along its longitudinal axis.

Point of daylight: The point where guy wire anchor assemblies or tower components meet the soil. Below this point, they are no longer exposed to daylight.

Power: The rate of doing work or the rate at which energy is used or dissipated. Electrical power is the rate of doing electrical work. Electrical power is measured in watts.

Power density: The means of quantifying the power available in wind per unit of area, generally expressed as watts per square meter (w/m^2); in English units, it is expressed as watts per square foot (w/ft^2).

Power-limited cable: Cable used for communication and control applications where the voltage is less than 30 volts and the current is less than 1,000 volt-amps. As a result, overcurrent protection for the wire is typically not required.

Programmable logic controller (PLC): A computer containing input and output modules that may be programmed to deliver specific output values, sequences, and other functions based on input values.

Prohibited approach boundary: An approach limit at a distance from exposed energized electrical conductors or components within which work is considered the same as making physical contact with the energized surface.

Protons: The smallest positively charged particles of an atom. Protons are contained in the nucleus of an atom.

Qualified worker: Electrically speaking, one who has the skills and knowledge related to the electrical equipment and installations, and has received safety training to recognize and avoid the hazards involved.

Rappel: Descent of a vertical surface by sliding down a rope, typically while facing the surface and performing a series of short backward leaps to control the descent.

Reeving: The passing of rope through one or more pulleys or openings. A significant length of rope can be reeved through pulleys and lifting blocks, adding to the necessary length needed for the task.

Relay: An electromechanical device consisting of a coil and one or more sets of contacts. Used as a switching device.

Resistance: An electrical property that opposes the flow of current through a circuit. Resistance (R) is measured in ohms.

Resistor: Any device in a circuit that resists the flow of electrons.

Restricted approach boundary: An approach limit at a distance from exposed energized electrical conductors or components within which there is an increased risk of electrical shock.

Schematic: A type of drawing in which symbols are used to represent the components in a system.

Series circuit (s): A circuit with only one path for current flow.

Series-parallel circuits: Circuits that contain both series and parallel current paths.

Site substation: A wind-site, outdoor electrical facility containing switchgear, transformers, and other electrical equipment designed to connect turbine output power to the commercial grid system. Also known as a collector substation.

Solenoid: An electromagnetic coil used to control a mechanical device such as a valve.

Solid-state relay: Relays that contain no moving parts, only integrated circuitry.

Stator windings: The stationary, non-moving winding within a generator. Stator windings deliver the generated power through wiring connections. The rotor rotates inside the stator winding, generating electrical power.

Step potential: The voltage between the feet (usually about 1m in length) of a person standing near an energized grounded object.

Supervisory control and data acquisition (SCADA) system: A computerized system used to supervise, control, monitor, and collect historical data from an individual wind turbine or a collection of turbine systems using real-time information and commands.

Suspension trauma: The development of physical symptoms due to blood accumulation in the lower extremities from being suspended in a harness. This trauma results from gravitational effects, lack of muscle movement to help pump blood back to the heart, and the restrictive nature of the groin straps; also known as orthostatic intolerance.

Swept area: The area that turbine rotor blades pass through.

Swing zone: The area in space where the momentum and inertia of a fall would cause the body or protected object to swing until a center of gravity is stabilized (hanging straight down).

Thermoplastic insulation: Conductor insulation made from plastic materials, that become elastic or melt when heated, and return to their rigid state at room temperature.

Thermoset insulation: A rubbery conductor insulation, made from plastic materials that maintains its form when exposed to high temperatures.

Tip brakes: Rotor blade tips designed to rotate independent of the rest of the blade, allowing it to spoil the aerodynamic characteristics and reduce rotor speed.

Touch potential: The voltage between the energized object being touched and ground.

Transformer: A device consisting of one or more coils of wire wrapped around a common core. It is commonly used to step voltage up or down.

Turbine transformer: An electromagnetic device that increases the turbine generator output voltage to the value required by the site substation, typically located adjacent to the tower base.

Valence shell: The outermost ring of electrons that orbit about the nucleus of an atom.

Vertical-axis wind turbine (VAWT): A wind turbine with a rotor that spins on a vertical or near-vertical axis. VAWTs are generally omnidirectional and allow the drive train and generators systems to be mounted at ground level.

Volt (V): The unit of measurement for voltage (electromotive force or difference of potential). One volt is equivalent to the force required to produce a current of one ampere through a resistance of one ohm.

Voltage: The driving force that makes current flow in a circuit. Voltage (E) is also referred to as electromotive force or difference of potential.

Voltage drop: The change in voltage across a component that is caused by the current flowing through it and the amount of resistance opposing it.

Voltmeter: An instrument for measuring voltage. The resistance of the voltmeter is fixed. When the voltmeter is connected to a circuit, the current passing through the meter will be directly proportional to the voltage at the connection points.

Watt (W): The basic unit of measurement for electrical power.

Wind rose: A circular graph that depicts the frequency at which winds blow from a given direction at a given location, generally reported as a percentage of time. Wind roses may also contain other information using color, such as how often the wind blows from a direction at a given speed.

Wind shear: Term used to describe the wind speed variations that occur at different heights above the Earth.

Yaw control: Management of a wind turbine's facing direction by rotation of the turbine assembly on its vertical axis.

Yaw deck: The final accessible deck located at the top of a tubular tower and just below the nacelle assembly.

Yaw pucks: Load-bearing surfaces for the turbine nacelle that provide a solid but slick surface for rotation of the turbine during yaw adjustments.

Index

A

AC power, (58101): 19
Adjustable resistors, (26103): 12
Aerial work platforms
  boom lifts, (58102): 30, 33–35
  powering, (58102): 30
  scissor lifts, (58102): 30, 31–33
  types of, (58102): 30
Aerodynamic brakes, (58101): 23, (58102): 8
Air density, (58101): 5
Air mass (m), (58101): 5
Air volume calculations, (58101): 5–6
Altitude, wind speed and, (58101): 6–8
Aluminum conductors, (26103): 8, (58104): 4
Ambient temperature, (58104): 2, 7–8, 32
American National Standards Institute (ANSI), (58101): 4
American National Standards Institute (ANSI) Standards
  fall protection industry, (58103): 27
  lanyards, (58103): 9
  personal protective equipment, (58103): 13, 14
  PFAS equipment, (58103): 2, 6
  rescue planning, (58103): 28
  Z89.1-2009 Type I, Class C, (58103): 13
American Society for Testing Materials (ASTM)
    International, (58101): 4
American Society of Mechanical Engineers (ASME), (58101):
    4, (58102): 25
American Wind Energy Association (AWEA), (58101): 4, 26,
    (58102): 1
American Wire Gauge (AWG) system, (58104): 2–3
Ammeters, (26103): 15, 16, 28, (26112): 5
Ampacity
  derating and correction, (58104): 9–10
  determining, (58104): 8
Ampacity rating
  defined, (58104): 8
  NEC® tables listing, (58104): 8
Ampere (A), (26103): 6, 19, 28
Ampere-hour, (26103): 7
Analog multimeter, (26103): 15
Anchor point, (58103): 1, 36
Anemometers, (58101): 8, 10, 16, 17, 35
ANSI (American National Standards Institute). see
    American National Standards Institute (ANSI)
Arc, (58102): 13, 46
Arc blast, (58102): 10, 13, 15, 46
Arc fault current, (58102): 13
Arc flash, (58102): 10, 13, 46
Arc flash protection boundary, (58102): 14, 15, 46
Ascent safety, (58103): 23–24, 25
ASME (American Society of Mechanical Engineers). see
    American Society of Mechanical Engineers (ASME)
ASTM (American Society for Testing Materials). see
    American Society for Testing Materials (ASTM)
    International
Atomic theory
  the atom, (26103): 1–2

conductors and insulators, (26103): 2–3
  magnetism, (26103): 3
Atoms, (26103): 1–2, 28
AWEA (American Wind Energy Association). see American
    Wind Energy Association (AWEA)
AWG (American Wire Gauge) system. see American Wire
    Gauge (AWG) system

B

Battery
  defined, (26103): 28
  ohmmeter, (26103): 19
  schematic, (26103): 9
  voltage from, (26103): 6–7
Betz, Albert, (58101): 12
Betz Limit, (58101): 11–12, 35
Big Horn wind farm, (58101): 24
Birds
  climbing safety and, (58103): 17
  wind turbines as danger to, (58101): 1, 24
Blade count, (58101): 13–14
Block and tackle systems, (58103): 28
Body harness, (58103): 1, 4–6, 36
Boom lifts
  applications, (58102): 33–34
  categories of, (58102): 33
  maintenance, (58102): 35
  operation of, (58102): 34
Brakes
  aerodynamic, (58101): 23, (58102): 8
  hydraulic, (58101): 23–24
  wind turbine
    overview, (58101): 23–24
    process, (58102): 8
    purpose and illustration, (58101): 17
    service systems, (58102): 8
Brush, Charles, (58101): 2
Brush Electric Company, (58101): 2
Bureau of Labor Statistics, U.S., (58101): 26

C

Cable connections. See also Conductors/cables
  compression connectors, (58104): 23–27
  importance of good, (58104): 14–15
  stripping and cleaning conductors prior to, (58104): 15–19
  tools for, (58104): 19–20, 21, 22–26
  wire connections under 600 volts
    control and sensor cables, (58104): 21–22
    dissimilar metals, (58104): 20
    heat-shrink insulators, (58104): 21
    low-voltage circuits, (58104): 22–23
Cable grabs, (58103): 11, 16, 36
Cable strippers, (58104): 16, 17, 18
Capacitors as shock hazard, (58102): 10
Cape Wind Project, (58101): 4
Carabiner, (58103): 5, 6, 16, 36
Carbon composition resistors, (26103): 12

Carbon fiber-reinforced plastic (CFRP) blades, (58101): 15
Case heaters, (58104): 1
Centrifugal force, (26103): 1
Centrifugal release units, (58101): 23, 35
CFR (Code of Federal Regulations). *see* Code of Federal Regulations (CFR)
CFRP (carbon fiber-reinforced plastic) blades. *see* Carbon fiber-reinforced plastic (CFRP) blades
Chain hoists, (58102): 23–24
Charge, (26103): 1, 6
Charts and maps, wind, (58101): 8–9
Circle area equation, (58101): 8
Circuits
    adding resistance to, (26103): 7
    defined, (26103): 28
    highlighted in text, (26103): 1
    language of, (26103): 8
    measuring power in, (26103): 19
    pictoral form, (26103): 9
    schematic diagram, (26103): 9
    work done in, (26103): 19
Circular mils, (58104): 3
Clamp-on ammeters, (26112): 5
Clipper Liberty, (58101): 22
Code of Federal Regulations (CFR), (58102): 1
Coil, (26112): 1, 16
Collection system, (58104): 1, 32
Color codes
    insulating material, (58104): 7, 23
    resistors, (26103): 12–13
    wire size, (58104): 7
Come-alongs, (58102): 22
Commercial power grid system, (58104): 5, 32
Commercial voltage, (26103): 4
Communications cables, (58104): 10–12
Company vehicles, driving, (58102): 27
Compression connectors, (58104): 23–27
Compression-type crimp connectors, (58104): 22–23
Compression-type terminators, (58104): 19
Conductivity, (58104): 4
Conductor material, value elements
    conductivity, (58104): 4
    workability, (58104): 4–5
Conductors
    current flow in, (26103): 6
    defined, (26103): 28, (58104): 2
    electrical characteristics, (26103): 2–3
    magnetic properties, (26103): 3
    resistance in, (26103): 7
    valence electrons in, (26103): 3
Conductors/cables
    applications, (58104): 10–12
    low-voltage, (58104): 12–13
    medium-voltage, (58104): 13–14
    properties of
        ampacity rating, (58104): 8
        conductive material, (58104): 4–5
        insulating material, (58104): 5–7
        physical size, (58104): 2–4
        temperature rating, (58104): 7–8
    stripping and cleaning
        copper communications conductors, (58104): 18
        data cables, (58104): 18
        faulty, results of, (58104): 15
        large cables/conductors, (58104): 15–16
        power cables, (58104): 15–16
        proper length for, (58104): 19
        small, (58104): 15

tools for, (58104): 15–16
Confined-space entry permit, (58102): 5–6
Confined-space entry program, (58102): 4
Confined spaces
    classifications, (58102): 4
    defined, (58102): 3
    permit-required, (58102): 4, 5–6
    in wind turbines
        nacelle, (58102): 1, 8–9, 10
        rotor hub, (58102): 9–10
    worker responsibilities in
        attendants, (58102): 4, 7
        entrants, (58102): 4
        rescue workers, (58102): 7
        supervisors, (58102): 7
Connecting devices, (58103): 2, 36
Construction standards, (58104): 2
Continuity tester, (26103): 17
Controller, (58101): 16, 35
Converter, (58101): 21
Cooling fans, (58104): 1
Copper conductors, (26103): 3, 8, (58104): 4, 18
Coulomb, (26103): 5–6, 28
Coulomb's law, (26103): 5–6
Craft instructor, (58101): 28
Cranes, (58102): 23, 25
Crimping tools, (58104): 23–26
Crimp-on wire lugs, (58104): 19
Crimp-type connectors, (58104): 19, 22–23
Crossing water safely, (58102): 29
Current (I)
    defined, (26103): 6, 28
    highlighted in text, (26103): 1
    measuring, (26103): 16, (26112): 5
    in resistor circuits, determining
        Kirchoff's laws for, (26104): 9–10
        Ohm's law for, (26104): 5–7
Current flow, (26103): 6, 7

D

Darrieus, D. G. M., (58101): 12
Data cables, (58104): 10–12, 18
DBI SALA, (58103): 28
DC power, (58101): 19
d'Arsonval, Arsene, (26112): 1
d'Arsonval meter movement, (26112): 1, 2, 16
de Coulomb, Charles, (26103): 7
De-energized equipment, (26112): 10, (58102): 18
Department of Energy (DOE), U.S., (58101): 3, 4, 8
Department of Labor (DOL), U.S., (58101): 26, (58103): 27
Descent safety, (58103): 23–24, 26
Digital multimeter, (26103): 15
Distribution conductors, (58104): 5, 32
DOE (Department of Energy, U.S.). *see* Department of Energy (DOE), U.S.
Drain, (58104): 11
D-ring, (58103): 4, 5, 8, 9, 11
Drive system components, HAWT systems
    braking and hydraulic systems, (58101): 23–24
    gearbox, (58101): 22–23
Driving safety
    company vehicles, (58102): 27
    ground clearance, (58102): 29
    in inclement weather, (58102): 29–30
    off-road, (58102): 27–30
Dynamo, (58101): 2, 35

E

Earth, resistance to current flow, (58102): 11–13
Earth burial of cables, (58104): 8
Effective ground level, (58101): 7, 35
Eggbeater turbine design, (58101): 12–13
Egress, (58102): 8, 46
Electrical charges, (26103): 2
Electrical circuit. *see* Circuit
Electrical grid expansion, historically, (58101): 3
Electrical hazards
   in nacelles, (58102): 10
   pre-climb meeting discussions on, (58103): 22
Electrical injuries
   arc flash/arc blast, (58102): 13
   hazard boundaries
      flash protection, (58102): 15
      minimum approach distance, (58102): 17–18
      shock protection, (58102): 15–16
      signage requirements, (58102): 15
      standards, (58102): 13–14
      types of, (58102): 14
   levels of, (58102): 10–11, 15
   shock following, (58102): 10–11
   statistics, (58102): 11
   step and touch potentials, (58102): 11–13
Electrically safe work conditions, (58102): 14
Electrical power
   commercial power grid system, (58104): 5, 32
   generation and distribution, (26103): 4
   overview, (26103): 19–21
   wind-generated, (58101): 1, 2–3
Electrical safety
   confined spaces and, (58102): 10
   de-energized equipment, (26112): 10, (58102): 18
   importance of, (26112): 10
   lockout/tagout
      procedure, (58102): 20–23
      purpose of, (58102): 18–20
   OSHA standards, (58102): 22
Electrical symbols, (26103): 10, 11, 12, 19
Electrical systems, (58104): 1
Electrical test equipment
   ammeters, (26103): 16, (26112): 5, 15, 28
   category hazard ratings, (26112): 9–10
   changes in, (26103): 18
   megohmmeter, (26112): 7
   meters, (26112): 1
   motor and phase rotation testers, (26112): 7–9
   multimeter, (26103): 15, (26112): 5–6
   ohmmeter, (26103): 16–17, 19, 28, (26112): 3–5
   primary purposes of, (26112): 1
   recording instruments, (26112): 9
   safety systems and standards, (26112): 9
   voltage testers, (26103): 17–18, 19, (26112): 2–3, 4
   voltmeter, (26103): 16, 28, (26112): 2
   volt-ohm-milliammeter (VOM), (26103): 15, (26112): 5–6
Electric charge
   current flow, (26103): 6
   measurement of, (26103): 5–6
   potential, (26103): 5–6
   resistance, (26103): 7–8
   voltage, (26103): 6–7
Electricity
   defined, (26103): 1
   usage, measuring, (26103): 20
   visual language of, (26103): 8
Electric power components, HAWT systems
   converter, (58101): 21
   generator, (58101): 19–20
   low-voltage distribution module (LVDM), (58101): 21
   pad-mounted transformer, (58101): 21
Electromagnet, (26103): 3
Electromotive force (emf), (26103): 5, 6, (26112): 2
Electron, (26103): 1, 2, 28
Electron flow (current), (26103): 6, 7
Electronics symbols, (26103): 10
Electrostatic field, (26103): 2
Electrostatic force, (26103): 1–2
Elevators, (58103): 13
Enercon E126, (58101): 14
Energy, unit of, (26103): 8
Equipotential plane, (58102): 11, 46
Eye protection, (58103): 14

F

FAA (Federal Aviation Administration). *see* Federal
   Aviation Administration (FAA)
FACE Program (NIOSH), (58103): 22
Fall arrest, (58103): 1, 36
Fall restraint, (58103): 1, 36
Federal Aviation Administration (FAA), (58101): 8
Fixed resistors, (26103): 12
Flash hazard analysis, (58102): 15, 46
Flash protection boundary, (58102): 14, 15, 46
Footwear, (58103): 14
Force, (26103): 6
Freestanding towers, (58101): 17–18
Frequency, (26112): 6, 16
Furling, (58101): 15, 35

G

Gearbox
   HAWT system, (58101): 22–23
   maintenance and repair, (58101): 27
   purpose and illustration, (58101): 16, 17
Gearbox-free generator, (58101): 21
Gedser turbine, (58101): 3
General Electric Company, (58101): 2
Generators
   gearbox-free, (58101): 21
   HAWT system, (58101): 19–20
   multiple, (58101): 22
   purpose and illustration, (58101): 16, 17
Generator stator (output) conductors, (58104): 12–13
Germanium, (26103): 3
Germany, (58101): 3, 14
Gold conductors, (58104): 4
Great Depression, (58101): 3
Groin straps, (58103): 5
Ground fault, (58102): 10
Ground wire, (58104): 11
Guyed towers, (58101): 17–18

H

Hailo elevator, (58103): 13
Hand protection, (58103): 13–14
HAWTs (horizontal-axis wind turbines). *see* Horizontal-axis
   wind turbines (HAWTs)
Head protection, (58103): 13
Hearing protection, (58103): 14, 17
Heat-shrink insulators, (58104): 21
Height, wind speed and, (58101): 6–8
High-speed shaft, (58101): 16, 17
High-temperature insulated conductors, (58104): 6–7
Hoists, (58102): 22, 23–24

Hooks, (58103): 7, 16
Horizontal-axis wind turbines (HAWTs)
   advantages of, (58101): 13
   basic components, (58101): 16–17
   confined spaces, (58102): 8–10
   converter, (58101): 21
   defined, (58101): 35
   design, (58101): 13
   drive system components
      braking and hydraulic systems, (58101): 23–24
      gearbox, (58101): 22–23
   electric power components
      converter, (58101): 21
      generator, (58101): 19–20
      low-voltage distribution module (LVDM), (58101): 21
      pad-mounted transformer, (58101): 21
   hoists, (58102): 23–24
   yaw and pitch in, (58101): 14–16
Horsepower (hp), (26103): 19
Household circuits, (26103): 13–14
Household voltage, (26103): 4, 10
Humans, as safety hazard, (58103): 18
Hutter, Ulrich, (58101): 3
Hydraulic brakes, (58101): 23–24
Hydraulic pump motors, (58104): 1
Hydroelectric power, (26103): 5
Hydrogen atom, (26103): 1

I

I^2R heating, (26103): 21
Iberdrola Renewables, (58101): 24
Ice
   climbing safety and, (58103): 18–19
   driving safely on, (58102): 29–30
IEA (International Energy Agency). *see* International Energy
   Agency (IEA)
IEC (International Electrotechnical Commission). *see*
   International Electrotechnical Committee (IEC)
IEC (International Electrotechnical Committee). *see*
   International Electrotechnical Commission (IEC);
   International Electrotechnical Committee (IEC)
IEEE (Institute of Electrical and Electronic Engineers). *see*
   Institute of Electrical and Electronic Engineers (IEEE)
Industrialization, wind power and, (58101): 2–3
In-line ammeter, (26103): 16
Insects, climbing safety and, (58103): 17–18
Institute of Electrical and Electronic Engineers (IEEE),
   (58101): 4
Insulating material
   color-coding, (58104): 7, 23
   high-temperature insulated conductors, (58104): 6–7
   thermoplastic- and thermoset-insulated conductors,
      (58104): 5–6
Insulators
   defined, (26103): 28
   electrical characteristics, (26103): 2–3
   heat-shrink, (58104): 21
   purpose of, (26103): 3
   resistance in, (26103): 7
   valence electrons, (26103): 3
International Electrotechnical Commission (IEC), (26112): 9
International Electrotechnical Committee (IEC), (58101): 4
International Energy Agency (IEA), (58101): 4
International Standards Organization (ISO), (58101): 4
Inverse Square Law, (26103): 7
Ions, (26103): 1
ISO (International Standards Organization). *see*
   International Standards Organization (ISO)

J

Jacobs, Joe, (58101): 3
Jacobs, Marcellus, (58101): 3
Jacobs Wind Electric Company, (58101): 3
Jibs, (58102): 25, 46
Job Hazard Analysis, (58102): 2
Job Safety Analysis (JSA)
   example, (58102): 41–46
   purpose, (58102): 2
   task analysis, (58102): 2
   using outside specialists, (58102): 2–3
Joule (J), (26103): 6, 19, 28
Joule, James, (26103): 8
Joule's Law, (26103): 8
JSA (Job Safety Analysis). *see* Job Safety Analysis (JSA)
Juul, Johannes, (58101): 3

K

Kilo, (26103): 28
Kilowatt-hour (kWh), (26103): 20
Kinetic energy, (58101): 5, 35
Kirchoff, Gustav, (26104): 10
Kirchoff's current law, (26104): 9, 18
Kirchoff's voltage law, (26104): 9–10, 18

L

La Cour, Poul, (58101): 3
Lad Saf® systems, (58103): 11, 12
Lamp (light bulb) schematic, (26103): 9
LAN (local area network). *see* Local area network (LAN)
Lanyards, fall arrest, (58103): 2, 8–10, 12, 15, 24, 25–26, 36
Lattice towers, (58101): 17–18, (58103): 20, 26
Law of Electrical Charges, (26103): 2
Law of Electrical Force, (26103): 7
Lighting fixtures, (58104): 1
Lightning safety, (58103): 19–20
Limited approach boundary, (58102): 16, 46
Load, (58104): 2, 32
Local area network (LAN), (58104): 11, 32
Lockout/tagout
   procedure, (58102): 20–23
   purpose, (58102): 18–20
Longitudinal, (58101): 15, 35
Low-speed shaft, (58101): 16, 17
Low-voltage cables/conductors, (58104): 1, 12–13
Low-voltage distribution module (LVDM), (58101): 19, 21,
   35, (58102): 22
LVDM (low-voltage distribution module). *see* Low-voltage
   distribution module (LVDM)

M

MAD (Minimum approach distance). *see* Minimum
   approach distance (MAD)
Magnetic flux lines, (26103): 3
Magnetism, (26103): 3, (26112): 1
Maintenance
   boom lifts, (58102): 35
   scissor lifts, (58102): 33
Maps and charts, wind, (58101): 8–9
Mass equation, (58101): 5
Matter, (26103): 3, 28
Mechanical braking, (58101): 23–24
Mechanical compression connectors, (58104): 19, 21
Mechanical power, (26103): 19
Medium-voltage conductors/cables
   single-phase collection system, (58104): 14
   three-phase, (58104): 13

voltages of, (58104): 1
Mega, (26103): 20, 28
Megawatts (MW), (26103): 20
Meggers®, (26112): 7
megohmmeter, (26112): 7
Messenger wire, (58104): 8, 32
Met (meteorological) towers, (58101): 8
Metallic electricity, (26103): 7
Meters, (26112): 1
Minimum approach distance (MAD), (58102): 14, 17–18, 46
Mobile cranes, (58102): 23
Monopole towers, (58101): 18
Motor and phase rotation testers, (26112): 7–9
Moving-coil meter, (26112): 1
Mud, driving in, (58102): 29
Multimeter, (26103): 15, (26112): 5–6
MW (Megawatts). *see* Megawatts (MW)

N

Nacelles
    case heaters voltage levels, (58104): 1
    confined spaces within, (58102): 1, 8–9, 10
    construction materials, (58104): 5
    deck location and, (58101): 19
    defined, (58101): 35
    electrical hazards, (58102): 10
    entry and escape routes, (58102): 8–9
    lifting loads to the, (58102): 23–24
    noise generated by, (58101): 1
    PLCs in, (58104): 10–11
    purpose and illustration, (58101): 16, 17
    safety guidelines, (58103): 24
    wiring materials, (58104): 4–5
    working in sub-zero weather, (58104): 5
National Center for Construction Education and Research
    (NCCER), (58101): 28–29
National Center for Construction Education and Research
    (NCCER) Accreditation Guidelines, (58101): 28
National Center for Construction Education and Research
    (NCCER) Accredited Training Sponsors and
    Assessment Centers, (58101): 28
National Center for Construction Education and Research
    (NCCER) credentials, (58101): 28
National Center for Construction Education and Research
    (NCCER) National Registry, (58101): 28
National Climatic Data Center (NCDC), (58101): 8
National Electrical Code® *(NEC*®)
    310.106(C), (58104): 4
    ampacity rating defined, (58104): 8
    Article 310, (58104): 8
    Article 725, (58104): 11
    requirements for electrical installation, (58102): 14
    Section 110.14, (58104): 19, 20
    Tables
        310.15(B)(2)(A), (58104): 35
        310.15(B)(3)(a), (58104): 8
        310.15(B)(16), (58104): 9, 33
        310.15(B)(16,17,18,19,20,21), (58104): 8
        310.15(B)(17), (58104): 9, 34
        310.15(B)(18), (58104): 9
National Fire Protection Association (NFPA)
    70, (58102): 14
    70E®, (58102): 14, 16
    electrical safety standards, (58102): 14
National Institute for Occupational Safety and Health
    (NIOSH), (58103): 3
National Renewable Energy Laboratory (NREL), (58101): 4
National Weather Service (NWS), (58101): 8

National Wind Technology Center (NWTC), (58101): 4
NCCER (National Center for Construction Education
    and Research). *see* National Center for Construction
    Education and Research (NCCER)
NCDC (National Climatic Data Center). *see* National
    Climatic Data Center (NCDC)
NEC® *(National Electrical Code*®). *see* National Electrical Code®
    (*NEC*®)
Neutrons, (26103): 2, 28
Newtons, (58103): 7, 8, 36
NFPA (National Fire Protection Association). *see* National
    Fire Protection Association (NFPA)
NIMBY. *see* Not in my backyard
NIOSH (National Institute for Occupational Safety and
    Health). *see* National Institute for Occupational Safety
    and Health (NIOSH)
Noise
    climbing safety and, (58103): 17
    rotor, (58101): 1
Nominal output voltage, (58104): 2, 32
Nonpermit-required confined space, (58102): 4
Not in my backyard (NIMBY), (58101): 1
NREL (National Renewable Energy Laboratory). *see*
    National Renewable Energy Laboratory (NREL)
Nucleus, (26103): 1, 2, 28
NWS (National Weather Service). *see* National Weather
    Service (NWS)
NWTC (National Wind Technology Center). *see* National
    Wind Technology Center (NWTC)

O

Occupational Safety and Health Administration (OSHA)
    defined, (58103): 36
    General Industry program, (58102): 1
    General Industry Safety Training, (58102): 1
    MADs established, (58102): 17
    personal protective equipment regulations, (58102): 15, 17
    purpose, (58102): 1, (58103): 1
    requirements
        electrical, (58102): 13–14
        mobile crane use, (58102): 25
        personal fall arrest systems, (58103): 1–2, 4
        personal protective equipment, (58103): 13, 14
        pre-climb meeting, (58103): 21–22
        use of aerial lifts, (58102): 30–31
    rescue planning regulations, (58103): 27
    safety analysis consultations, (58102): 3
    Standards
        29, CFR 1910.180, (58102): 25
        29, CFR 1926, Subpart N, (58102): 25
        1910,269 Electric Power Generation, Transmission, and Dis-
            tribution, (58102): 22
        1926, Subpart N, (58102): 30–31
        1926.100(a), (58103): 13
        1926.502, Subpart M, Section (d), (58103): 1
        working at heights, (58103): 1
Ocean winds, (58101): 4
Off-road driving
    crossing water, (58102): 29
    first rule of safety in, (58102): 27–28
    mud, (58102): 29
    steep inclines, (58102): 28–29
Offshore turbines, (58103): 21
Offshore wind farms, (58101): 4, 19–20
Off-slab boom lifts, (58102): 33–34
Ohm (Ω), (26103): 1, 7, 28
Ohm, George, (26104): 10
Ohmmeter, (26103): 16–17, 19, 28, (26112): 3–5

Ohm's law, (26103): 8–9, 28, (26104): 5–9
Ohm's law circle, (26103): 21
Ohm's law for power, (26103): 20–21
Oil crisis (1973), (58101): 3
½ power law, (58101): 8
Optical fiber cable, (58104): 11, 12
OSHA (Occupational Safety and Health Administration).
    see Occupational Safety and Health Administration
    (OSHA)
Owatonna, MN, (58101): 23

P

Pad-mounted transformer, (58101): 21
Parallel circuits
    calculating resistance in, (26104): 1–3
    defined, (26104): 18
    finding voltage and current in, (26104): 6–7
    highlighted in text, (26104): 1
    overview, (26103): 13–15
Parallel-tap connector, (58104): 20
Paved/slab boom lifts, (58102): 33–34
Pelican style hooks, (58103): 7
Permit-required confined space, (58102): 4, 46
Personal fall arrest system (PFAS)
    anchor points, (58103): 2–4
    carabiners and hooks, (58103): 7
    chest straps, (58103): 5
    connecting devices, (58103): 5–7
    defined, (58103): 36
    D-ring, (58103): 4, 5
    full body harness, (58103): 4–6, 24
    groin straps, (58103): 5
    illustration, (58103): 3
    inspecting, (58103): 14–16
    modifying, (58103): 5
    saddles, (58103): 5
    standards, (58103): 1–2
    suspension trauma strap, (58103): 5, 7
    waist/tool belt, (58103): 5
Personal protective equipment
    for flash protection, (58102): 15
    OSHA requirements, (58102): 15, 17
    tower climbers, (58103): 13–14
Personnel lifts. see Scissor lifts
Phase rotation testers, (26112): 7–9
Phi turbine design, (58101): 12–13
Pitch, (58101): 16, 17
Pitch control, (58101): 15, 35
Plastic insulator, (26103): 3
Point of daylight, (58103): 21, 36
Porcelain insulator, (26103): 3
Post mill, (58101): 2
Potential, (26103): 5
Power. See also Wind power
    defined, (26103): 1, 19, 28
    electrical
        commercial power grid system, (58104): 5, 32
        generation and distribution, (26103): 4
        overview, (26103): 19–21
        wind-generated, (58101): 1, 2–3
Power density, (58101): 10–11, 35
Power equation, (26103): 20–21
Power grid system, commercial, (58104): 5, 32
Power Industry Fundamentals, Basic Rigging (NCCER),
    (58102): 23
Power Industry Fundamentals, Basic Safety (NCCER), (58102): 1
Power-limited cable, (58104): 11, 32

Programmable logic controller (PLC), (58104): 5, 32
Prohibited approach boundary, (58102): 16, 46
Protons, (26103): 1, 2, 28
Pump motors, hydraulic, (58104): 1

Q

Qualified worker, (58102): 14, 46

R

Raceway installations, (58104): 8
Rain safety, (58103): 18–19
Rappel, (58103): 26, 36
Ratchet lever hoists, (58102): 22
Recording instruments, (26112): 9
Record-keeping
    Job Safety Analysis (JSA) forms, (58102): 41–46
    pre-climb meeting, (58103): 21
    wind turbine maintenance, (58101): 26–27
Reeving, (58102): 25, 46
Relay, (26103): 3, 28
Rescue
    confined space, (58102): 7
    at height
        equipment for, (58103): 28
        planing for, (58103): 27–28
        suspension trauma, (58103): 26–27
Resistance (R)
    current flow and, (26103): 1
    defined, (26103): 28
    measuring, (26103): 16–17, (26112): 3–5
    overview, (26103): 7–8
Resistive circuits
    finding voltage and current in
        Kirchoff's laws for, (26104): 9–10
        Ohm's law for, (26104): 5–7
    loop equations, (26104): 10–11
    in parallel, (26104): 1–3, 6–7
    in series, (26104): 1, 5–6
    series-parallel, (26104): 2–5, 7–9
Resistors
    color-coding, (26103): 12–13
    defined, (26103): 7, 28
    function of, (26103): 10
    power rating of, (26103): 21
    types of, (26103): 11–12
Restricted approach boundary, (58102): 16, 46
Rigging superintendent, (58102): 26
Rollgliss® R350 System, (58103): 28
Rollgliss® R500 System, (58103): 28
Rope, in lift plans, (58102): 25
Rope grabs, (58103): 12, 16
Rotor blades
    carbon fiber-reinforced plastic, (58101): 15
    damage, undetectable, (58102): 26
    diameter, (58102): 26
    ice buildup on, (58103): 18–19
    purpose and illustration, (58101): 16, 17
    size and configuration, (58101): 14
    stress failures, (58101): 15
Rotor braking, (58101): 23
Rotor hub, (58102): 1, 9–10, (58103): 24
Rotors
    Betz Limit for, (58101): 11–12
    purpose and illustration, (58101): 17
    swept area, (58101): 8–10

S

Saddles, (58103): 5
Safety. *See also* Job Safety Analysis (JSA); Wind tower
    climbing: safety; Wind turbine safety
  ascent and descent, (58103): 18–24
  in confined spaces, (58102): 3–4, 6–7
  electrical test equipment systems, (26112): 9–10
  fall statistics, (58103): 1
  lockout/tagout for, (58102): 18–20
  off-road driving, (58102): 27–28
Safety hazards
  category ratings, (26112): 9
  wind turbines, (58102): 1, 8
SCADA (supervisory control and data acquisition). *see*
    Supervisory control and data acquisition (SCADA)
Schematic, (26103): 9, 28
Scissor lifts
  controls and indicators, (58102): 32
  heights and capacities, (58102): 31
  maintenance, (58102): 33
  operating
    precautions, (58102): 31–32
    procedure, (58102): 32–33
  powering, (58102): 31
Self-retracting lanyards, (58103): 10, 15
Semiconductors, (26103): 3
Series circuit
  calculating resistance in, (26104): 1
  defined, (26103): 28, (26104): 18
  finding voltage and current in, (26104): 5–6
  highlighted in text, (26103): 10, (26104): 1
Series-parallel circuits
  calculating resistance in, (26104): 2–5
  defined, (26104): 18
  finding voltage and current in, (26104): 7–9
  highlighted in text, (26104): 1
  overview, (26103): 15
Shadow flicker, (58101): 1
Shielding cables, (58104): 11
Shock, electrical, (58102): 10–11
Shock-absorbing lanyards, (58103): 9–10
Shock protection boundary, (58102): 15, 16–17
Siemens, (58101): 26
Signage requirements, hazard boundaries, (58102): 15
Silicon, (26103): 3
Silver conductors, (26103): 3, (58104): 4
Single-phase collection system cables, (58104): 14
Site assessment, pre-climb, (58103): 20–21
Site substation
  defined, (58104): 32
  highlighted in text, (58104): 1, 2
  terrain, (58102): 27–30
  typical, (58104): 1
  wiring, (58104): 1–2
Slings, in lift plans, (58102): 25
Small wind market, (58101): 14
Smith-Putnam turbine, (58101): 3
Snakes and climbing safety, (58103): 18
Snap-On®, (58103): 8
Snow
  climbing safety and, (58103): 18–19
  driving safely in, (58102): 29–30
Solenoid, (26103): 3, 28
Solid-state relay, (58104): 1, 32
Standards development, (58101): 4
Standards organizations, (58101): 4
Stator (output) conductors, (58104): 12–13
Stator windings, (58104): 5, 32

Steep inclines, driving on, (58102): 28–29
Step potential, (58102): 10, 11–13, 46
Stranded wire, (58104): 4, 19
Supervisory control and data acquisition (SCADA), (58101):
    16, 23, 35, (58104): 11
Suspension trauma, (58103): 5, 26–27, 36
Suspension trauma strap, (58103): 5, 7
Swept area, (58101): 8, 10–12, 35
Swing zone, (58103): 4, 36
Switch schematic, (26103): 9
Symbols, electrical, (26103): 10, 11, 12, 19

T

Task safety analysis, (58102): 2, 3
Temperature, resistance and, (26103): 8
temperature rating, conductors, (58104): 7–8
Thermoplastic insulation, (58104): 5–6, 32
Thermoset insulation, (58104): 5–6, 32
Three-phase medium-voltage conductors/cables, (58104): 13
Tip brakes, (58101): 23, 35
Tool lanyards, (58103): 8
Touch potential, (58102): 10, 11–13, 46
Tower mills, (58101): 2
Transformer, (26103): 3, 4, 6, 28
Truss towers, (58101): 17–18
Tubular towers, (58101): 17–19, (58103): 20, 21
Turbines, (26103): 4
Turbine transformer, (58104): 1, 13, 32

V

Valence electrons, (26103): 2–3
Valence shell, (26103): 2, 28
Variable resistors, (26103): 12
VAWTs (vertical-axis wind turbines). *see* Vertical-axis wind
    turbines (VAWTs)
Velocity (V), (58101): 5–6
Vertical-axis wind turbines (VAWTs), (58101): 12–13, 35
Volt (V), (26103): 1, 28
Volta, Alessandro, (26103): 7
Voltage
  in commercial buildings, (26103): 4
  creating, (26103): 6–7
  defined, (26103): 1, 6, 28
  household, (26103): 4, 10
  in long-distance transmission lines, (26103): 4
  measuring, (26103): 16, 28, (26112): 2
  in resistor circuits, determining
    Kirchoff's laws for, (26104): 9–10
    Ohm's law for, (26104): 5–7
Voltage drop, (26103): 10, 28
Voltage testers, (26103): 17–18, 19, (26112): 2–3, 4
Voltaic pile, (26103): 7
Voltmeter, (26103): 16, 28, (26112): 2
Volt-ohm-milliammeter (VOM), (26103): 15, (26112): 5–6

W

Waist/tool belt, (58103): 5
Water, driving across safely, (58102): 29
Watt (W), (26103): 1, 19, 28
Weather
  arc potential and, (58102): 13
  climbing safety and
    lightning, (58103): 19–20
    rain, snow, and ice, (58103): 18–19
    sunshine, (58103): 18
    temperature extremes, (58103): 19
    wind, (58103): 20

Weather (*continued*)
  hub entry and, (58102): 9
  inclement, driving safely in, (58102): 29–30
  sub-zero, working in, (58104): 5
Weight element in lift planning, (58102): 24–25
Wind
  climbing safety and, (58103): 20
  creation of, (58101): 5
  lifting loads and, (58102): 24
Wind and Water Program (DOE), (58101): 4
Wind data, acquisition and use
  instruments, (58101): 10
  SCADA systems for, (58101): 16
  wind maps and charts, (58101): 8–9
Wind energy
  calculating, (58101): 5–6
  factors determining, (58101): 5
  intercepting, (58101): 8, 9–10
  research on, (58101): 4
Wind energy equation, (58101): 5
Wind farms
  aesthetic appeal of, (58101): 1, 17
  Big Horn, (58101): 24
  development criteria, (58101): 24–25
  locating, (58101): 3, 24
  maintenance, (58101): 25–27
  noise generated by, (58101): 1
  offshore, (58101): 4, 19–20
Wind industry
  present day, (58101): 3–5
  standards, (58101): 4
Wind industry workers
  confined spaces responsibilities, (58102): 4, 7
  training
    online programs, (58102): 1
    OSHA General Industry program, (58102): 1
    OSHA General Industry Safety Training, (58102): 1
Wind maps and charts, (58101): 8–9
Wind power
  advantages of, (58101): 1
  available, determining
    averaging for, (58101): 6, 8–9
    Betz Limit, (58101): 11–12
    power density for, (58101): 10–11
  history of, (58101): 1–3
  international use of, (58101): 3
  statistics, (58101): 1
  US and
    capacity growth, (58101): 1, 3
    goal for use of, (58101): 3–4, 4
  wind speeds effect on, (58101): 6
Wind Powering America Program (DOE), (58101): 8
Wind rose, (58101): 8, 35
Wind shear, (58101): 7–8, 35
Wind speed
  factors affecting, (58101): 5
  height and, (58101): 6–8
  impact on power, (58101): 6
Wind Technologies Market Report (DOE), (58101): 4
Wind tower climbers
  basic skills for
    ascent, (58103): 25
    descent, (58103): 26
    maneuvering and positioning, (58103): 25–26
    safe work habits, (58103): 24–25
  demand for, (58103): 1
  injuries to, types of
    fatalities, (58103): 22, 27

  impact, (58103): 5
  suspension trauma, (58103): 26–27, 36
  swing zone, (58103): 4
  personal protective equipment, (58103): 13–14, 17
  physical demands on, (58103): 25
  rescue of, (58103): 26–28
Wind tower climbing
  first aid, (58103): 13, 21–22, 23
  preparations
    emergency medical facility location, (58103): 22
    equipment inspection, (58103): 22
    policy and guideline review, (58103): 22, 23–24
    pre-climb meeting, (58103): 20, 21–22
    rescue planning, (58103): 22, 23, 27–28
    site and tower assessment, (58103): 20–21, 22
  safety
    ascent and descent, (58103): 18–25, 26
    OSHA requirements, (58103): 1–2, 4, 13, 21–22
    qualified workers, (58102): 10, (58103): 23
    responsibility for, (58103): 24–25
    vigilance in, (58103): 24
  safety equipment
    assistance devices, (58103): 10–12
    cable grabs, (58103): 11, 16
    inspecting, (58103): 14–16, 22
    lanyards, (58103): 8–10, 12, 15, 24, 25–26
    modifying, (58103): 5, 8
    personal fall arrest systems, (58103): 1–7, 14–16, 24
    personal protective equipment, (58103): 13–14
    powered climb assist system, (58103): 12
    regulations, (58103): 1–2, 4, 6
    rope grabs, (58103): 12, 16
Wind tower climbing hazards
  electrical, (58103): 22
  environmental
    birds, insects, snakes, (58103): 17–18
    communication tools, (58103): 17
    humans, (58103): 18
    noise, (58103): 17
    weather, (58103): 18–19
  swing zone, (58103): 4
Wind towers
  confined spaces within, (58102): 10
  heights of, (58102): 23
  purpose and illustration, (58101): 16, 17
  types of, (58101): 17–19
Wind turbine assembly
  lattice towers, (58101): 17
  simple pole towers, (58101): 18
Wind turbine assembly lifts
  aerial work platforms
    boom lifts, (58102): 30, 33–35
    powering, (58102): 30
    scissor lifts, (58102): 30, 31–33
    types of, (58102): 30
  complexity of, (58102): 23
  critical classification, (58102): 24–25
  critical lifts, (58102): 25–26
  lift categorization, (58102): 25
  lift plan implementation, (58102): 26–27
  lift planning, (58102): 24–25
  rigging
    hoists, (58102): 22, 23–24
    mobil cranes, (58102): 23
  rigging superintendents role, (58102): 26
Wind turbine design
  blade count in, (58101): 13–14
  blade size and configuration, (58101): 14

HAWTs and VAWTs, (58101): 12–13
HAWT yaw and pitch, (58101): 14–16
standards, (58104): 2
Wind Turbine Maintenance Technician Climbing Wind
Towers, (58102): 1
Wind turbine maintenance technicians
demand for, (58101): 4, 26, (58103): 1
driving records, importance of, (58102): 27
fatal injuries to, (58102): 21
NCCER credentials, (58101): 37–40
qualified worker, (58102): 14, 46
rescue of, (58102): 7
responsibilities, (58104): 1
skills requirements, AWEA, (58101): 27
training, (58101): 26, 28–29, (58102): 1, (58104): 15
Wind turbines. *See also specific types of*
construction materials, (58101): 18
development of, historically, (58101): 2–3
inspection checklist, (58101): 26
maintenance of, (58101): 16, 25–27
placement, (58101): 6
size of, (58101): 1, 2
transportation of, (58102): 26
warranty periods, (58101): 25
Wind turbine safety
aerial work platforms
boom lifts, (58102): 34
scissor lifts, (58102): 31–32
critical lifts, (58102): 24–25
driving
company vehicles, (58102): 27
in inclement weather, (58102): 27–30
off-road, (58102): 27–30
introduction, (58102): 1
repetitive tasks and, (58102): 27
rigging, (58102): 22
use of aerial lifts, (58102): 30–31
using hoists, (58102): 24

Wind Turbine Technician (DOL), (58101): 26
Wind vane, (58101): 16, 17
Wire, measuring resistance in, (26103): 7–8
Wire connections under 600 volts
control and sensor cables, (58104): 21–22
dissimilar metals, (58104): 20
heat-shrink insulators, (58104): 21
methods and tools, (58104): 19–20
Wire lugs, (58104): 19
Wire size
AWG identification system, (58104): 2–3
circular mils, (58104): 3
color-coding, (58104): 7
markings, (58104): 3
Wire strippers, (58104): 15–16, 18
Wire-wound resistors, (26103): 11
Work
electrical, (26103): 19–20
Joule's Law and, (26103): 8
potential, (26103): 5
useful and wasted, (26103): 19
Workability of conductor materials, (58104): 4–5
World Wind Energy Association (WWEA), (58101): 4
WWEA (World Wind Energy Association). *see* World Wind
Energy Association (WWEA)

Y
Yaw and pitch, HAWTs, (58101): 14–16
Yaw control, (58101): 15, 35
Yaw deck, (58101): 19, 35
Yaw drive, (58101): 16, 17
Yaw motors, (58101): 16, 17
Yaw pucks, (58101): 1, 35